Introduction to Computer-Intensive Methods of Data Analysis in Biology

This guide to the contemporary toolbox of methods for data analysis will serve graduate students and researchers across the biological sciences. Modern computational tools, such as Bootstrap, Monte Carlo and Bayesian methods, mean that data analysis no longer depends on elaborate assumptions designed to make analytical approaches tractable. These new 'computer-intensive' methods are currently not consistently available in statistical software packages and often require more detailed instructions. The purpose of this book therefore is to introduce some of the most common of these methods by providing a relatively simple description of the techniques. Examples of their application are provided throughout, using real data taken from a wide range of biological research. A series of software instructions for the statistical software package S-PLUS are provided along with problems and solutions for each chapter.

DEREK A. ROFF is a Professor in the Department of Biology at the University of California, Riverside.

Introduction to Computer-Intensive Methods of Data Analysis in Biology

DEREK A. ROFF
Department of Biology
University of California

CAMBRIDGE UNIVERSITY PRESS
Cambridge, New York, Melbourne, Madrid, Cape Town, Singapore, São Paulo

Cambridge University Press
The Edinburgh Building, Cambridge CB2 2RU, UK

Published in the United States of America by Cambridge University Press, New York

www.cambridge.org
Information on this title: www.cambridge.org/9780521846288

First published 2006

Printed in the United Kingdom at the University Press, Cambridge

A catalogue record for this publication is available from the British Library

Library of Congress Cataloging in Publication data

Roff, Derek A., 1949–
    Introduction to computer-intensive methods of data analysis in biology / Derek A. Roff.
        p. cm.
    Includes bibliographical references.
    ISBN-13: 978-0-521-84628-8 (hardback)
    ISBN-13: 978-0-521-60865-7 (pbk.)
    ISBN-10: 0-521-84628-5 (hardback)
    ISBN-10: 0-521-60865-1 (pbk.)
1. Biology–Data processing.   I. Title.

    QH324.2 R62 2006
    570.285–dc22   2006001857

ISBN-13: 978-0-521-84628-8 hardback
ISBN-10: 0-521-84628-5 hardback

ISBN-13: 978-0-521-60865-7 paperback
ISBN-10: 0-521-60865-1 paperback

Contents

Preface

Easy access to computers has created a revolution in the analysis of biological data. Prior to this easy access even "simple" analyses, such as one-way analysis of variance, were very time-consuming. On the other hand, statistical theory became increasingly sophisticated and far outstripped the typical computational means available. The advent of computers, particularly the personal computer, and statistical software packages, changed this and made such approaches generally available.

Much of the development of statistical tools has been premised on a set of assumptions, designed to make the analytical approaches tractable (e.g., the assumption of normality, which underlies most parametric methods). We have now entered an era where we can, in many instances, dispense with such assumptions and use statistical approaches that are rigorous but largely freed from the straight-jacket imposed by the relative simplicity of analytical solution. Such techniques are generally termed "computer-intensive" methods, because they generally require extensive numerical approaches, practical only with a computer. At present, these methods are rather spottily available in statistical software packages and very frequently require more than simple "point and click" instructions. The purpose of the present book is to introduce some of the more common methods of computer-intensive methods by providing a relatively simple mathematical description of the techniques, examples from biology of their application, and a series of software instructions for one particular statistical software package (S-PLUS). I have assumed that the reader has at least an introductory course in statistics and is familiar with techniques such as analysis of variance, linear and multiple regression, and the χ^2 test. To relieve one of the task of typing in the coding provided in an appendix to this book, I have also made it available on the web at http://www.biology.ucr.edu/people/faculty/Roff.html.

1

An introduction to computer-intensive methods

What are computer-intensive data methods?

For the purposes of this book, I define computer-intensive methods as those that involve an iterative process and hence cannot readily be done except on a computer. The first case I examine is maximum likelihood estimation, which forms the basis of most of the parametric statistics taught in elementary statistical courses, though the derivation of the methods via maximum likelihood is probably not often given. Least squares estimation, for example, can be justified by the principle of maximum likelihood. For the simple cases, such as estimation of the mean, variance, and linear regression analysis, analytical solutions can be obtained, but in more complex cases, such as parameter estimation in nonlinear regression analysis, whereas maximum likelihood can be used to define the appropriate parameters, the solution can only be obtained by numerical methods. Most computer statistical packages now have the option to fit models by maximum likelihood but they typically require one to supply the model (logistic regression is a notable exception).

The other methods discussed in this book may have an equally long history as that of maximum likelihood, but none have been so widely applied as that of maximum likelihood, mostly because, without the aid of computers, the methods are too time-consuming. Even with the aid of a fast computer, the implementation of a computer-intensive method can chew up hours, or even days, of computing time. It is, therefore, imperative that the appropriate technique be selected. Computer-intensive methods are not panaceas: the English adage "you can't make a silk purse out of a sow's ear" applies equally well to statistical analysis. What computer-intensive methods allow one to do is to apply a statistical analysis in situations where the more "traditional" methods fail. It is important to remember that, in any investigation, great efforts should be put into making the experimental design amenable to traditional methods, as these have both well-understood statistical properties and are easily carried out, given the available statistical programs. There will, however, inevitably be circumstances in which the assumptions of these methods cannot be met. In the next section, I give several examples that illustrate the utility of computer-intensive methods discussed in this book. Table 1.1 provides an overview of the methods and comments on their limitations.

Why computer-intensive methods?

A common technique for examining the relationship between some response (dependent) variable and one or more predictor (independent) variables is linear and multiple regression. So long as the relationship is linear (and satisfies a few other criteria to which I shall return) this approach is appropriate. But suppose one is faced with the relationship shown in Figure 1.1, that is highly nonlinear and cannot be transformed into a linear form or fitted by a polynomial function. The fecundity function shown in Figure 1.1 is typical for many animal species and can be represented by the four parameter (M,k,t_0,b) model

$$F(x) = M(1 - e^{-k(x-t_0)})e^{-bx} \qquad\qquad (1.1)$$

Using the principle of maximum likelihood (Chapter 2), it can readily be shown that the "best" estimates of the four parameters are those that minimize the residual sums of squares. However, locating the appropriate set of parameter values cannot be done analytically but can be done numerically, for which most statistical packages supply a protocol (see caption to Figure 1.1 for S-PLUS coding).

In some cases, there may be no "simple" function that adequately describes the data. Even in the above case, the equation does not immediately "spring to mind" when viewing the observations. An alternative approach to curve fitting for such circumstances is the use of local smoothing functions, described in Chapter 6. The method adopted here is to do a piece-wise fit through the data, keeping the fitted curve continuous and relatively smooth. Two such fits are shown in Figure 1.2 for the *Drosophila* fecundity data. The loess fit is less rugged than the cubic spline fit and tends to de-emphasize the fecundity at the early ages. On the other hand, the cubic spline tends to "over-fit" across the middle and later ages. Nevertheless, in the absence of a suitable function, these approaches can prove very useful in describing the shape of a curve or surface. Further, it is possible to use these methods in hypothesis testing, which permits one to explore how complex a curve or a surface must be in order to adequately describe the data.

Table 1.1 *An overview of the techniques discussed in this book*

Method	Chapter	Parameter estimation?	Hypothesis testing?	Limitations
Maximum likelihood	2	Yes	Yes	Assumes a particular statistical model and, generally, large samples
Jackknife	3	Yes	Yes	The statistical properties cannot generally be derived from theory and the utility of the method should be checked by simulation for each unique use
Bootstrap	4	Yes	Possible[a]	The statistical properties cannot generally be derived from theory and the utility of the method should be checked by simulation for each unique use. Very computer-intensive.
Randomization	5	Possible	Yes	Assumes difference in only a single parameter. Complex designs may not be amenable to "exact" randomization tests
Monte Carlo methods	5	Possible	Yes	Tests are usually specific to a particular problem. There may be considerable debate over the test construction.
Cross-validation	6	Yes	Yes	Generally restricted to regression problems. Primarily a means of distinguishing among models.
Local smoothing functions and generalized additive models	6	Yes	Yes	Does not produce easily interpretable function coefficients. Visual interpretation difficult with more than two predictor variables
Tree models	6	Yes	Yes	Can handle many predictor variables and complex interactions but assumes binary splits.
Bayesian methods	7	Yes	Yes	Assumes a prior probability distribution and is frequently specific to a particular problem

[a]"Possible"=Can be done but not ideal for this purpose.

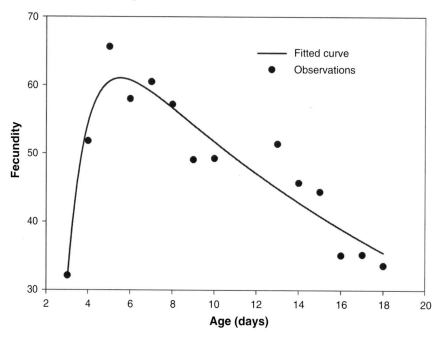

Figure 1.1 Fecundity as a function of age in *Drosophila melanogaster* with a maximum likelihood fit of the equation $F(x)=M(1-e^{k(x-t_0)})e^{-bx}$. Data are from McMillan *et al.* (1970).

Age (x)	3	4	5	6	7	8	9	10	13	14	15	16	17	18
F	32.1	51.8	66	58	60.5	57.2	49.1	49.3	51.4	45.7	44.4	35.1	35.2	33.6

S-PLUS coding for fit:

```
# Data contained in data file D
# Initialise parameter values
    Thetas <- c(M=1, k=1, t0=1, b=.04)
# Fit model
    Model <- nls(D[,2]~M*(1-exp(-k*(D[,1]-t0)))*exp(-b*D[,1]), start=Thetas)
# Print results
    summary(Model)
OUTPUT
Parameters:
          Value     Std. Error    t value
   M  82.9723000   7.52193000   11.03070
   k   0.9960840   0.36527300    2.72696
  t0   2.4179600   0.22578200   10.70930
   b   0.0472321   0.00749811    6.29920
```

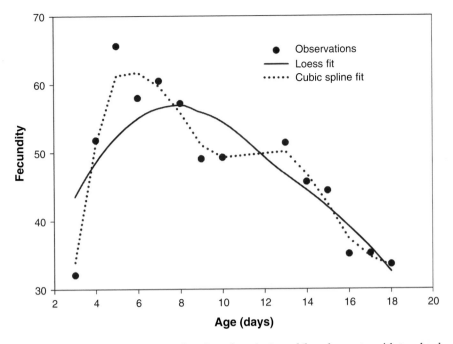

Figure 1.2 Fecundity as a function of age in *Drosophila melanogaster* with two local smoothing functions. Data given in Figure 1.1.
S-PLUS coding to produce fits:

```
# Data contained in file D. First plot observations          # Plot points
    plot (D[,1], D[,2])
    Loess.model <- loess(D[,2]~D[,1], span=1, degree=2)       # Fit loess model
# Calculate predicted curve for Loess model
    x.limits <- seq(min(D[,1]), max(D[,1]), length=50)        # Set range of x
    P.Loess <- predict.loess(Loess.model, x.limits, se.fit=T) # Prediction
    lines(x.limits, D.INT$fit)                                # Plot loess prediction
    Cubic.spline <- smooth.spline(D[,1], D[,2])               # Fit cubic spline model
    lines(Cubic.spline)                                       # Plot cubic spline curve
```

An important parameter in evolutionary and ecological studies is the rate of increase of a population, denoted by the letter r. In an age-structured population, the value of r can be estimated from the Euler equation

$$1 = \sum_{x=0}^{\infty} e^{-rx} l_x m_x \tag{1.2}$$

where x is age, l_x is the probability of survival to age x and m_x is the number of female births at age x. Given vectors of survival and reproduction, the above equation can be solved numerically and hence r calculated. But having an estimate of a parameter is generally not very useful without also an estimate of the variation about the estimate, such as the 95% confidence interval. There are two computer-intensive solutions to this problem, the jackknife (Chapter 3) and the bootstrap (Chapter 4). The jackknife involves the sequential deletion of a single observation from the data set (a single animal in this case) giving n (= number of original observations) data sets of $n-1$ observations whereas the bootstrap consists of generating many data sets by random selection (with replacement) from the original data set. For each data set, the value of r is calculated; from this set of values, each technique is able to extract both an estimate of r and an estimate of the desired confidence interval.

Perhaps one of the most important computer-intensive methods is that of hypothesis testing using randomization, discussed in Chapter 5. This method can replace the standard tests, such as the χ^2 contingency test, when the assumptions of the test are not met. The basic idea of randomization testing is to randomly assign the observations to the "treatment" groups and calculate the test statistic: this process is repeated many (typically thousands) times and the probability under the null hypothesis of "no difference" estimated by the proportion of times the test statistic from the randomized data sets exceeded the test statistic from the observed data set. To illustrate the process, I shall relate an investigation into genetic variation among populations of shad, a commercially important fish species.

To investigate geographic variation among populations of shad, data on mitochondrial DNA variation were collected from 244 fish distributed over 14 rivers. This sample size represented, for the time, a very significant output of effort. Ten mitochondrial haplotypes were identified with 62% being of a single type. The result was that almost all cells had less than 5 data points (of the 140 cells, 66% had expected values less than 1.0 and only 9% had expected values greater than 5). Following Cochran's rules for the χ^2 test, it was necessary to combine cells. This meant combining the genotypes into two classes, the most common one and all others. The calculated χ^2 for the combined data set was 22.96, which just exceeded the critical value (22.36) at the 5% level. The estimated value of χ^2 for the uncombined data was 236.5, which was highly significant ($P<0.001$) based on the χ^2 with 117 degrees of freedom. However, because of the very low frequencies within many cells, this result was suspect. Rather than combining cells and thus losing information, we (Roff and Bentzen 1989) used randomization (Chapter 5) to test if the observed χ^2 value was significantly larger than the expected value under the null hypothesis of homogeneity among the rivers. This analysis showed that the probability of obtaining a χ^2 value as large or larger than that observed for the ungrouped data was less than one in a thousand. Thus, rather than being merely marginally significant the variation among rivers was highly significant.

Most of the methods described in this book follow the frequentist school in asking "What is the probability of observing the set of n data $x_1, x_2, \ldots, x_n$ given the set of k parameters $\theta_1, \theta_2, \ldots, \theta_k$?" In Chapter 7 this position is reversed by the Bayesian perspective in which the question is asked "Given the set of n data $x_1, x_2, \ldots, x_n$, what is the probability of the set of k parameters $\theta_1, \theta_2, \ldots, \theta_k$?" This "reversal" of perspective is particularly important when management decisions are required. For example, suppose we wish to analyze the effect of a harvesting strategy on population growth: in this case the question we wish to ask is "Given some observed harvest, say x, what is the probability that the population rate of increase, say θ, is less than 1 (i.e., the population is declining)?" If this probability is high then it may be necessary to reduce the harvest rate. In Bayesian analysis, the primary focus is frequently on the probability statement about the parameter value. It can, however, also be used, as in the case of the James–Stein estimator, to improve on estimates. Bayesian analysis generally requires a computer-intensive approach to estimate the posterior distribution.

Why S-PLUS?

There are now numerous computer packages available for the statistical analysis of data, making available an array of techniques hitherto not possible except in some very particular circumstances. Many packages have some computer-intensive methods available, but most lack flexibility and hence are limited in use. Of the common packages, SAS and S-PLUS possess the breadth of programming capabilities necessary to do the analyses described in this book. I chose S-PLUS for three reasons. First, the language is structurally similar to programming languages with which the reader may already be familiar (e.g., BASIC and FORTRAN. It differs from these two in being object oriented). In writing the coding, I have attempted to keep a structure that could be transported to another language: this has meant in some cases making more use of looping than might be necessary in S-PLUS. While this increases the run time, I believe that it makes the coding more readable, an advantage that outweighs the minor increase in computing time. The second reason for selecting S-PLUS is that there is a version in the public domain, known as R. To quote the web site (http://www.r-project.org/), "R is a language and environment for statistical com-puting and graphics. It is a GNU project which is similar to the S language and environment which was developed at Bell Laboratories (formerly AT&T, now Lucent Technologies) by John Chambers and colleagues. R can be considered as a different implementation of S. There are some important differences, but much code written for S runs unaltered under R." The programs written in this book will, with few exceptions, run under R. The user interface is definitely better in

S-PLUS than R. My third reason for selecting S-PLUS is that students, at present, can obtain a free version for a limited period at http://elms03.e-academy.com/splus/.

Further reading

Although S-PLUS has a fairly steep learning curve there are several excellent text books available, my recommendations being:

Spector, P. (1994). *An Introduction to S and S-PLUS*. Belmont, California: Duxbury Press.

Krause, A. and Olson, M. (2002). *The Basics of S-PLUS*. New York: Springer.

Crawley, M. J. (2002). *Statistical Computing: An Introduction to Data Analysis using S-PLUS*. UK: Wiley and Sons.

Venables, W. N. and Ripley, B. D. (2002). *Modern Applied Statistics with S*. New York: Springer.

An overview of the language with respect to the programs used in this book is presented in the appendices.

2

Maximum likelihood

Introduction

Suppose that we have a model with a single parameter, θ, that predicts the outcome of an event that has some numerical value y. Further, suppose we have two choices for the parameter value, say θ_1 and θ_2, where θ_1 predicts that the numerical value of y will occur with a probability p_1 and θ_2 predicts that the numerical value of y w'ill occur with a probability p_2. Which of the two choices of θ is the better estimate of the true value of θ? It seems reasonable to suppose that the parameter value that gave the highest probability of actually observing what was observed would be the one that is also closer to the true value of θ. For example, if p_1 equals 0.9 and p_2 equals 0.1, then we would select θ_1 over θ_2, because the model with θ_2 predicts that one is unlikely to observe y, whereas the model with θ_1 predicts that one is quite likely to observe y. We can extend this idea to many values of θ by writing our predictive model as a function of the parameter values, $\varphi(\theta_i)=p_i$, where i designates particular values of θ. More generally, we can dispense with the subscript and write $\varphi(\theta)=p$, thereby allowing θ to take on any value. By the **principle of maximum likelihood** we select the value of θ that has the highest associated probability, p.

The important element of maximum likelihood estimation (often contracted to MLE) is that there is a definable probability function that can be used to generate the **likelihood** of the observed event. The most frequently used probability functions are the **normal distribution** and the **binomial distribution**.

There are three areas to be considered:

(1) **Point estimation.** Given some statistical model with k unknown parameters $\theta_1, \theta_2, \ldots, \theta_k$ how do we use MLE to obtain estimates of these parameters, denoted as $\hat{\theta}_1, \hat{\theta}_2, \ldots, \hat{\theta}_k$?

(2) **Interval estimation.** Having the set of estimates $\hat{\theta}_1, \hat{\theta}_2, \ldots, \hat{\theta}_k$ is only marginally useful, because we have no idea whether the estimates are

likely to be close to or far from the true values. In conjunction with point estimation we must, therefore, also estimate a confidence region for the estimates, typically 95%.

(3)   **Hypothesis testing.** In many instances, we are interested in testing hypotheses about the parameter values: for example, given two data sets we could test the hypothesis that they have a common mean. Maximum likelihood provides a mechanism to both compare different parameter values and to compare different statistical models.

Point estimation

Why the mean?

The underlying distribution of much of statistical estimation is the normal distribution (Figure 2.1). Under this distribution, the probability of observing a value, say x, is given by

$$\varphi(x) = \frac{1}{\sigma\sqrt{2\pi}} e^{-\frac{1}{2}\left(\frac{x-\mu}{\sigma}\right)^2} \tag{2.1}$$

where $\varphi(x)$ is called the **probability density function of x**. This function is symmetrical and characterized by two parameters μ and σ. Anyone who has had a first course in statistics will recognize these two as the "mean" and the "standard deviation," respectively. The mean is a measure of central tendency, and the standard deviation a measure of spread of the distribution (Figure 2.1). We typically estimate the parameter μ as the **arithmetic average**

$$\hat{\mu} = \frac{1}{n}\sum_{i=1}^{n} x_i \tag{2.2}$$

where n is the number of observations and x_i is the ith observation. The "hat" over μ indicates that this is an estimate of the true value of μ: this is a general symbol for the estimate of a parameter, but in the case of the average, we frequently use the symbol $\bar{x}$.

There are actually three measures of central tendency, the arithmetic average, the **mode** (the most commonly occurring value), and the **median** (the value that divides the sample into two equal portions). Why should we use the arithmetic average as the estimate of μ? The use of the arithmetic average as the preferred estimate of μ can be justified by the fact that it is the maximum likelihood estimate of μ. Suppose we have a sample of n observations

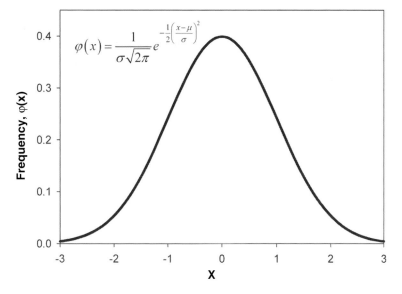

$$\varphi(x) = \frac{1}{\sigma\sqrt{2\pi}} e^{-\frac{1}{2}\left(\frac{x-\mu}{\sigma}\right)^2}$$

Figure 2.1 The normal distribution with $\mu=0$ and $\sigma=1$.

$x_1, x_2, x_3, \ldots, x_i, \ldots, x_n$: the probability of observing this sequence assuming a normal probability density function is

$$L = \varphi(x_1)\varphi(x_2)\varphi(x_3)\ldots\varphi(x_i)\ldots\varphi(x_n) = \prod_{i=1}^{n} \varphi(x_i) \tag{2.3}$$

where L is the likelihood of observing this particular sequence. We could consider all possible arrangements of the set of observations, but, as will become obvious, this does not change the final answer; so for notational convenience, we shall ignore this minor complication. Writing out the probability density function in full we have

$$L = \prod_{i=1}^{n} \varphi(x_i) = \prod_{i=1}^{n} \frac{1}{\sigma\sqrt{2\pi}} e^{-\frac{1}{2}\left(\frac{x_i-\mu}{\sigma}\right)^2} \tag{2.4}$$

Now, according to the maximum likelihood principle we should choose μ such that the likelihood, L, is maximized. To find this value, we could simply vary $\hat{\mu}$ and calculate the likelihood, selecting that value at which L is a maximum; how close we can get to the "best" value depends only upon the step size of our iteration. In many cases, and this is the reason for the computer-intensive nature of maximum likelihood estimators, this numerical approach is the only one available. However, in the present case, we can arrive at an exact solution by means of the calculus. Recall that to find the maximum or minimum of a function, we set the derivative of the function equal to zero. It is very inconvenient to work with derivatives of multiplicative functions. It is much easier if we take the logarithm of the likelihood. This does not change the result, because the turning point of the log transform is exactly the same as the untransformed value, so, taking the natural logs we have

$$\ln(L) = n \ln\left(\frac{1}{\sigma\sqrt{2\pi}}\right) - \sum_{i=1}^{n} \frac{1}{2}\left(\frac{x_i - \mu}{\sigma}\right)^2 \tag{2.5}$$

Differentiating

$$\frac{d\ln(L)}{d\mu} = 0 + \sum_{i=1}^{n}\left(\frac{1}{2}\right)\left(\frac{2}{\sigma^2}\right)(x_i - \mu) = \sum_{i=1}^{n}\frac{1}{\sigma^2}(x_i - \mu) \tag{2.6}$$

Setting the derivative equal to zero

$$\frac{d\ln(L)}{d\mu} = 0 \quad \text{when} \quad \frac{1}{\sigma^2}\sum_{i=1}^{n}(x - \mu) = 0 \tag{2.7}$$

After some simple algebraic rearrangement, we arrive at

$$\mu = \frac{1}{n}\sum_{i=1}^{n} x_i \tag{2.8}$$

which is the arithmetic average or mean (note that σ is irrelevant). At this point you may be concerned that we have μ exactly equal to the arithmetic average, whereas previously we asserted that it was only an estimate of μ (i.e., $\hat{\mu}$). The reason for the discrepancy is that we have treated the likelihood function as if it were exactly an algebraic relationship, whereas in any finite sample, the actual probability of the observed sequence will not be invariably maximal when μ is set equal to the arithmetic average. Suppose we take the extreme lower limit of a sample, that is, a sample of one; according to the above derivation the parameter μ (mean) is equal to the sample value, which is clearly nonsense, in general. Consider what happens as the number of observations in the sample increases. It is intuitively obvious that as n becomes larger and larger, the difference between the arithmetic average and μ becomes smaller and smaller, and in the limit, when n equals infinity, the arithmetic average is equal to μ (i.e., $\hat{\mu} \to \mu$ as $n \to \infty$). This resolves the problem: the derivation above implicitly assumes that the sample is very large, and, hence, for small samples the arithmetic average is only an estimate of μ ($\hat{\mu}$ or $\bar{x}$, depending upon your symbolic preference). This is a very important result, because it means that we cannot ignore the size of the sample. We shall return to this issue in the next section, "Interval estimation."

Nuisance parameters don't always disappear

In the previous example, there were two parameters, μ and σ, and we were interested only in μ. The parameter σ in this case is called a **nuisance parameter**, because it is of unknown quantity and could make the estimate uncertain if it does not drop out in the analysis. For the estimation of the mean, the nuisance parameter does drop out and is thus irrelevant in this instance. However, this is frequently not the case and we can be left with a joint estimation problem. Such is the problem when we use maximum likelihood to derive the best estimator for the second parameter of the normal distribution, the standard deviation, σ, or its square, the variance (σ^2).

Recall that the log-likelihood, $\ln(L)$ of the normal is

$$n \ln\left(\frac{1}{\sigma\sqrt{2\pi}}\right) - \sum_{i=1}^{n} \frac{1}{2}\left(\frac{x_i - \mu}{\sigma}\right)^2$$

(Eqn. (2.5)). Expanding this to make the differentiation more obvious gives

$$\ln(L) = -n\ln(\sigma) - n\ln(\sqrt{2\pi}) - \sum_{i=1}^{n} \frac{1}{2}\left(\frac{x_i - \mu}{\sigma}\right)^2 \tag{2.9}$$

As before, we differentiate $\ln(L)$ and set the result to zero

$$\frac{d\ln(L)}{d\sigma} = -\frac{n}{\sigma} + \sum_{i=1}^{n} \frac{2(x_i - \mu)^2}{2\sigma^3} = \frac{1}{\sigma}\left(-n + \sum_{i=1}^{n} \frac{(x_i - \mu)^2}{\sigma^2}\right) \tag{2.10}$$

$$= 0 \quad \text{when} \quad \frac{\sum_{i=1}^{n}(x - \mu)^2}{\sigma^2} = n$$

Upon rearrangement we have

$$\sigma^2 = \frac{1}{n}\sum_{i=1}^{n}(x_i - \mu)^2 \tag{2.11}$$

which is the readily recognized formula for the variance. As previously, the left-hand side should be indicated as an estimator, $\hat{\sigma}^2$, as it approaches the true value only as n becomes large. Annoyingly, to estimate the variance we have to know the exact value of the mean, which we do not. Thus, in this case, the nuisance parameter, μ, inconveniently remains in the estimation formula for σ. What can we do? One possibility is to substitute our estimate of the mean, giving $\hat{\sigma}^2 \approx (1/n)\sum_{i=1}^{n}(x_i - \hat{\mu})^2$. Because we know that, unless n is infinitely large, $\hat{\mu}$ is not exactly equal to μ, I have denoted this estimate as an approximation. In fact, it is a biased estimate, which could be problematical for small samples. Fortunately, this bias can be readily removed by rewriting the formula as

$$\hat{\sigma}^2 = \frac{1}{n-1}\sum_{i=1}^{n}(x_i - \hat{\mu})^2 \tag{2.12}$$

In many cases, the log-likelihood functions cannot be resolved into such simple, single-parameter formulae and then one must use numerical methods to locate the combination of point estimates that maximize the likelihood, and hence are the maximum likelihood estimators.

These two examples show that our use of the "standard" formulae for the mean and standard deviation are appropriate from the perspective of maximum likelihood *when the distribution is normal*. If the distribution is not normal, then we can still estimate the mean and standard deviation using these formulae, but we have no guarantee that they correctly estimate any particular parameter in the true probability density function.

Why we use least squares so much

Throughout a first course in statistics one comes across the use of "least-squares" estimation. It is, for example, the foundation of estimation in linear regression and analysis of variance. As with the commonly used estimators of mean and variance of a normal distribution, we can justify the use of least-squares by reference to maximum likelihood. To illustrate this, consider the simple linear regression equation (Figure 2.2)

$$y = \theta_1 + \theta_2 x + \varepsilon \tag{2.13}$$

The parameters θ_1 and θ_2 are the intercept and slope, respectively. They are frequently denoted as α and β, with estimated values of a and b, respectively.

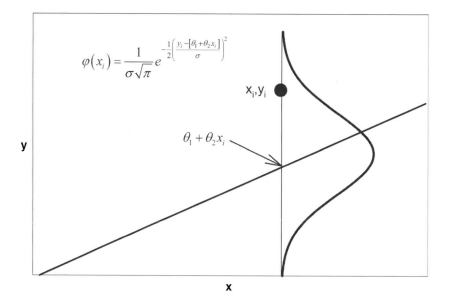

Figure 2.2 A regression line. The line is described by the equation $y = \theta_1 + \theta_2 x$. At each point along the line the data are distributed normally as $N(\theta_1 + \theta_2 x_i, \sigma)$.

To avoid the proliferation of confusing symbols (particularly as α and β are also used in context of type 1 and 2 errors), I shall use the **symbol "θ" as the general symbol for a parameter to be estimated**, noting if in specific cases the parameter typically has another symbol. The term ε refers to the essential assumption of linear regression, namely that the error about the line is normally distributed with a mean of zero (i.e., $\mu=0$) and an unspecified standard deviation (σ). I shall denote the normal distribution with mean μ and standard deviation σ as $N(\mu,\sigma)$. Thus, in the present example, we would say that ε is distributed as $N(0,\sigma)$. Using this notation, we can write that y is distributed as $N(\theta_1+\theta_2x,\sigma)$, which is to say that y is a normally distributed variable with mean $\theta_1+\theta_2x$ (the value of the line) and standard deviation σ (Figure 2.2). We can now assign a probability to observing some particular value y_i:

$$\varphi(y_i) = \frac{1}{\sigma\sqrt{2\pi}}e^{-\frac{1}{2}\left(\frac{y_i-[\theta_1+\theta_2x_i]}{\sigma}\right)^2} \tag{2.14}$$

Hence the probability or likelihood of observing some sequence $y_1,y_2,y_3,\ldots,y_i,\ldots,y_n$ is

$$L = \prod_{i=1}^{n}\varphi(y_i) = \prod_{i=1}^{n}\frac{1}{\sigma\sqrt{2\sigma}}e^{-\frac{1}{2}\left(\frac{y_i-[\theta_1+\theta_2x_i]}{\sigma}\right)^2} \tag{2.15}$$

As before, and as in general, it is more convenient to work with the natural logarithms

$$\ln(L) = -n\ln(\sigma\sqrt{2\pi}) - \frac{1}{2\sigma^2}\sum_{i=1}^{n}(y_i - [\theta_1+\theta_2x_i])^2 \tag{2.16}$$

According to the principle of maximum likelihood, the best estimates of θ_1 and θ_2 are those that maximize the log-likelihood, which will be those values at which the summation $SS = \sum_{i=1}^{n}(y_i - [\theta_1+\theta_2x_i])^2$ is minimized. This, of course, is the **least squares procedure**. Note that, as with the estimation of the mean of the normal distribution, the variance about the regression line does not enter into the estimation of the slope (θ_1) or intercept (θ_2) of the line. For the sake of completeness, let us calculate the least squares estimation equations for both parameters. We have to carry out two differentiations, one with respect to θ_1 and another with respect to θ_2

$$\frac{dSS}{d\theta_1} = -2\sum_{i=1}^{n}(y_i - [\theta_1+\theta_2x_i])$$

$$\frac{dSS}{d\theta_2} = -2\sum_{i=1}^{n}(y_i - [\theta_1+\theta_2x_i])x_i \tag{2.17}$$

Setting these to zero, and noting that $\sum_{i=1}^{n} \theta_1 = n\theta_1$, we have the simultaneous equations

$$\sum_{i=1}^{n} y_i = \theta_1 n + \theta_2 \sum_{i=1}^{n} x_i$$

$$\sum_{i=1}^{n} x_i y_i = \theta_1 \sum_{i=1}^{n} x_i + \theta_2 \sum_{i=1}^{n} x_i^2$$

(2.18)

Multiplying the first by $\sum_{i=1}^{n} x_i$, the second by n and subtracting gives the estimate for θ_2

$$\hat{\theta}_2 = \frac{\sum_{i=1}^{n} (x_i - \bar{x})(y_i - \bar{y})}{\sum_{i=1}^{n} (x_i - \bar{x})^2}$$

(2.19)

Note that the estimate is a function of the arithmetic means of x and y (for convenience I have used their usual bar notation, but could have written them as $\hat{\mu}_x$ and $\hat{\mu}_y$, respectively). We can use the two simultaneous equations to estimate θ_1 or more simply make use of the relationship $\theta_1 = \bar{y} - \theta_2 \bar{x}$. These equations point out the fact that, because they use the same data, the estimates of the two parameters are not independent of each other. This is a general statement about multiple estimates from the same statistical model and usually poses no problem. But suppose we estimate the same regression, say fecundity on body weight, from a number of different populations and then calculate the correlation between $\hat{\theta}_1$ and $\hat{\theta}_2$, on finding a significant correlation between the two, we might be persuaded to interpret this to be due to some biologically meaningful cause, when it probably arises as a statistical artifact.

More on least squares

The maximum likelihood approach to the estimation of the parameters of the simple linear regression indicates that the least squares approach is justified. Further, in this instance, exact estimation equations for the two parameters can be found. Here, I present a more complex example in which there are several least squares solutions, depending upon the assumption of the error structure.

Many organisms, particularly fish, grow continuously throughout their lives but the rate of growth slows with age (Figure 2.3). A growth curve that captures this behavior is known as the von Bertalanffy growth model (though the physiological basis von Bertalanffy actually used to derive the curve is fallacious). Ignoring any source of error or variation, the length at some age t is given by the relationship

$$l_t = \theta_1(1 - e^{-\theta_2(t - \theta_3)})$$

(2.20)

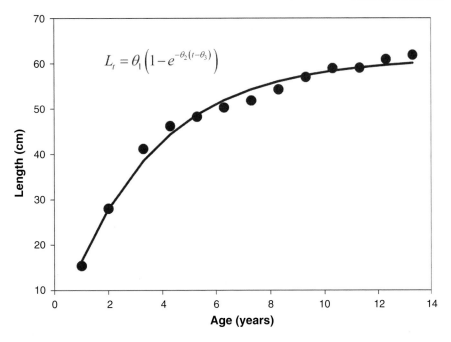

$$L_t = \theta_1 \left(1 - e^{-\theta_2(t-\theta_3)}\right)$$

Figure 2.3 Plot of average length at each age for female Pacific hake, with the estimated von Bertalanffy curve (Data from Kimura 1980).

Age	1.0	2.0	3.3	4.3	5.3	6.3	7.3	8.3	9.3	10.3	11.3	12.3	13.3
Length	15.4	28.0	41.2	46.2	48.2	50.3	51.8	54.3	57.0	58.9	59.0	60.9	61.8

where θ_1 is the asymptotic length (generally denoted as L_∞), θ_2 is the rate of growth (generally denoted as k), θ_3 is the hypothetical age at which the length is zero (generally denoted as t_0). Now the above equation can also be written in terms of the length at two consecutive ages, say t and $t+1$

$$l_{t+1} = \theta_1(1 - e^{-\theta_2}) + e^{-\theta_2}l_t \tag{2.21}$$

which suggests a regression method of estimating θ_1 and θ_2. The natural logarithm of the slope of the regression of l_{t+1} on l_t is an estimate of $-\theta_2$ and the intercept is an estimate of $\theta_1(1 - e^{-\theta_2})$. This method is known as the Ford–Walford method and provides excellent initial estimates of the parameter values (except for θ_3). The statistical problem with this method is that the length at each age successively changes from being the dependent variable (l_{t+1}) to being the independent variable (l_t), which invalidates the basic assumption of linear regression.

How can we use the maximum likelihood method to obtain parameter estimates of the von Bertalanffy equation (Eq. (2.20))? To use this method we have to decide how to incorporate ε, the normally distributed error term. The simplest method is to tack it onto the equation in the same manner as in the linear regression model (for alternate models see Kimura 1980)

$$l_t = \theta_1(1 - e^{-\theta_2(t-\theta_3)}) + \varepsilon \tag{2.22}$$

Now following the method discussed in the previous section, the log-likelihood function is

$$\ln(L) = -n\ln(\sigma\sqrt{2\pi}) - \frac{1}{2\sigma^2}\sum_{t=1}^{n}(l_t - \theta_1[1 - e^{-\theta_2(t-\theta_3)}])^2 \tag{2.23}$$

It should be readily apparent that the log-likelihood function will be maximized when, as with the linear regression model, the sum of squares between the observed (l_t) and predicted ($\theta_1[1 - e^{-\theta_2(t-\theta_3)}]$) values are minimized, i.e., the least squares solution. Assuming that the variance, σ^2, is constant, there are two scenarios to be considered. First, as assumed in the linear regression model, we can posit that the distribution of individual observations, l_t, is $N(0,\sigma)$. Second, we can focus upon the mean values and posit that the distribution of mean lengths at a given age is $N(0,\sigma)$. With either scenario, we estimate the parameters using least squares, but in the first case, we use the individual values, whereas in the second we use the means. In the above derivation, the variance dropped out: we obtain the MLE of σ^2 in the usual manner by taking the partial derivative of the log-likelihood function (Eq. (2.23)) with respect to σ^2 and setting this equal to zero. It is left to the reader to verify that

$$\hat{\sigma}^2 = \sum_{t=1}^{n}\left(l_t - \hat{\theta}_1[1 - e^{-\hat{\theta}_2(t-\hat{\theta}_3)}]\right)^2 \Big/ n \tag{2.24}$$

i.e., the mean residual sum of squares of the best fitting model (the MLE). However, as with the variance estimate of the normal distribution previously discussed, the above estimate is biased by the use of the parameter estimates rather than their true values. To remove this bias, we divide not by n but by n minus the number of estimated parameters (i.e., $n-3$ in the present case).

Unlike the linear regression model, there is no exact analytical solution for the above model and hence one must resort to numerical methods. Fortunately, virtually every statistical package has a nonlinear fitting routine using least squares. The curve shown in Figure 2.3 was obtained using the "regression wizard" in SigmaPlot. The routine failed to converge to a solution when I used Eq. (2.20), but changing the equation to $l_t = \theta_1(1 - e^{-\theta_2 t+\theta'_3})$, which in no way alters the structure of the equation (the parameter θ_3' is merely the product $\theta_2\theta_3$), produced the fit shown in Figure 2.3. A similar failure occurred with the nonlinear routine in SYSTAT but not in S-PLUS (the equation can be fitted using dialog boxes or the command: `nls(LENGTH~b1*(1-exp(-b2*(AGE-b3))),data=D,` `start=list(b1=50,b2=.1,b3=.1)))`. The parameter estimates from S-PLUS when using the altered equation were identical to those from SigmaPlot, indicating that when convergence is obtained both packages do get the same solution, which is not guaranteed, although solutions should always be similar. I could not get convergence in the SYSTAT routine unless I deleted θ_3 entirely from the equation. Because θ_3 is a very small value (-0.057), its deletion from the equation changes little, but the message to draw from the three analyses is that different statistical packages perform differently (and I do not mean to imply that the ranking in performance can be judged from a single example) and that small changes to the equation to be fitted can make big differences in the ability of the routine to converge to a solution.

Generalizing the MLE approach in relation to least squares estimation

Let $y = \varphi(\theta_1, \theta_2, \ldots, \theta_k, x)$ be a model comprising k parameters (e.g., linear regression has two, the von Bertalanffy function has three) and an independent variable, x. Assume that the error term, ε, is $N(0, \sigma)$ and the observed value y_i can be predicted from

$$y_i = \varphi(\theta_1, \theta_2, \ldots, \theta_k, x_i) + \varepsilon \tag{2.25}$$

The log-likelihood function is thus

$$\ln(L) = -n \ln(\sigma\sqrt{2\pi}) - \frac{1}{2\sigma^2} \sum_{i=1}^{n} (y_i - \varphi(\theta_1, \theta_2, \ldots, \theta_k, x_i))^2 \tag{2.26}$$

The maximum likelihood estimates of the k parameters are obtained by minimizing the sum of the squared difference between the observed and predicted values, i.e.,

$$\text{Minimize} \sum_{i=1}^{n} (\text{Observed value} - \text{Predicted value})^2 \tag{2.27}$$

(see, for example, exercise 2.5). But remember, you are making an assumption about how the variability about the predicted value is distributed! Under the assumption above, the residuals should be normally distributed with mean zero and variance σ^2. Checking this assumption is standard practice in linear regression analysis and the same methods apply to the general case.

Leaving normality

The fundamental assumption of the maximum likelihood approach is that the data are distributed according to a known distribution. This need not be the normal distribution. Consider the situation in which there are two outcomes: for example, in a mate selection experiment, a female might be presented with two choices (e.g., in acoustically orienting animals, two different songs might be played and the female choice recorded), or in a genetical study there might be two alleles at a particular locus of interest. Let the probability of the first outcome be p, in which case the probability of the second is $1-p$. Given two outcomes, the resulting distribution can be described by the **binomial probability function**: the probability or likelihood of observing the first outcome r times in n trials is thus

$$L = \frac{n!}{r!(n-r)!}p^r(1-p)^{n-r} \tag{2.28}$$

Taking the natural logarithms gives a more easily handled equation

$$\ln(L) = \ln\left(\frac{n!}{r!(n-r)!}\right) + r\ln(p) + (n-r)\ln(1-p) \tag{2.29}$$

To find the maximum likelihood estimate of p we differentiate $\ln(L)$ with respect to p and set the result to zero:

$$\frac{d\ln(L)}{dp} = \frac{r}{p} - \frac{n-r}{1-p}$$

$$\frac{d\ln(L)}{dp} = 0 \quad \text{when} \quad \frac{r}{p} = \frac{n-r}{1-p} \tag{2.30}$$

$$\text{i.e., } \hat{p} = \frac{r}{n}$$

which is the intuitively obvious estimate. Note that I have substituted $\hat{p}$ for p in the final equation; it is very important to remember that our estimate approaches the true value only as the sample size increases.

Multiple likelihoods: estimating heritability using the binomial and MLE

Heritability is a parameter that defines the degree to which offspring resemble their parents due to the additive effect of genes. It is a parameter that can be used to predict how much a quantitatively varying trait such as body weight changes when selection is applied to a population. The predictive equation, known as the breeder's equation, is $R=\theta S$, where R is the response to selection (the difference between the means of the population in each

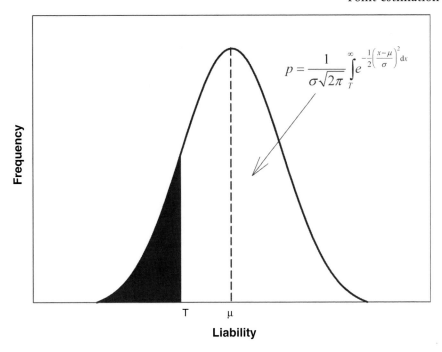

$$p = \frac{1}{\sigma\sqrt{2\pi}} \int\limits_{T}^{\infty} e^{-\frac{1}{2}\left(\frac{x-\mu}{\sigma}\right)^2} dx$$

T   μ

Liability

Figure 2.4 Illustration of the threshold model. The underlying trait, called the liability, is normally distributed. Individuals above the threshold, T, display one morph, whereas individuals below the threshold display the alternate.

generation), θ is the heritability (invariably designated as h^2: despite this, for consistency, I shall still use θ), and S is the selection differential (the difference between the population and parental means). Heritability can be estimated after one generation of selection by rearranging the breeder's equation to give $\theta = R/S$. In general, a single generation of selection is insufficient to produce a reliable estimate, but this problem is ignored here for the purpose of clarity.

There is a class of traits, known as threshold traits, which are peculiar in that they are manifested as dichotomous traits, but breeding experiments show them to be determined by the action of many genes. Examples of threshold traits include twins versus singletons in sheep, certain diseases such as schizophrenia, the phenomenon of "jacking" in salmon (early maturation at a very reduced size), wing and horn dimorphism in certain insect species. To account for the quantitative inheritance pattern in these traits, a threshold model has been proposed: according to this model there is a continuously distributed trait termed the liability and a threshold of sensitivity. If the value of the liability lies above the threshold, one outcome results, whereas if the value lies below the threshold, the alternate outcome results (Figure 2.4). The liability is assumed to be normally distributed and thus the proportion, p, lying above the threshold is given by

$$p = \frac{1}{\sigma\sqrt{2\pi}} \int_T^\infty e^{-\frac{1}{2}\left(\frac{x-\mu}{\sigma}\right)^2} dx \tag{2.31}$$

where the liability is distributed as $N(\mu,\sigma)$ and T is the threshold value. Without loss of generality, we can rescale the above by setting $\sigma=1$ and $T=0$, leaving us only a single parameter, the mean liability (μ) to calculate. If $p=0.5$ the mean liability equals 0, whereas if $p=0.8$ the mean liability is equal to -0.84. Suppose we subject a threshold trait to selection by taking as parents only those of a designated morph (e.g., only winged individuals in a wing-dimorphic species). We can arbitrarily designate these individuals as lying above the threshold, in which case their mean liability is the mean of a truncated normal distribution and is

$$\mu + \frac{e^{-\frac{1}{2}\mu^2}}{p\sqrt{2\pi}} = \mu + \frac{\varphi(\mu)}{p}$$

Letting the number of the designated morph in the parental generation be r_0 and the total sample size be n_0, we can write the likelihood using the binomial formula given in Eq. (2.28), which I shall designate as L_0.

Using the breeder's equation we can predict the mean liability in the offspring generation as

$$\mu_1 = \mu_0(1-\theta) + \theta\left(\mu_0 + \frac{\varphi(\mu_0)}{p}\right) \tag{2.32}$$

and hence the predicted proportion of the designated morph. From the selection experiment, we observe the proportion of the designated morph in each generation but directly observe neither the mean liability nor the parameter we wish to estimate, the heritability of the liability. We thus have two parameters to estimate, the one of interest (heritability) and a "nuisance" parameter (mean liability of the initial population). As in the parental generation, the likelihood of obtaining the observed number, r_1, of the designated morph from the observed offspring sample, n_1, is given by the binomial example (Eq. (2.28)). Designate this likelihood L_1. We have two likelihoods, L_0 and L_1. The overall likelihood, L_{01}, is simply the product of the two likelihoods (sum of the log-likelihoods). Therefore, we find the combination of μ_0 and θ that maximize $\ln(L_{01})$.

Suppose in the first sample we observe 50 of the designated morph out of a total of 100 individuals. We then use the designated morph as parents for the next generation obtaining 68 offspring of the designated morph out of a total sample of 100 offspring. Thus $r_0=50$, $n_0=100$, $r_1=68$, $n_1=100$. Estimates of p and

θ can be obtained using the S-PLUS routine `nlminb`, which allows for restriction of parameter values, in this case between 0 and 1 (it is possible to use the unrestricted minimization function `nlmin` but a warning is issued as the search routine takes parameter values below zero causing an error in the routine `qnorm`). Most packages minimize a function rather than maximize and hence we use the negative log-likelihood function. Because constants in the likelihood function do not affect the value of the estimates, as a simplification, these are generally dropped from the analysis. Suitable S-PLUS coding to estimate p and θ is given in Appendix C.2.1.

The heritability used to generate the numbers for the offspring generation was 0.6 (giving an expected number of 68 as actually used): the output from S-PLUS is 0.500000 for p (which is simply r_0/n_0) and 0.5861732 for the heritability (θ or h^2).

Logistic regression: connecting the binomial with regression

Suppose that the probability of an event occurring is a function of some other variable x. For example, the proportion of insects killed by an insecticide would be expected to increase with the dosage to which they are exposed (Figure 2.5). We might be inclined to model this relationship using the simple additive model, $p = \theta_1 + \theta_2 x$. The problem with this model is that it does not restrict p within the required range of 0 to 1, and as can be seen in the example, the shape of the curve is sigmoidal. This shape is a natural form (but not the only possibility) when the upper and lower limits are bounded. What we require is a model in which there is a lower bound to p that is equal to zero and an upper bound to p of 1. A model that satisfies these requirements is the logistic equation

$$p_i = \frac{e^{\theta_1 + \theta_2 x_i}}{1 + e^{\theta_1 + \theta_2 x_i}} \tag{2.33}$$

This equation can be linearized by the transformation

$$\ln\left(\frac{p_i}{1 - p_i}\right) = \theta_1 + \theta_2 x_i \tag{2.34}$$

The left-hand side is termed the **logit**, and is a contraction of the phrase "**logistic unit.**" It is also known as the **log odds**. Equation (2.34) provides a simple means of graphically representing the data and crudely estimating the parameter values. For obvious reasons, the method of maximum likelihood is to be preferred. To obtain the log-likelihood function, we first note that Eq. (2.29) can be rearranged as

$$\ln(L) = \ln\left(\frac{n!}{r!(n-r)!}\right) + n\ln(1 - p) + r\ln\left(\frac{p}{1 - p}\right) \tag{2.35}$$

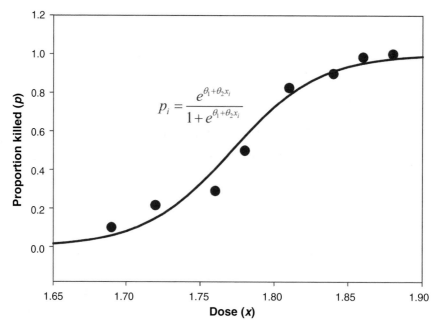

Figure 2.5 Plot of beetle mortality vs. dose of gaseous carbon disulphide. Solid line shows fitted curve (see Appendix C.2.2 for coding. Data from Dobson 1983 after Bliss 1935).

Dose	1.69	1.72	1.76	1.78	1.81	1.84	1.86	1.88
N	59	60	62	56	63	59	62	60
R	6	13	18	28	52	53	61	60

The log-likelihood function for the logistic is then

$$\ln(L) = \sum_{i=1}^{N} \left[\ln\left(\frac{n_i!}{r_i!(n_i - r_i)!}\right) - n_i \ln(1 + e^{\theta_1 + \theta_2 x_i}) + r_i(\theta_1 + \theta_2 x_i) \right] \tag{2.36}$$

The summation sign is required because we have N observations, the ith observation consisting of r_i "successes" in n_i "trials" (e.g., r_i is the number of individuals killed when n_i individuals are subjected to dose i of insecticide). Estimates of the parameters θ_1 and θ_2 are obtained by minimizing $\ln(L)$. Logistic regression can include more than one explanatory variable (e.g., in the insecticide example we might include body size as a second variable), and is so widely used that most statistical packages include **logistic regression** as a specific option and it is only necessary to give the linear component of the model (note that in SYSTAT the dialog box for logistic regression is named "logit").

But most programs expect the data to be in the form of one row per individual with the dependent variable (e.g., mortality) being categorical: in the case of the beetle data, we would code an individual as 0 if alive and 1 if dead. This is a convenient method of coding if there are several explanatory variables (e.g., dose and body size) but is definitely a nuisance for the present example. The data can still be analyzed in the tabulated form but it is necessary to specify the log-likelihood function. Programs frequently minimize rather than maximize functions and hence it is necessary to use minus log-likelihood. SYSTAT calls the function to be minimized the LOSS function. Because the term $\sum_{i=1}^{N} \ln(n_i!/(r_i!(n_i - r_i)!))$ is a constant it can be omitted. In SYSTAT one supplies the model function (e.g., `DEATHS=SAMPLE*exp(b1+b2*DOSE)/(1+exp(b1+b2*DOSE))`, where DEATHS is r_i, SAMPLE is n_i, DOSE is x_i, and b1, b2 are θ_1, θ_2, respectively) and the loss function (e.g., `LOSS = -(DEATHS*(b1+b2*DOSE)-SAMPLE*LOG(1+exp(b1 +b2*DOSE))))`), both of which can be done via dialog boxes. In S-PLUS it is necessary to write a function and use the nonlinear minimizing routine (see Appendix C.2.2 for coding and for an alternative approach that uses 0,1 data and the general linear model routine `glm`, see Appendix C.2.8).

From binomial to multinomial

In many cases, there are more than two possible outcomes (e.g., several alleles at a locus). Such a distribution is said to be multinomial. Suppose we have a sample consisting of a set of categories, such as an age sample of animals: using the multinomial distribution the likelihood for the sample is

$$L = \frac{n!}{x_1 x_2 \ldots x_k} p_1^{x_1} p_2^{x_2} \ldots p_k^{x_k} = \frac{n!}{\prod_{i=1}^{k} x_i} \prod_{i=1}^{k} p_i^{x_i} \tag{2.37}$$

where x_i is the number of observations in the ith category (e.g., age class) and p_i is the true proportion in the ith class. The log-likelihood function is

$$\ln(L) = \ln(n!) - \sum_{i=1}^{k} \ln(x_i) + \sum_{i=1}^{k} x_i \ln(p_i) \tag{2.38}$$

To find the maximum likelihood estimates $(\hat{p}_1, \hat{p}_2, \ldots, \hat{p}_k)$, we can proceed by differentiating and setting the result to zero, but an easier approach is as follows: the probability of an animal being in age class i is p_i and hence the probability that it is not in age class i is $1 - p_i$. Thus from this perspective we have a simple binomial distribution, all age classes except age class i being collapsed into one. Hence the maximum likelihood estimator for age class i is simply x_i/n.

Combining simulation and MLE to estimate population parameters

Hooded seals are commercially harvested and hence it is essential to be able to predict the result of particular harvesting strategies on population rates of increase or decline. To do this analysis, five population parameters are required:

(1)   The number of pups produced at the starting point of the projection (1945),

(2)   The instantaneous rate of natural mortality, M (i.e., probability of surviving each year is e^{-M}), and

(3)   The proportions of 4, 5, and 6-year-old females that breed (none breed earlier than 3 years and all breed by 7 years).

The available data on hooded seals consisted of age distributions collected each year from 1972 to 1978. Jacobsen (1984) constructed a simulation model of hooded seal population dynamics which generated age distributions for the years 1945–1986. Now consider the year 1972: for this year there is an observed distribution of ages. *Taking the simulation to be the model* we can construct the log-likelihood function using the predicted and observed age distributions

$$\ln(L_{1972}) = \ln(n_{1972}!) - \sum_{i=1}^{k} \ln(x_{i,1972}) + \sum_{i=1}^{k} x_{i,1972} \ln(p_{i,1972}) \tag{2.39}$$

where n_{1972} is the number of seals sampled in 1972, $x_{i,1972}$ is the number of seals of age i in the 1972 sample and $p_{i,1972}$ is the proportion of seals of age i in 1972 predicted by the simulation model. The first two terms are constant and hence maximizing the log-likelihood is accomplished by maximizing the third term $\sum_{i=1}^{k} x_{i,1972} \ln(p_{i,1972})$. Each simulation run with a given combination of the five unknown parameters will produce a likelihood value. The preferred combination of parameter values is that which maximizes the log-likelihood.

As previously noted, likelihood values are multiplicative: thus the likelihood for all the years 1972 to 1978 is $L_{1972-1978} = L_{1972}L_{1973}\ldots L_{1978}$, and the preferred set of population parameters is the set that maximizes $\sum_{j=1972}^{1978} \sum_{i=1}^{k} x_{i,j} \ln(p_{i,j})$.

Whereas the general principle set out above is correct, the execution proved troublesome, because different combinations of initial pup production and natural mortality rate produce almost identical age distributions but widely divergent population projections (Figure 2.6, Table 2.1). The reason is that population size is not itself constrained: the observed changing age distribution can be modeled by an initially small population (1945 pup production) and a low mortality rate or by a very large initial population and a high mortality rate (Table 2.1). There is thus a ridge of likelihood values corresponding to a positive relationship between 1945 pup production and natural mortality. Unfortunately,

Table 2.1 *Maximum likelihood estimates of the five parameters (pup production in 1945, natural mortality, partial recruitment to the breeding stock of females aged 4, 5, and 6 years). Note that the partial recruitment values change little but that there is a positive correlation between pup production and natural mortality. Taken from Jacobsen (1984)*

| Pup production in 1945 | Natural mortality | Partial recruitment at | | | Log-likelihood |
		4 years	5 years	6 years	
64350	0.08	0.40	0.74	0.98	−8308
79250	0.10	0.40	0.74	0.97	−8306
98900	0.12	0.41	0.75	0.96	−8304
125500	0.14	0.42	0.75	0.95	−8302
162500	0.16	0.42	0.74	0.94	−8301

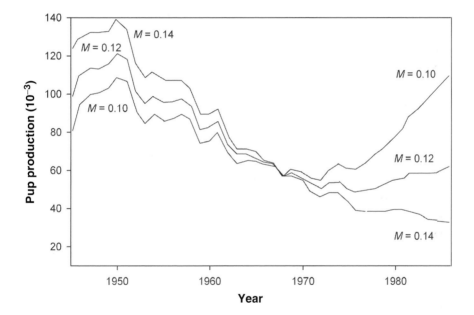

Figure 2.6 Predicted pup production estimates of hooded seals in the West Ice for the years 1945–1986 for three combinations of natural mortality (M) and initial pup production that have virtually identical, and maximum, likelihood (Table 2.1). Redrawn from Jacobsen (1984).

future population trajectories vary very widely (Figure 2.6), from increasing (hence a sustainable catch rate) to decreasing (hence an unsustainable catch rate). To constrain the estimates we need at least one population estimate either in the early years (e.g., between 1945 and 1960) or the later years (e.g., between 1975 and

1985): but note that an estimate around 1968 would be of no use, because all the trajectories converge at that year. Although the maximum likelihood method in this instance cannot give good estimates of the parameter values, it does very clearly demonstrate the problem with the available data and indicates the type of information that is required.

Interval estimation

Method 1: an exhaustive approach

Having an estimate is of little use if we do not also have a measure of confidence in the estimate. The usual measure is the 95% confidence interval. How can we calculate this interval for the likelihood estimate? Consider the likelihood function for a single variable, θ, plotted against θ. Each likelihood represents the relative support for the particular value of θ that generates this likelihood. Thus, for example, if the likelihood at θ_i is 0.1 and that at θ_j is 0.05, we can state that θ_i has twice the support of θ_j. If we divide throughout by the area under the distribution, we arrive at a distribution whose total area is equal to 1, and the 95% confidence region can be approximated by the values of θ that cut off the tails of the distribution at 0.025 and 0.975. Suppose, for example, we wished to estimate the confidence limits for the mean, the likelihood for a given value of μ, say μ_j, is

$$L(\mu_j) = \prod_{i=1}^{n} \frac{1}{\sigma\sqrt{2\pi}} e^{-\frac{1}{2}\left[\frac{x_i - \mu_j}{\sigma}\right]^2} \tag{2.40}$$

Because σ is the same for all values of μ_j, we can assign it any value we choose (1 is the simplest value). Similarly, $\sqrt{2\pi}$ is a constant and hence can be dropped. Therefore, our modified function, $L^*(\mu_j)$ is

$$L^*(\mu_j) = \prod_{i=1}^{n} e^{-\frac{1}{2}(x_i - \mu_j)^2} \tag{2.41}$$

Now we calculate $L^*(\mu_j)$ between limits that enclose most of the probability distribution (this can be done by trial and error; it is better to have a very large range rather than a small range) and iterate using some pre-assigned step length, so we have a series of equally spaced parameter values, $\mu_1, \mu_2, \ldots, \mu_N$, where μ_1 is the smallest value and μ_N is the largest. We then divide each $L^*(\mu_j)$ by the sum of all values, giving

$$L^*_S(\mu_j) = \frac{L^*(\mu_j)}{\sum_{i=1}^{N} L^*(\mu_i)} \tag{2.42}$$

We calculate the cumulative sum of the above $L_{\text{cum},k} = \sum_{j=1}^{k} L_S^*(\mu_j)$, where k ranges from 1 to N. The upper and lower confidence values are then the values of μ at which $L_{\text{cum},k}=0.025$ and $L_{\text{cum},k}=0.975$, respectively.

The above method is numerically intensive but rigorous in that it makes no assumption about the actual distribution of the likelihood. It can be extended to any number of parameters. For example, if there are two parameters to be estimated (θ_1, θ_2), we would vary both parameters and the result would be a bivariate confidence region.

Method 2: the log-likelihood ratio approach

For large samples, we can make use of the fact that the sampling distribution of log-likelihood function is approximately

$$2(LL(\hat{\theta}_1, \hat{\theta}_2, \ldots, \hat{\theta}_k) - LL(\theta_1, \theta_2, \ldots, \theta_k)) \sim \chi_k^2 \tag{2.43}$$

where $LL(\hat{\theta}_1, \hat{\theta}_2, \ldots, \hat{\theta}_k)$ is the log-likelihood at the maximum likelihood estimators and $LL(\theta_1, \theta_2, \ldots, \theta_k)$ is the log-likelihood at the true parameter values. Confidence regions can be approximated by the set of parameter combinations that lie $\chi_k^2/2$ units distant from $LL(\hat{\theta}_1, \hat{\theta}_2, \ldots, \hat{\theta}_k)$. Thus, for a model with one parameter, the confidence range is given by the two log-likelihoods that give a value of $\frac{1}{2}\chi_1^2 = 1.92$. To illustrate the procedure, consider the problem of estimating the heritability of a threshold trait discussed previously. For simplicity, we shall assume that the initial proportion is 0.5 and we have a single parameter, θ (=h^2), for which to provide confidence limits. To obtain the lower and upper confidence values, a simple approach is as follows (see Appendix C.2.3 for S-PLUS coding):

Step 1: Find the MLE for θ $(= \hat{\theta})$

Step 2: Calculate the log-likelihood (i.e., $LL(\hat{\theta})$)

Step 3: Iterate over a range of θ_i (e.g., 0.01–0.99) and for each calculate $LL(\theta_i)$

Step 4: Calculate Diff $= LL(\hat{\theta}) - LL(\theta_i) - 0.5\chi_1^2$

Step 5: Find value of θ_i at which Diff is equal to zero. There will be two values, corresponding to the upper and lower confidence limits. These values can be found graphically, as shown in Figure 2.7, or numerically, as shown in Appendix C.2.3 (also see exercise 2.8).

For the von Bertalanffy equation, there are four parameters (three θs and σ). With more than two parameters the confidence region cannot be visualized. To obtain a visual picture, we can proceed as follows, using the von Bertalanffy

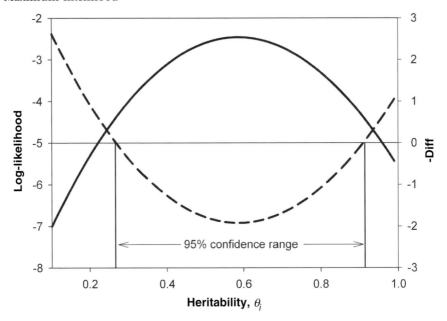

Figure 2.7 Plot of log-likelihood vs. heritability. Dotted line shows the negative of Diff = $LL(\hat{\theta}) - LL(\theta_i) - 0.5\chi_1^2$, where θ_i is a heritability estimate. The values at which Diff = 0 demark the lower and upper 95% confidence interval.

function as an example (see Appendix C.2.4 for coding). The two parameters of most interest are θ_1, the asymptotic length, and θ_2, the growth rate.

Step 1: Find the MLEs for all four parameters, which I shall refer to as the **global MLEs**

Step 2: Iterate over a range of values of the two parameters of interest (θ_1 and θ_2)

Step 3: At each iteration find the MLE for θ_1 and θ_2, keeping the other two parameters at their global MLE. Designate these new MLE values as θ_1^*, θ_2^*.

Step 4: Calculate $2(LL(\hat{\theta}_1, \hat{\theta}_2, \hat{\theta}_3, \hat{\sigma}) - LL(\theta_1^*, \theta_2^*, \hat{\theta}_3, \hat{\sigma})) - \chi_2^2$, where, for the 95% region, $\chi_2^2 = 5.991$

Step 5: Use a numerical method to construct the contour line corresponding to zero. This line demarcates the 95% confidence region (Figure 2.8).

Method 3: a standard error approach

A simple but approximate method of assessing the variability in a parameter value is to examine the standard errors of the estimates; roughly speaking, the 95% confidence limits are ±2 the standard errors. The variance of

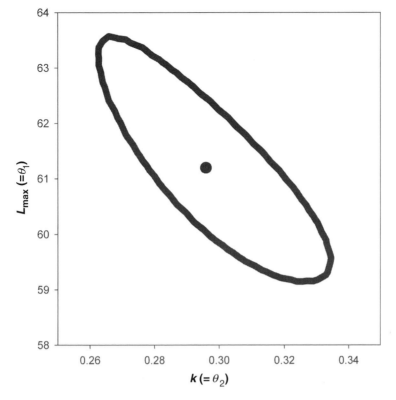

Figure 2.8 The 95% confidence ellipse for the estimates L_{max} ($=\theta_1$) and k ($=\theta_2$) generated by conditioning on t_0 ($=\theta_3$) and σ. Dot shows MLE combination. Coding that generated the data matrix from which the contour was estimated is given in Appendix C.2.4.

a parameter is approximately equal to the negative of the inverse of the second derivative at the maximum likelihood point

$$\sigma_\theta^2 = -\left(\frac{\partial^2 LL}{\partial \theta^2}\right)^{-1} \tag{2.44}$$

For example, in the case of the mean of a normal distribution, the variance of the mean, σ_μ^2 (obtained by taking the derivative of Eq. (2.6)) is

$$\sigma_\mu^2 = -\left(\sum_{i=1}^{n} -\frac{1}{\sigma^2}\right)^{-1} = \frac{\sigma^2}{n} \tag{2.45}$$

and the standard error is thus $\sigma/\sqrt{n}$, which is the well-known formula. When there are several estimated parameters, the estimation is somewhat more tricky as it is necessary to invert the matrix of second partial derivatives. Fortunately, the standard errors of parameter estimates are typically given in the output from statistical packages, along with an approximate t-test for $\theta = 0$, computed as $\hat{\theta}/\hat{\sigma}_\mu$. The output from S-PLUS for fitting the von Bertalanffy growth function is shown in Appendix C.2.5. In addition to the standard errors, the output also includes the correlation matrix: this can be useful to examine the independence of the parameter estimates. In the present example, the parameter $k(=\theta_2)$ is highly correlated with both $L_\infty(=\theta_2)$, and $t_0(=\theta_3)$ and thus variation in L_∞ and k cannot be considered independently, a point made clear by the bivariate contour interval (Figure 2.8).

Hypothesis testing

There are two basic questions we have to answer having fitted a model: first, is the model a poor fit to the data, and, second, does the model explain significantly more variation than a model with fewer parameters? In both cases, we make use of the chi-square distribution introduced in the previous section.

Testing model fit

The adequacy of a model is defined in relation to a model that has the same number of parameters as observations and thus completely describes the data. This model is known as the **maximal** or **saturated model**. For example, consider the log-likelihood function for the mean

$$\ln(L) = n \ln\left(\frac{1}{\sigma\sqrt{2\pi}}\right) - \sum_{i=1}^{n} \frac{1}{2}\left(\frac{x_i - \mu}{\sigma}\right)^2 \tag{2.46}$$

In the saturated model, there are n parameters; that is, each observation has a different mean equal to the observation ($\mu_i = x_i$). The log-likelihood for this model is thus

$$\ln(L) = LL_{\text{Sat}} = n \ln\left(\frac{1}{\sigma\sqrt{2\pi}}\right) \tag{2.47}$$

More generally, define LL_{Sat} as the log-likelihood of the saturated model with n observations and LL_{MLE} as the log-likelihood at the maximum likelihood estimated values. Now

$$D = 2(LL_{\text{Sat}} - LL_{\text{MLE}}) \sim \chi^2_{n-k} \tag{2.48}$$

where k is the number of parameters estimated. D is known as the **scaled deviance** (or simply the **deviance**). If the model fits the data well D will be smaller than the critical value of χ^2_{n-k}.

To illustrate, D for the von Bertalanffy function is

$$D = \frac{2}{2\sigma^2} \sum_{t=1}^{n} \left(l_t - \hat{\theta}_2\left[1 - e^{-\hat{\theta}_2(t-\hat{\theta}_3)}\right]\right)^2 \sim \chi^2_{n-3} \tag{2.49}$$

Because σ^2 is unknown, we cannot use D directly to test for **lack of fit**. For the present data set, in which we have only a single observation per age, we could go no further. Of course, we can always examine the residual sums of squares to assess how much variation is accounted for by the fit.

If there are several observations per age, we can approximately test for lack of fit using a method suggested by Draper and Smith (1981), which is exact for linear models. We divide the residual sums of squares into a **pure error component** (SS_{PE}) and a **lack of fit component** (SS_{LOF})

$$SS_{PE} = \sum_{t=1}^{n} (m_t - 1)\hat{\sigma}_t^2$$

$$SS_{LOF} = SS(\hat{\theta}_1, \hat{\theta}_2, \hat{\theta}_3) - SS_{PE}$$

(2.50)

where m_t is the number of observations in age group t, $\hat{\sigma}_t^2$ is the estimated variance within age group t, and $SS(\hat{\theta}_1, \hat{\theta}_2, \hat{\theta}_3)$ is the sums of squares at the maximum likelihood estimators. With no lack of fit

$$\frac{SS_{LOF}/n - 3}{SS_{PE}/N - n} \sim F_{n-3, N-n}$$

(2.51)

where N is the total sample size ($= \sum_{t=1}^{n} m_t$). Thus if the right-hand side exceeds the critical point of $F_{n-3, N-n}$ the model is deemed a poor fit, even though it may still be a better fit than competing models.

For the models based on the binomial, we are generally in a better situation. For the logistic model, D can be shown to be (Dobson 1983, p. 77)

$$D = 2\sum_{i=1}^{N}\left[\text{Obs}_i \ln\left(\frac{\text{Obs}_i}{\text{Exp}_i}\right)\right] \sim \chi_{N-k}^2$$

(2.52)

where Obs_i is the observed numbers in the ith category (i.e., observed r_i and $n_i - r_i$), Exp_i is the expected numbers and $N-k$ is the degree of freedom, which is equal to the number of subgroups minus the number of estimated parameters. For the beetle data, $N=8$ and $k=2$, and hence $\chi_{N-k}^2 = 12.59$. The estimated value of D is 13.66, which indicates that the model does not fit the data very well, although visually the fit certainly appears quite adequate (Figure 2.5, see Appendix C.2.6 for S-PLUS coding), and clearly better than the simpler alternative of a constant proportion (an issue discussed in the next section). When the expected value is equal to zero, D is undefined (ln(0) undefined), and in the present example I added a very small number to avoid this problem (Appendix C.2.6), though the existence of the problem suggests that the sample sizes are too small or the model is inadequate.

Comparing models

I shall only consider models that have the same structure but differ in the number of parameters. For example, in the case of the von Bertalanffy function, we might wish to compare a model with θ_3 with one without θ_3

$$l_t = \theta_1(1 - e^{-\theta_2(t-\theta_3)}) \quad \text{vs.} \quad l_t = \theta_1(1 - e^{-\theta_2 t}) \tag{2.53}$$

To do so we can make use of the chi-square property of the deviance. For these two models the deviances are

$$D_{n-3} = \frac{1}{\sigma^2}\sum_{t=1}^{n}\left(l_t - \hat{\theta}_{F,2}\left[1 - e^{-\hat{\theta}_{F,2}(t-\hat{\theta}_{F,3})}\right]\right)^2 \sim \chi^2_{n-3}$$

$$D_{n-2} = \frac{1}{\sigma^2}\sum_{t=1}^{n}\left(l_t - \hat{\theta}_{R,2}\left[1 - e^{-\hat{\theta}_{R,2}t}\right]\right)^2 \sim \chi^2_{n-2} \tag{2.54}$$

where the subscripts F and R stand for "Full" and "Reduced" model, respectively. Now, by the additive nature of chi-square, we have $D_{n-2} - D_{n-3} \sim \chi^2_1$. But we still have the problem of the nuisance σ^2. By construction of the following ratio, we can both eliminate this parameter and produce the F-statistic

$$\frac{D_{n-2} - D_{n-3}}{D_{n-3}/(n-3)} \sim F_{1,n-3} \tag{2.55}$$

For models in which the maximum likelihood estimates are found by minimizing the sums of squares, we can write a general formula for comparing two models:

$$\frac{(SS_R - SS_F)/(F - R)}{SS_F/(n - F)} \sim F_{F-R,n-F} \tag{2.56}$$

where SS_F is the sums of squares of the full model with F parameters, SS_R is the sums of squares of the reduced model with R parameters ($F > R$). Appendix C.2.7 shows coding to compare the von Bertalanffy model with three vs. two parameters. The analysis shows that the three parameter model does not fit significantly better than the two parameter model ($F_{1,10}=0.09$, $P=0.767$). This is also evident from the standard error of θ_3 (t_0) given in Appendix C.2.5.

For models other than those for which the maximum likelihood estimates are obtained by least squares, we can employ the deviances directly

$$\frac{D_R - D_F}{F - R} \sim \chi^2_{F-R} \tag{2.57}$$

Suppose, for example, we wished to compare the logistic fit with a constant proportion model. The latter model is equivalent to the logistic model in which θ_2 is set equal to zero (i.e., $p_i = e^{\theta_1}/(1 + e^{\theta_1}) = $ constant), that is, a one parameter model (see Appendix C.2.8 for coding to compare these models). The deviance for the two parameter model is 13.63 and for the one parameter model it is 287.22: thus $D_1 - D_2 = 273.59$, to be compared to $\chi_1^2 = 3.84$, which is obviously highly significantly different ($P > 0.0001$).

In some instances one might wish to compare several samples: for example, do the means from two separate populations come from the same statistical population (i.e., the null hypothesis of $\mu_1 = \mu_2$ versus the alternate hypothesis of $\mu_1 \neq \mu_2$). This is conceptually and mathematically equivalent to comparing a two parameter model with a one parameter model:

$$\text{One parameter model} \quad \mu_i = \mu$$
$$\text{Two parameter model} \quad \mu_i = \mu + d_i \theta \tag{2.58}$$

where d_i is a "dummy" variable that takes the value 0 for population 1 and 1 for population 2. The statistic μ is estimated by minimizing sums of squares and hence we can use Eq. (2.56) to compare the two models. The two deviances are

$$D_1 = \frac{1}{\sigma^2} \sum_{j=1}^{2} \sum_{i=1}^{n} (x_{ij} - \bar{x})^2 = \sigma^2 SS_1$$
$$D_2 = \frac{1}{\sigma^2} \sum_{j=1}^{2} \sum_{i=1}^{n} (x_{ij} - \bar{x}_j)^2 = \sigma^2 SS_2 \tag{2.59}$$

where, for simplicity, I have assumed equal sample sizes. In the two parameter model there are $2n$ data points and 2 parameters and thus "$n-F$" is equal to $2n-2$, and we test the hypothesis that the two parameter model explains significantly more variance than the one parameter model (i.e., is a better fit to the data) with the F-statistic

$$\frac{D_1 - D_2}{D_1/(2n - 2)} = \frac{SS_1 - SS_2}{S_1/(2n - 2)} \sim F_{1,2n-2} \tag{2.60}$$

which the reader will no doubt recognize as the calculation used in a one-way analysis of variance.

A more complex example is the comparison of two functions that have several parameters. Consider the problem of comparing two growth curves fitted using the von Bertalanffy function (Figure 2.9), the two shown corresponding to male and female curves. The curves could differ in several ways; all three parameters

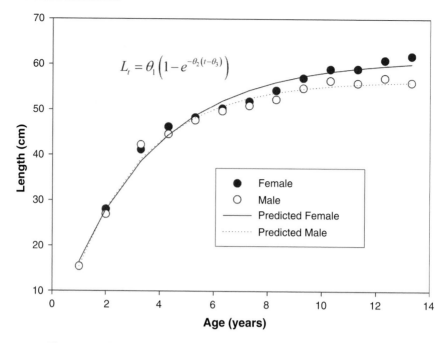

$$L_t = \theta_1\left(1 - e^{-\theta_2(t-\theta_3)}\right)$$

Figure 2.9 Plot of average length at each age for male and female Pacific hake, with the estimated von Bertalanffy curves. Data modified from Kimura (1980).

Age	1.0	2.0	3.3	4.3	5.3	6.3	7.3	8.3	9.3	10.3	11.3	12.3	13.3
Female	15.4	28.0	41.2	46.2	48.2	50.3	51.8	54.3	57.0	58.9	59.0	60.9	61.8
Male	15.4	26.9	42.2	44.6	47.6	49.7	50.9	52.3	54.8	56.4	55.9	57.0	56.0

might differ between the populations or only one. Suppose we wish to test the hypothesis that a model in which all three parameters differ fits the data better than one in which none differ between populations: we proceed in the same manner as above (Appendix C.2.9 shows the coding using the fitting function nlmin and C.2.10 shows the coding using the supplied function nls, which fits a function using least squares. Both methods give identical results and are presented to illustrate that several routes may be taken to achieve the same test. Interestingly, nlmin failed to fit the function with dummy variables). First, we fit the curves separately for the two samples and calculate the combined sums of squares. Second, we fit a single curve using the combined data. Third, we apply Eq. (2.60). For the example data set, we have $F_{3,20}=4.7$, $P=0.01$, indicating that a model with all three parameters different is to be preferred over a model with common parameters. This does not indicate that a model with only one or two common parameters does not fit the data equally as well. From the previous analyses, we might suspect that the parameter θ_3 $(=t_0)$ does not differ between populations. Therefore, it is reasonable to compare the full model with one that incorporates a dummy variable for sex but not θ_3. The full model has six parameters and the reduced model has four parameters (coding in Appendix C.2.11). The full model is not significantly better than the reduced model ($F_{2,22}=0.48$, $P=0.63$).

Summary

(1) The method of maximum likelihood presumes that one can assign to a set of observations a probability, or likelihood (L) that is a function of one or more parameters $\theta_1, \theta_2, \ldots$, the values of which are to be estimated. The parameter values that maximize the probability of obtaining the observed data are the maximum likelihood estimates. It is frequently most convenient to work with the log-likelihood.

(2) In many cases, the probability distribution is based on the normal distribution, leading to the maximum likelihood estimates being obtained by minimizing the residual sums of squares, that is Minimize $\sum_{i=1}^{n}$ (Observed value $-$ Predicted value)^2. Another commonly occurring situation is that in which there are two outcomes (e.g., alive or dead) and the likelihood is based on the logistic model.

(3) The two most commonly used methods of estimating confidence limits are the log-likelihood ratio approach and the standard error approach. The first method involves five steps:

Step 1: Find the MLE for θ

Step 2: Calculate the log-likelihood (i.e., $LL(\hat{\theta})$)

Step 3: Iterate over a range of θ (e.g., 0.01 to 0.99) and for each calculate $LL(\theta)$

Step 4: Calculate Diff $= LL(\hat{\theta}) - LL(\theta) - 0.5\chi_1^2$

Step 5: Find value of θ at which Diff is equal to zero. There will be two values, corresponding to the upper and lower confidence limits.

(4) The second method of assessing the variability in a parameter value is to examine the standard errors of the estimates; roughly speaking, the 95% confidence limits are ± 2 the standard error. The variance of a parameter is approximately equal to the negative of the inverse of the second derivative at the maximum likelihood point $\sigma_\theta^2 = -(\partial^2 LL/\partial\theta^2)^{-1}$. When several parameters are estimated, the matrix of second derivatives must be inverted to obtain the variance–covariance matrix.

(5)   To examine a model for how well it conforms to the observed data, define LL_{Sat} as the log-likelihood of the saturated model with n observations and LL_{MLE} as the log-likelihood at the maximum likelihood estimated values. The saturated model is that in which the log-likelihood is equal only to the constant component of the log-likelihood (e.g., for a normal distribution it would be $-n\ln(\sigma\sqrt{2\pi})$). Now $D = 2(LL_{Sat} - LL_{MLE}) \sim \chi^2_{n-k}$, where k is the number of parameters estimated. D is known as the scaled deviance (or simply the deviance). If the model fits the data well D will be smaller than the critical value of χ^2_{n-k}.

(6)   To compare two models that have the same structure but differ in the number of parameters we construct either an F or χ^2 statistic. For models in which the maximum likelihood estimates are found by minimizing the sums of squares, a general formula for comparing two models is

$$\frac{(SS_R - SS_F)/(F - R)}{SS_F/(n - F)} \sim F_{F-R,n-F}$$

For models other than those for which the maximum likelihood estimates are obtained by least squares, we can generally employ the deviances directly

$$\frac{D_R - D_F}{F - R} \sim \chi^2_{F-R}$$

Further reading

Cox, D. R. and Hinkley, D. V. (1974). *Theoretical Statistics*. London: Chapman and Hall.

Cox, D. R. and Snell, E. J. (1989). *Analysis of Binary Data*. London: Chapman and Hall.

Dobson, A. J. (1983). *An Introduction to Statistical Modelling*. London: Chapman and Hall.

Eliason, S. R. (1993). *Maximum Likelihood Estimation*. Newbury Park: Sage Publications.

Kimura, D. K. (1980). Likelihood methods for the von Bertalanffy growth curve. *Fishery Bulletin*, **77**, 765–76.

Stuart, A., Ord, K. and Arnold, S. (1999). *Kendall's Advanced Theory of Statistics: Classical Inference and the Linear Model*, Vol. 2A. London: Arnold.

Exercises

(2.1) Using the 10 values of x given below, and assuming a normal distribution with $\sigma=1$, plot the log-likelihoods from -3 to $+3$, using a step interval of 0.1. Compare the estimate of μ with the arithmetic average.

−0.793	0.794	−0.892	0.112	1.371	1.417	1.167	−0.531	0.921	−0.577

Hint: if using S-PLUS, consider the following routines: mean, seq, length, for, max, plot

(2.2) Show that $(1/n)\sum_{i=1}^{n}(x_i - \bar{x})^2$ is a biased estimator of σ^2 and that an unbiased estimator is $(1/n - 1)\sum_{i=1}^{n}(x_i - \bar{x})^2$. Hint: There is no loss in generality in assuming that $\mu = 0$, which makes the proof simpler.

(2.3) A frequently used distribution used for sparsely spatially distributed data (e.g., the distribution of a rare organism) is the Poisson distribution, which has probability density function $p(r) = e^{-\theta}(\theta^r/r!)$, where $p(r)$ is the probability of r events occurring (e.g., probability of a sampling unit containing r individuals). Show that the maximum likelihood estimate of θ is equal to (Total number of individuals counted)/(Total number of sampling units).

(2.4) Generate 20 regression lines using the same probability distribution and estimate the correlation between $\hat{\theta}_1$ and $\hat{\theta}_2$. Assume $\theta_1 = 0$, $\theta_2 = 1$, the error term is $N(0,1)$, and there are 10 x values evenly spaced from 1 to 10 (i.e., 1,2,3,...,9,10). Hint for coding: consider using the following routines: for, seq, rnorm, lm, cor.test

(2.5) Egg production in many organisms follows a triangular pattern, first increasing with age and then decreasing. A function suggested by McMillan *et al.* (1970) to describe this pattern in the fruitfly is $y = \theta_1(1 - e^{-\theta_2(x-\theta_3)})e^{-\theta_4 x}$, where y is eggs laid on day x. Assuming that errors are normally distributed (as discussed for the von Bertalanffy function), estimate the four MLE parameters for the following data set

Day	1.5	3	4	5	6	7	8	11	14
Eggs laid	21.6	63.7	61.6	59.9	53.8	55.5	50.8	31.5	24.4

(2.6) A mate choice experiment is run twice, the first time with a sample size n_1 and the second with a sample size n_2. Show that the maximum likelihood estimate is $(r_1 + r_2)/(n_1 + n_2)$ rather than $\frac{1}{2}[r_1/n_1 + r_2/n_2]$.

(2.7) Using the 10 data points from $N(0,1)$ given in question 1 construct a 95% confidence interval using the exhaustive approach method for the mean. Compare the results with those obtained by the usual method (i.e., $\pm t^*SE = \pm 2.262^*SE$). Use a range from -2 to $+2$ and a step length of 0.01. Hint for coding: consider using the following routines: rnorm, mean, var, seq, length, prod, for, sum, cumsum.

(2.8) Using the above data, estimate the 95% confidence limits using the log-likelihood ratio approach. Hint: check Appendix C.2.3.

(2.9) Consider the von Bertalanffy function $l_t = \theta(1 - e^{-k(t-t_0)})$, where l_t is length at age t, k and t_0 are known constants, and θ is an unknown parameter to be estimated. Show that the maximum likelihood estimate of the standard error of θ is $1/\sigma^2 \sum_{t=1}^{n} (1 - e^{-k(t-t_0)})^2$. Hint: make use of the second derivative.

(2.10) The following mean length at age are measured for a particular species of fish.

Age	1	2	3	4	5	6	7	8	9	10
Length	23.61	43.10	57.54	68.24	76.16	82.03	86.38	89.60	91.99	93.76

Assuming a von Bertalanffy growth curve as in question 7, with $k=0.3$ and $t_0=0.05$, estimate θ and the nuisance parameter σ^2 using the nls routine in S-PLUS (or other statistical package). Estimate the standard error of θ using the result from question 7 (Note that σ^2 is estimated as described in the main text).

(2.11) The table below shows egg production in a second strain of Drosophila melanogaster. Fit the function $y = \theta_1(1 - e^{-\theta_2(x-\theta_3)})e^{-\theta_4 x}$ to these data and test the hypothesis $\theta_3 = 0$

Day	1	2	3	4	5	6	7	8	9	10	11	12	13	14
Eggs laid	54.8	73.5	78	71.4	75.6	73.2	65.4	61.9	61.7	60.1	55.1	50.4	44.3	42.3

List of symbols used in Chapter 2

Symbols may be subscripted

ε	Error term
θ	Parameter to be estimated
$\hat{\theta}$	Estimate of θ
$\varphi(\theta)$	Function of θ
σ	Standard deviation
σ^2	Variance
μ	Mean
$\hat{\mu}$	Estimate of μ
π	Pi ($=3.14\ldots$)
D	Deviance
F	Number of parameters in the full model
L	Likelihood

LL	Log-likelihood
L_∞	Asymptotic length (von Bertalanffy equation)
MLE	Maximum likelihood estimate(s)
N	Total number of observations $\left(= \sum_{i=1}^{n} m_i\right)$
$N(\mu, \sigma)$	Normal distribution with mean μ and standard deviation σ
P	Probability
R	Response to selection or number of parameters in reduced model
S	Selection differential
SS	Residual sums of squares
T	Threshold value in heritability model
d	Dummy variable
h^2	Heritability
k	Number of parameters or growth rate (von Bertalanffy equation)
l	Length (von Bertalanffy equation)
m	Number of observations in a subgroup
n	Number of observations or number of subgroups
p	Probability
$\hat{p}$	Estimate of p
r	Number of "successes" in a set of binomial trials
t	Age (von Bertalanffy equation)
t_0	Hypothetical length at age 0 (von Bertalanffy equation)
x	Observed value
$\bar{x}$	Mean value of x
y	Observed value (typically a function of x)
$\bar{y}$	Mean of y

3

The jackknife

Introduction

The jackknife was invented by Quenouille (1949) as a means of eliminating bias in an estimate. Tukey (1958) suggested that Quenouille's method could be used as a non-parametric means of estimating the mean and variance of an estimate, and coined the term "jackknife," to signify an all-purpose statistical tool. The jackknife has proven to be invaluable in the estimation of parameters for which standard techniques are unsatisfactory. However, at the outset it must be recognized that this method is not without assumptions and should not be used without justification, either from a theoretical or numerical analysis. In this chapter, I shall describe the jackknife method, first in a very general sense and then by a series of examples taken from the biological literature.

The jackknife: a general procedure

Point estimation

Suppose we wish to estimate some parameter θ. To do so using the jackknife method, we first estimate θ according to the appropriate algorithm (e.g., we might be estimating the coefficients in a linear regression, in which case the algorithm could be the least squares regression method): let this estimate be $\hat{\theta}$. Next we delete a single datum from the data set. This datum could be a single observation or it could be a group of observations (e.g., in a genetical analysis there might be n families, each consisting of m individuals, and the datum to be dropped is a family rather than an individual). Using the remaining $n-1$ observations, we recalculate the estimate of θ: let this estimate be denoted as $\hat{\theta}_{-1}$. Now we calculate the quantity called the **pseudovalue**

$$S_1 = n\hat{\theta} - (n-1)\hat{\theta}_{-1} \tag{3.1}$$

We return the deleted datum back into the data set, delete the next observation and calculate the second pseudovalue (S_2). This operation is repeated until each datum has been deleted and the corresponding set of n pseudovalues are calculated. The jackknife estimate, $\tilde{\theta}$, is the mean of the pseudovalues

$$\tilde{\theta} = \frac{1}{n}\sum_{i=1}^{n} S_i = \hat{\theta} - \frac{n-1}{n}\sum_{i=1}^{n}\hat{\theta}_{-i} \qquad (3.2)$$

The above is called the **delete-one jackknife**. By deleting more than one observation each time, higher order jackknife estimates can be constructed: these have rarely been used and I shall restrict discussion to the delete-one jackknife and refer to it simply as the jackknife.

Interval estimation

An estimate of the standard error (SE) of θ is given by the standard error of the pseudovalues

$$\text{SE}(\tilde{\theta}) = \sqrt{\frac{1}{n(n-1)}\sum_{i=1}^{n}(S_i - \tilde{\theta})^2} \qquad (3.3)$$

Assuming that the pseudovalues are normally distributed, confidence limits on θ can be computed as

$$\tilde{\theta} \pm t_{\alpha/2,\, n-1}\text{SE}(\tilde{\theta}) \qquad (3.4)$$

where $t_{\alpha/2,\, n-1}$ is the value that is exceeded with probability $\alpha/2$ for the t distribution with $n-1$ degrees of freedom.

Hypothesis testing

Hypothesis testing using the jackknife derives immediately from the assumption that the pseudovalues are normally distributed. For two samples one can use the t distribution or analysis of variance. For several samples analysis of variance is appropriate. For example, as described below, one can use the jackknife to estimate a population rate of increase, r (in this case the symbol r is so well entrenched in the literature that I shall use it instead of θ). Suppose we had several populations for which we had jackknife estimates: we might wish to ask if there is an overall difference among the populations, which could be done by a oneway ANOVA of the pseudovalues. What is particularly useful about the analysis of the pseudovalues is that one can introduce several independent variables in the analysis. In the above example, the rate of increase might have been estimated for several species at each geographic location. We can then ask if there are differences with respect to population, species or an interaction between the two.

Several parameters might be estimated simultaneously using the jackknife. An example, discussed below, is the estimation of the components of a variance–covariance matrix. Rather than doing multiple tests on the individual estimates, we can take a multivariate approach by using multivariate analysis of variance on the pseudovalues. In short, the jackknife is potentially a highly flexible and general approach to both estimation and hypothesis testing.

Examples of the use of the jackknife

The jackknife of the mean

To more fully understand the rationale of the jackknife, it is instructive to examine the pseudovalues of the jackknife estimate of the mean, $\bar{x}$. Deletion of the ith observation gives

$$S_i = n\bar{x} - (n-1)\hat{\theta}_{-i} \tag{3.5}$$

Now, the mean value after deletion of the ith observation is

$$\hat{\theta}_{-i} = \frac{1}{n-1}\left(\sum_{j=1}^{n} x_j - x_i\right) \tag{3.6}$$

Rearranging gives

$$x_i = \sum_{j=1}^{n} x_j - (n-1)\hat{\theta}_{-i} \tag{3.7}$$

But

$$\sum_{j=1}^{n} x_j = n\bar{x} \tag{3.8}$$

and hence the pseudovalue is

$$x_i = n\bar{x} - (n-1)\hat{\theta}_{-i} = S_i \tag{3.9}$$

Thus, in this case we recover the value of the ith observation from the mean and the ith value removed. This may not impress the reader since x_i was what we deleted: but if there is not the simple linear relationship between the ith observation and $\tilde{\theta}_{-i}$, then this will not be the case. In essence, the jackknife turns the estimation problem into a problem of the estimation of mean and variance of a normal distribution, for which we have well established methods. The trick is to ensure that the transformation actually does this. In some cases, theoretical justification can be advanced but in many, if not most, it will be necessary to resort to simulation modeling to check the method. The strength of the method is its extreme simplicity.

Variance components

A common objective, particularly in genetic studies, is the estimation of variance components. Three questions that should be answered to justify the use of the jackknife in this case and in general are: first, does the jackknife give a correct estimate for the variance; second, is the estimated 95% confidence interval correct in that the true value lies above the upper value 2.5% of the time and the true value lies below the lower confidence value 2.5% of the time; and third, can the pseudovalues be used in hypothesis testing?

The jackknife estimate of confidence limits of the variance can be shown to be incorrect but a log transformation does produce appropriate limits, at least for large sample sizes (Miller 1974; Manly 1997). Whether it does so for small sample sizes can only be decided by simulation. Before examining the question of transformations, let us examine the jackknife on the untransformed statistic. Many statistical packages now have routines for doing the delete-one jackknife where each row represents a datum to be deleted (the question of jackknifing when the unit consists of multiple lines is addressed below). Appendix C.3.1 provides the coding for the S-PLUS case in which there are 10 observations per sample. The first set of lines generate 1000 sets of 10 random normal values with mean zero and unit standard deviation: these data are stored in a matrix **X**, with each column representing a single sample. The second group of lines iterate over these replicates calculating the jackknife mean and SE of the estimated variance and storing the results in the matrix Output. The final group of lines calculates the upper and lower confidence limits for each sample and determines in how many cases the true value lies outside the limits. Not unexpectedly, the grand mean of the jackknife estimates, 1.009637, is an unbiased estimate of the true variance. However, the confidence limits are incorrect. We require that the upper confidence limit be less than the true value in 2.5% of cases, whereas the jackknife estimated upper limit is less than the true value in 13.3% of the replicates. Thus the jackknife estimated upper limit is too low. The opposite problem occurs with the lower confidence limit. We require that the lower confidence limit be greater than the true value in 2.5% of cases, whereas the jackknifed estimated lower limit is greater than the lower value in only 0.1% of the replicates. The confidence limits estimated from the jackknife are too small (86.6% rather than 95%) and shifted downwards.

To test the efficacy of the log transformation, we need to alter three lines: replace `var(x)` with `log(var(x))` and instead of testing the limits against 1, test them against 0 (because log(1)=0). The result is better in that the lower and upper percentiles are now 1.5 and 95.9%, respectively (instead of the required 2.5 and 97.5%). Thus, the correct value is included in 93.4% of the replicates but the limits are not symmetrical. This exercise illustrates the very important point that an analysis of the performance of an estimation procedure should not look simply at the overall confidence region but should examine both the upper and lower limits.

The jackknife does appear to provide an unbiased estimate of the variance in the above example. We now consider whether it can be used to test for a difference between two variances given that the null hypothesis of equal variances is true (coding in Appendix C.3.2). Given a type 1 error of 5% we expect, as the samples are drawn from the same population, that a difference should be declared significant in 5% of the replicates. The actual percentage of significant differences obtained was 3.5%, which is close to but still significantly different from the expected 5% (χ^2=4.737, degrees of freedom [df]=1, P=0.03). Using the log transformation produced 6.3% significant differences, which is marginally non-significantly different from the expected 5% (χ^2=3.558, df=1, P=0.0593). Using the untransformed data gives a test that is conservative, whereas using the transformed data gives a test that for all practical purposes is quite satisfactory, though slightly too liberal in declaring significance.

We now move to a more useful task for the jackknife, that of estimating and testing variation among variance–covariance matrices. The estimation of variance–covariance matrices is of particular interest to evolutionary biologists, because two such matrices are required to predict the evolution of multiple traits. As discussed in Chapter 2, the response to selection, R, for a single trait is given by $R=h^2S$, where h^2 is the heritability of the trait and S is the selection differential (the difference between the mean of the population and the mean of the parents contributing to the next generation). Heritability is the ratio of additive genetic variance to the total phenotypic variance, $h^2 = \sigma_A^2/\sigma_P^2$. Selection on two traits must take into account not only selection acting directly upon a trait but also indirect selection resulting from genes that contribute to the expression of both traits, the latter selection producing a correlated response. Using matrix notation, the response to selection for two traits (R_1, R_2) can be written as

$$\begin{pmatrix} R_1 \\ R_2 \end{pmatrix} = \begin{pmatrix} \sigma_{A11}^2 & \sigma_{A12} \\ \sigma_{A21} & \sigma_{A22}^2 \end{pmatrix} \begin{pmatrix} \sigma_{P11}^2 & \sigma_{P12} \\ \sigma_{P21} & \sigma_{P22}^2 \end{pmatrix}^{-1} \begin{pmatrix} S_1 \\ S_2 \end{pmatrix} \tag{3.10}$$

where the first matrix contains the additive genetic variances (diagonal) and covariances (off-diagonal) and the second contains the phenotypic variances and covariances. Extension to more than two traits is immediate and obvious. The above equation is written in shorthand as $\mathbf{R}=\mathbf{GP}^{-1}\mathbf{S}$, where $\mathbf{G}$ is the genetic variance–covariance matrix and $\mathbf{P}$ is the phenotypic variance–covariance matrix. The prediction of evolutionary response requires not only estimates of the variances and covariances, but also the associated standard errors to place confidence bounds errors on the predicted response. Knapp *et al.* (1989) investigated the utility of the jackknife in predicting the response to selection for a single trait (i.e., $R=h^2S$). The data comprised a set of simulated families. The response to selection was predicted first using all families to estimate heritability (the estimation of heritability is discussed in the following section); pseudovalues for the response to selection were then created by deleting sequentially entire families, recomputing heritability, and thence the predicted response from which the pseudovalue of the response could be calculated. Note that, in this case, the basic unit of deletion is not an individual observation but a group of observations: it is important to consider carefully what is the unit of deletion, because use of the incorrect unit will generate incorrect standard errors. The families simulated here represent full-sib families (i.e., each family has a single sire and dam that contribute to no other families. Contrast this with a half-sib structure in which a sire is mated to several dams, leading to both full-sib and half-sib relationships). Both untransformed and log transformed values were used. In both cases, the confidence regions about the predicted response (both 80 and 95% were examined) were not significantly different from the stated coverages. So far as I know, the analysis of multivariate response to selection has not been done, though the results for the single trait case are encouraging.

A commonly considered issue is whether two or more covariance matrices are the same. Differences in the genetic covariance matrix, the $\mathbf{G}$ matrix, could arise as a result of selection or genetic drift (i.e., variation resulting from random sampling of genes in small populations). Several methods have been proposed to test for differences [reviewed in (Roff 1997, 2000)]. The jackknife provides a possible solution (Roff 2002). The procedure, illustrated in Table 3.1, is as follows: for each group, calculate the $\mathbf{G}$ matrix, which typically can be done using standard statistical methods. Next, delete in turn one sampling unit (e.g., for a full-sib design it would be a single family and for a half-sib design it would be a single sire group), and calculate the pseudovalues according to the usual jackknife procedure. As shown in Table 3.1, the final data matrix can be arranged such that the columns comprise the pseudovalues of each covariance (note that the variance is included as it is the covariance of a variable with itself) and the rows are the results for the deletion of a given family (so the ith row, jth column

Table 3.1 *A sample output data file of pseudovalues calculated for a full-sib experiment involving two species (GF, GP). Two traits are shown, femur length and head width, which generate two variances and one covariance. In the first sample (GF) there are 43 families and hence 43 pseudovalues for each component covariance. The second sample (GP) contains 39 families and hence 39 pseudovalues per component covariance*

| | Pseudovalues for | | | |
| | Variance | | Covariance | |
Omitted Family[a]	Femur	Head	Femur × Head	Species
1	1.1621	0.1588	0.4342	GF
2	−0.0157	−0.0241	−0.0627	GF
.	0.1549	0.0174	0.0564	GF
.	.	.	.	.
43	1.0667	0.2455	0.5204	GF
1	−0.0760	−0.0113	−0.0267	GP
2	0.5166	0.2222	0.3307	GP
3	0.0987	0.1981	.	GP
	.	.	−0.0708	.
39	−0.1624	−0.0419	0.07 (0.03)	GP
Means[b] (SE) by species	0.17 (0.06)	0.04 (0.01)	0.09 (0.03)	GF
	0.23 (0.07)	0.05 (0.01)		GP

Reproduced from (Roff 2002).

[a]The ith pseudovalue, S_i, is computed as $S_i = n\hat{\theta} - (n-1)\hat{\theta}_{-i}$, where $\hat{\theta}$ is the statistic calculated using the full data set, $\hat{\theta}_{-i}$, is the statistic calculated using the full data set minus the ith data point (family), and n is the sample size (number of families: $n=43$ or 39).

[b]The jackknife estimate, $\tilde{\theta}$, is calculated as $\tilde{\theta} = \sum_{i=1}^{n} S_i/n$ and the standard error, SE, as $SE = \sqrt{[\sum (S_i - \tilde{\theta})]/[n(n-1)]}$. The means from the jackknife were the same (to two decimal places) as the estimates obtained using the entire data set ($n=43$ or 39). The results from the MANOVA analysis are as follows:

Source	SS	df	MS	F	P
Univariate F-Tests					
Femur	0.083	1	0.083	0.454	0.502
Error	14.617	80	0.183		
Head	0.010	1	0.010	0.310	0.579
Error	2.650	80	0.033		
FXH	0.003	1	0.003	0.418	0.520
Error	0.575	80	0.007		

Statistic	Value	F-Statistic	df	Prob
Multivariate Test Statistics				
Wilkes' Lambda	0.985	0.391	3, 78	0.760
Pillai trace	0.015	0.391	3. 78	0.760
Hotelling–Lawley trace	0.015	0.391	3. 78	0.760

Both univariate and multivariate tests indicate a non-significant difference.

is the pseudovalue for the *j*th covariance for the sample with the *i*th family deleted. This procedure is readily implemented in S-PLUS. It is important to remember that the grouping variable (FAMILY in this instance) is categorical, and so is best coded as a character variable to avoid accidentally entering it into the analysis as a continuous variable (this is true for a number of statistical packages and a good rule to follow is to use character variables for any categorical variable). The general steps for jackknifing when the basic datum to be deleted is a group are as follows:

(1) Assign a character variable to each group: for example, if there are three families in the data, these could be coded as 1, 2, 3 provided that they are designated as character variables and not numeric variables (so A, B, C would be better). Coding does not have to be sequential: so 1, 7, 3 would do as well.

(2) Calculate statistic for the entire data set.

(3) Iterate through the groups, deleting one group at each iteration. Suppose that there are *n* groups with the group designators stored in a vector called "Group.Designator." The data are stored in a file (a dataframe in S-PLUS) called "Data" and the group designator is in a column labeled "Group." Let the data minus one of the groups be called "Data.minus.one." A simple coding sequence for S-PLUS is

```
for (i in 1:n)
{
Ith.Group <- Group.Designator[i]
Data.minus.one <- Data[Data$Group!=Ith.Group,]
Insert Lines that calculate and store the pseudovalue
}
```

Coding for the estimation of the genetic variance–covariance matrix is shown in Appendix C.3.3. Having calculated the set of pseudovalues for two or more data sets, one can investigate whether there are significant differences among the matrices and if these are statistically associated with other factors, such as environmental variables. To illustrate this approach, I shall consider data on the amphipod, *Gammarus minus*.

Gammarus minus is a common amphipod species of karst areas throughout the central and eastern United States. Fong (1989) and Jernigan *et al.* (1994) compared the **G** matrices of four populations in west Virginia: (1) a population from Benedict's Cave; (2) a population from Davis Spring, which resurges from Benedict's Cave; (3) a population from Organ Cave; and (4) a population from Organ Spring, which resurges from Organ Cave. The two caves and their

associated springs are in two separate drainage basins and the cave populations are assumed to have been derived from the stream populations. For the
present analysis, I use eight morphological characters, comprising one measure
of overall size (head width), three measures of eye structure and four measures
of antennal length. The populations do not differ much in overall size, but
individuals from cave populations have a reduced number of ommatidia, smaller
eyes, and larger antennal components than those from the spring populations.
These differences are consistent with the hypothesis of adaptive evolutionary
change in the cave populations, selection favoring an increase in tactile sensory
organs and a loss of ocular sensory organs.

With respect to the genetic correlations, Fong (1989) found a significant correlation between the two cave populations and between the two spring populations
but not between populations in the same drainage basin. This result suggests
that genetic correlations are more related to habitat (cave or spring) than to
history (common ancestry). This hypothesis can be addressed by using the **G**
matrix pseudovalues calculated for the four populations in a two-way MANOVA.

Differences among **G** matrices can arise simply as a function of scale effects
if the groups being compared differ markedly in size. In the present analysis,
I transformed all variables to a log scale, which eliminated the large differences
in phenotypic variances. The two-way MANOVA shows a highly significant effect
of habitat and drainage basin but no significant interaction (Table 3.2).
To determine which trait covariances contribute most to the variation we can
examine the results of univariate tests, keeping in mind that individual significance levels are inflated because of multiple tests. The univariate tests are used
to see if there is a pattern of variation rather than isolating specific covariances.
A clear pattern does emerge for the differences associated with habitat, with
covariances involving eye length or antennal components generally being
"significant" ($P < 0.05$; Table 3.3). In contrast, there are only three "significant"
($P < 0.05$) univariate tests for the effect of drainage basin, all involving eye length.
This difference between habitat and drainage basin indicates that much greater
differences arise between habitats than between populations within the same

Table 3.2 *A two-way MANOVA examining the influence of habitat (Cave, Spring) and drainage basin*
on the G matrices of four populations of the amphipod Gammarus minus. *From (Roff 2002)*

	Wilkes λ	Approx F	df	P
Habitat	0.538	3.770	36, 158	<0.0005
Basin	0.671	2.154	36, 158	0.001
Habitat × Basin	0.766	1.341	36, 158	0.114

Table 3.3 *Univariate tests of the* **G** *matrix elements in the amphipod* Gammarus minus. *Probabilities associated with habitat are shown above those associated with drainage basin. Probabilities less than 0.05 are shown in bold*

		Head	Ommat	EyeL	EyeW	Ped1	Ped2	Flag2	Nf1
Head	(Habitat)	0.679	0.236	0.055	0.396	**0.045**	0.051	0.150	0.285
	(Basin)	0.370	0.744	0.209	0.776	0.302	0.237	0.373	0.593
Ommat	(Habitat)		0.151	0.457	0.797	0.168	0.320	0.361	0.351
	(Basin)		0.577	0.087	0.122	0.124	0.367	0.594	0.929
EyeL	(Habitat)			0.562	**0.029**	**0.016**	**0.012**	**0.020**	**0.021**
	(Basin)			**0.008**	**0.048**	**0.023**	0.062	0.162	0.292
EyeW	(Habitat)				**0.000**	0.801	0.789	0.888	0.998
	(Basin)				0.271	0.087	0.143	0.416	0.831
Ped1	(Habitat)					**0.011**	**0.007**	**0.02**	**0.029**
	(Basin)					0.085	0.103	0.258	0.418
Ped2	(Habitat)						**0.011**	**0.021**	**0.030**
	(Basin)						0.143	0.277	0.436
Flag2	(Habitat)							**0.011**	**0.021**
	(Basin)							0.565	0.749
Nf1	(Habitat)								0.141
	(Basin)								0.928

From Roff (2002).

Head=head width, Ommat=number of ommatidia in the compound eye, EyeL=eye length, EyeW=eye width, Ped1=length of the peduncle of the first antennae, Ped2=Length of the peduncle of the second antennae, Flag2=length of the flagellum of the second antennae, Nf1=number of flagellar segments of the first antennae.

drainage basin. These results support Fong's conclusion that habitat is more important than history in *G. minus* in molding genetic variation.

The above example illustrates the method of using the jackknife to address questions of variation among variance–covariance matrices, but it is still necessary to demonstrate that the method is valid. Under the null model of no difference between two **G** matrices, the Jackknife-MANOVA method should declare the two matrices to be different with a probability of 5%. To test this, I generated populations (see Appendix C.3.4 for simulation) by using identical covariances for each population (but differing among traits) and subjected these to the MANOVA method of analysis. To roughly match typical sample sizes (i.e., those used in Begin *et al.* 2004) I used populations of 50 full sib families with 10 offspring per family. The actual values of the heritabilities and correlations should make no difference under the null hypothesis: thus, under the null hypothesis, 5% of simulations should produce significant probabilities.

I used a range of parameter values: as all gave the same result, I combined the data to estimate the probability. Out of 2100 simulations, 4.7% were found to be significant, which is not significantly different from the predicted 5% ($\chi^2 = 0.49$, df $= 1$, $P = 0.48$). Thus, with regard to type I errors, the Jackknife-MANOVA method appears to be valid, at least for this data set. Further simulations are necessary to determine if the method breaks down with smaller sample sizes and to determine the power of the test.

The estimation of ratios: variances and covariances

The estimation of ratios is a particular area in which the jackknife can be very useful. Arvesen and Schmitz (1970) studied the use of the jackknife for the estimation of variance component ratios, specifically the F statistic. They showed both from theory and simulation that the jackknife could be used provided the estimate was log-transformed. Expanding on this result, they suggested that the standard error of heritability could also be estimated using the jackknife, though again they suggested that the estimate be transformed. Finally, Arvesen and Schmitz (1970) considered the estimation of the genetic correlation for which they could not derive an appropriate transformation but suggested that Fisher's z transformation might be appropriate. Knapp *et al.* (1989) and Simons and Roff (1994) checked the efficacy of the jackknife in estimating the mean and variance of the heritability (Appendix C.3.4 describes the model used by Simons and Roff).

As explained previously, in the case of the genetic analysis of full sibs, the jackknife is performed on the individual families rather than the individuals. This presents some technical difficulties in terms of implementation. One approach is outlined in Appendix C.3.3, in which the jackknife is performed "manually" (i.e., not using the routine "jackknife"). An alternate method is to make each row a separate family with individuals occupying different columns. Suppose, for example, the data set consisted of 20 families and 5 individuals per family. The data matrix would be set up with 20 rows and 10 columns, the first 5 columns containing the family code and the last 5 columns containing the data. This data set can then be converted into a two column data set in which the first column contains the family codes and the second column contains the data. Coding for this is illustrated in Appendix C.3.5.

Using the transformation gives confidence intervals somewhat closer to the correct values than the untransformed values (Figure 3.1). The performance of the untransformed estimate improves with the number of families (the lowest sample size is lower than would typically be recommended). In contrast to these results, the untransformed estimate performs better with respect to bias than the transformed estimate (Figure 3.1). The bias is particularly bad when the true

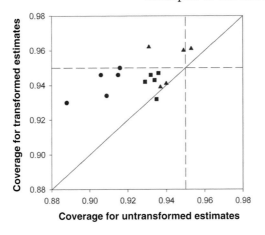

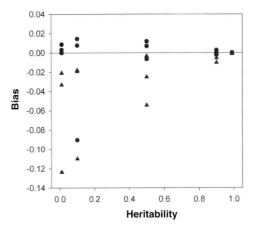

Figure 3.1 Simulation analysis of jackknife estimation of heritability in a full-sib design. Top panel shows the probability (=coverage) that the true value of heritability lies within 95% confidence limits estimated by using the jackknife; ●=20 families per sample, ■=60 families per sample, ▲=100 families per sample. Bottom panel shows the bias in the estimates (●=untransformed, ▲=transformed) as a function of the true heritability. Data from Knapp *et al.* (1989).

heritability is less than 0.2. Overall, it would appear that the untransformed estimator is better, provided that a sufficient number of families is used. Simons and Roff (1994) compared the performance of the jackknife using untransformed estimators with the approximate parametric estimate of the standard error. The jackknife estimate was closer to the true heritability than the ANOVA estimate but the difference was minor (the third decimal place). However, there was a difference in the coverage, with the jackknife estimate giving an overall coverage generally closer to the desired 95% (Figure 3.2). Further, the parametric estimate gave a lower limit that was consistently less than the jackknife method

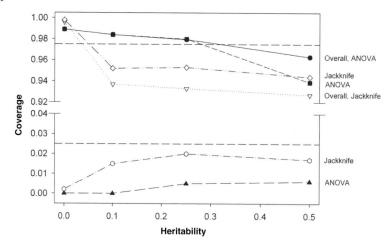

Figure 3.2 A comparison of the estimated confidence 95% limits using either a parametric estimator (ANOVA) of heritability or the jackknife method. The overall coverage should be 0.95, the upper limit at 0.975 (upper dashed line) and the lower limit at 0.025 (lower dashed line). Data from Simons and Roff (1994).

and well below the required 0.025. Contrarily, the jackknife estimate gave an upper limit consistently lower than the parametric estimate.

Roff and Preziosi (1994) tested the jackknife method for the estimation of phenotypic and genetic correlations. We found that the method produced excellent results for both the estimation of the parameter values and the estimated 95% confidence limits.

The estimation of ratios: ecological indices

Frequently, ecologists describe population processes or states with an index. Not untypically, there are a number of proposed indices and, equally, not untypically, the statistical properties of these indices are not clear. The jackknife may provide a means of estimating these indices and their associated standard errors, though this should not be assumed. As an example, I shall consider measures of niche overlap. We desire to calculate the niche overlap of two sympatric species that utilize a set of n resources, the utilization of the ith resource by the first species being p_i (= the proportion of total resources used) and the utilization by the second species being q_i. Four proposed indices are:

(1) Coefficient of community (C_1): $C_1 = \sum_{i=1}^{n} \min(p_i, q_i)$
(2) Morisita's index (C_2): $C_2 = 2 \sum_{i=1}^{n} p_i q_i / (\sum_{i=1}^{n} p_i^2 + \sum_{i=1}^{n} q_i^2)$
(3) Horn's index (C_3): $C_3 = (\sum_{i=1}^{n} (p_i + q_i) \log(p_i + q_i) - \sum_{i=1}^{n} p_i \log p_i - \sum_{i=1}^{n} q_i \log q_i) / 2 \log 2$
(4) Euclidean distance (C_4): $C_4 = 1 - \sqrt{\sum_{i=1}^{n} (p_i - q_i)^2 / 2}$.

Mueller and Altenberg (1985) compared the jackknife with the delta method, which is a commonly used analytical method to approximate expected values and their variances (the method is outlined in detail in Lynch and Walsh [1998, pp. 807–21]). To compare these two methods (and also the bootstrap, which is discussed in the next chapter), Mueller and Altenberg carried out a number of different simulations. First, they considered two resource categories (i.e., species 1 utilized one resource with a probability p_1 and the alternate with a probability $1 - p_1$, whereas species 2 utilized one resource with a probability q_1 and the alternate with a probability $1 - q_1$). Twenty-five different combinations were used (all pairwise combinations of $p_1 = 0.1, 0.2, 0.3, 0.4$, or 0.5, and $q_1 = 0.15$, 0.35, 0.55, 0.75, or 0.95). For each combination 200 replicates were made using the following protocol: (1) Draw a sample of n observations ($n = 20, 60$ or 200) from which p_1 and q_1 are calculated; and (2) calculate the estimates and their associated confidence region using the jackknife and delta methods. There are two questions to be addressed: first, which method has a smaller bias and a standard error closer to the correct value, and second, is either useful? Being a "better" estimator doesn't make it an adequate or acceptable estimator!

In all cases, the jackknife estimate had a smaller bias and generally a more accurate confidence region than the delta estimate. Similar results were obtained for two multiple category examples tested (e.g., set of p is 0.1, 0.1, 0.1, 0.1, 0.1, 0.1, 0.1, 0.1, 0.1, 0.1 and set of q is 0.4, 0.3, 0.1, 0.1, 0.05, 0.01, 0.01, 0.01, 0.01). However, whereas the performance of the jackknife was better than the delta method it still produced confidence limits somewhat smaller than required (Table 3.4).

The estimation of ratios: population parameters

An important parameter in both ecology and evolution is the rate of increase, r. In an unstructured population (or one with a stable age distribution)

Table 3.4 *Summary of analysis of niche overlap statistics with the multiple categories given in the text. Data from Mueller and Altenberg (1985)*

Index	% Bias		Coverage[a]	
	Delta	Jackknife	Delta	Jackknife
C1	19.5	6.5	0.85	0.93
C2	16.4	2.6	0.91	0.87
C3	22.1	3.5	0.61	0.92
C4	8.0	0.4	0.86	0.93

[a]Proportion of times true value fell inside the estimated 95% confidence limits.

in which there is no density dependence, population growth is exponential and can be described by the equation, $N_t = N_0 e^{rt}$, where N_t is population size at time t. With density-dependence, it is necessary to introduce a term that causes the growth rate to decline with density. One such formulation is the logistic equation, $N_t = K/(1 + e^{c-rt})$, where K is the carrying capacity and c is a constant. In both models, the parameter r plays a prominent role. The estimation of r from an age-structured population is done by solving the Euler equation

$$1 = \sum_{i=0}^{\Omega} e^{-ri} l_i m_i \tag{3.11}$$

where Ω is the last age, l_i is the probability of surviving to age i and m_i is the number of female births at age i. Equation (3.1) can be solved numerically but estimating confidence limits is problematical. Meyer *et al.* (1986) proposed the jackknife and tested its utility with simulation of two different hypothetical populations (Table 3.5). In the first population, there was no mortality until

Table 3.5 *The life histories of five females taken from the computer simulations of two hypothetical cladoceran populations. Underlined values indicate that the individual died in the next time interval*

	Number of offspring produced									
	Hypothetical Population 1					Hypothetical Population 2				
Age	1	2	3	4	5	1	2	3	4	5
1	0	0	0	0	0	0	0	0	0	0
2	0	0	0	0	0	0	0	0	0	0
3	0	0	0	0	0	0	0	0	0	0
4	0	0	0	0	0	0	0	0	0	0
5	0	0	0	0	0	0	0	0	0	0
6	0	0	0	0	0	0	0	0	0	0
7	0	0	12	8	0	0	0	0	0	<u>0</u>
8	4	0	0	0	0	10	<u>0</u>	0	11	
9	0	0	10	0	12	0		8	0	
10	11	8	0	8	0	0		0	7	
11	0	0	9	0	0	<u>10</u>		<u>0</u>	0	
12	9	0	0	9	12				0	
13	0	11	8	0	0				<u>8</u>	
14	0	0	0	0	0					
15	15	0	9	14	10					
.	.	.	.	.	.					
28	0	0	0	10	0					

Modified from Meyer *et al.* (1986).

age 28 at which time all individuals died, whereas in the second there was a relatively heavy mortality. Offspring (males and females are not distinguished as the simulation was of a clonal species) were assigned to females according to the following rules: (1) no reproduction before age 7; (2) a brood period of 2 or 3 days (equal probability); and (3) a brood size that was normally distributed about the mean brood size of 10 offspring and a coefficient of variation of 0.25. Each population consisted of 100 animals from which 10 females were taken at random and the rate of increase, r, calculated using Eq. (3.1). A jackknife estimate was made by sequential deletion of individual females from the sample. For each population, 1000 replicates were made. For the long-lived population, the true population value was 0.374 as was also the mean of the jackknifed estimates. The estimated 95% confidence limits included the true value in 94.4% of the replicates. Similarly, excellent results were obtained for the short-lived population, the true value of r being 0.313 compared to the mean jackknife estimate of 0.311. The estimated 95% confidence limits included the true value in 96.5% of the replicates.

Estimating ratios or products from two separate samples

In ecology, there are cases in which a statistic is the ratio or product of two variables from different samples: for example, population growth rate is the ratio of population size at two times. The jackknife can be used to generate estimates and confidence intervals but the technique differs from the usual implementation as there are two different samples. It is termed a **weighted jackknife estimator**. Let the true ratio of the two random variables from samples X and Y be θ_R and the estimate be $\hat{\theta}_R$, with n_X and n_Y observations in X and Y, respectively. The weighted jackknife proceeds by first creating pseudovalues by deleting one observation from sample X,

$$S_{Xi} = \hat{\theta}_R - \frac{(n_X + n_Y)}{n_X}(n_X - 1)(\tilde{\theta}_{-Xi} - \hat{\theta}_R) \qquad (3.12)$$

where $\tilde{\theta}_{-Xi}$ is the estimate of θ_R obtained by deleting the ith observation from sample X. Similarly, we create pseudovalues by single deletions from sample Y: let the pseudovalue generated by deletion of the jth observation from Y be S_{Yj}. We now have a total of $n_X + n_Y$ pseudovalues (n_X values of S_X and n_Y values of S_Y), which, for simplicity, I shall label as S_i where i goes from 1 to N ($= n_X + n_Y$). The jackknife estimate of θ_R is the mean of S, which is also algebraically equivalent to

$$\tilde{\theta}_R = n_Y\hat{\theta}_R - \frac{(n_Y - 1)}{n_Y}\sum_{j=1}^{n_Y} S_{Yj} \qquad (3.13)$$

The jackknife estimate of the standard error is

$$
\mathrm{SE}(\tilde{\theta}_R) = \sqrt{\frac{1}{N(N-2)} \sum_{i=1}^{N} S_i}
\tag{3.14}
$$

Note the 2 in the denominator, rather than 1 as given for the usual standard error estimate (Eq. (3.3)). For further discussion on the estimation of ratios and products using the jackknife, see Buonaccorsi and Liebhold (1988).

Estimating parameters of nonlinear models

In the previous chapter, the problem of estimating the parameter values in a nonlinear model such as the von Bertalanffy equation was tackled using the maximum likelihood approach. An alternative is the jackknife. To illustrate its use in this situation, I shall consider the simplified von Bertalanffy equation

$$
l_t = \theta_1(1 - e^{-\theta_2 t})
\tag{3.15}
$$

where l_t is length at age t (I have simplified the equation by assuming that the initial length can be set at zero). To generate suitable data, I simulated a population consisting of five age groups (1, 2, 3, 4, and 5) and drew from this population five individuals from each age group, the length of the ith individual in age group t, $l_{t,i}$, being generated by the expression $l_{t,i} = l_t + \varepsilon_{t,i}$, where $\varepsilon_{t,i}$ is a random variable specific to this individual (see Appendix C.3.6 for coding). I set the true values at $\theta_1 = 100$ and $\theta_2 = 1$ and used three error distributions, all with zero mean: (1) a random normal with a mean of zero and standard deviation of 10 (this is the one shown in Appendix C.3.6); this error distribution satisfies the assumption for the maximum likelihood method; (2) a uniform distribution from -5 to $+5$ (in S-PLUS, `Error <- runif(n, min=-5, max=5)`; and (3) a random normal in which the standard deviation increased proportionally with age (in S-PLUS, `Error <- rnorm(n, 0, Age*2)`).

The jackknife estimates were estimated as follows (Appendix C.3.7). First, the parameters were estimated by minimizing the residual sums of squares (i.e., the method of least squares), which is the same approach as taken for the maximum likelihood method. Next, one individual was deleted and the parameters estimated by least squares and thence the pseudovalue computed. This process was repeated for all 25 individuals to produce the required 25 pseudovalues from which the jackknifed estimates were made. The results shown in Appendix C.3.7 for the single simulated data set indicate that the jackknife estimates closely match those from maximum likelihood.

For each error distribution, I generated 1000 replicates (Appendix C.3.8). The mean values of the parameter estimates were very close to the true

Table 3.6 *Results of parameter estimation by the jackknife (Jack.) and maximum likelihood estimate (MLE) methods for 1000 replicate simulations of the von Bertalanffy growth function. See text for details of simulation*

Parameter	Mean MLE	Mean Jack.	P<LC[a] MLE	P<LC[a] Jack.	P>UC[b] MLE	P>UC[b] Jack.	Coverage[c] MLE	Coverage[c] Jack.
Random normal error								
θ_1	100.14	99.89	0.014	0.013	0.033	0.037	0.953	0.950
θ_2	1.011	0.998	0.006	0.004	0.037	0.044	0.957	0.952
Uniform error								
θ_1	99.99	99.97	0.022	0.027	0.019	0.029	0.959	0.944
θ_2	1.002	1.001	0.023	0.027	0.031	0.037	0.946	0.936
Proportionally increasing error								
θ_1	100.03	99.91	0.049	0.018	0.064	0.038	0.887	0.944
θ_2	1.003	1.001	0.005	0.014	0.013	0.023	0.982	0.963

[a]Proportion of replicates in which the true value fell below the lower 95% confidence limit.
[b]Proportion of replicates in which the true value fell above the upper 95% confidence limit.
[c]Proportion of replicates in which the true value fell within the 95% confidence limits.

values, showing no evidence of significant bias (Table 3.6). However, whereas the jackknife 95% confidence limits closely approximated the correct value (i.e., enclosed the true value in 95% of cases), the maximum likelihood estimates (MLE) performed relatively poorly for the proportionally increasing error (last two columns in Table 3.6). The jackknife limits tended to be somewhat skewed with the upper limit being higher than 2.5% and the lower limit lower than 2.5%: this was true for the MLE with random normal error, not with uniform error, and varied with the parameter being estimated with proportionally increasing error (Table 3.6). Overall, the performance of the jackknife estimator was superior to that of the maximum likelihood estimator. Of course, if one knew that the variance increased in proportion to the age, one could include this in the maximum likelihood procedure. The superiority of the jackknife is that it is relatively robust to the underlying error distribution.

The next question to address is whether the jackknife is a suitable method for comparing parameter values. To answer this question, I generated two samples from the same distribution and compared the parameter values in two ways: first, by pairwise comparison of the parameters (e.g., in the present case I compared the two θ_1 values and the two θ_2 values) using either a *t*-test or ANOVA, and second I compared all parameters simultaneously using MANOVA. The latter test is important because the estimates are likely to be correlated, as indeed is indicated by the maximum likelihood output in Appendix C.3.7.

The proportion of significant differences obtained from 1000 replicated pairs of data sets was 0.041 for all three comparisons, which is not significantly different from the expected 0.05 ($\chi^2 = 1.71$, df$=1$, $P=0.1916$). Using the proportionally increasing error variance produced 4.5% significant differences, which is also not significantly different from the expected 5%.

In conclusion, for nonlinear models the jackknife method appears to be an excellent alternative to the maximum likelihood method and is superior when the assumptions of MLE are not met. It cannot be assumed that the method will be valid for all nonlinear models and such models should be subject to the simulation analysis outlined above. The method is valid for the von Bertalanffy function (this study) and has also been shown to work for the Michaelis–Menten function (Oppenheimer *et al.* 1981; Matyska and Kovar 1985), suggesting that it will be generally applicable.

Checking the jackknife by bootstrapping the data

In all the cases so far investigated, the efficacy of the jackknife was examined by specifying the statistical model and producing simulated data. For example, suppose we wish to simulate data to test the estimation of parameter values in a function such as the von Bertalanffy growth curve. One way to proceed is to generate a set of random ages between the minimum and maximum ages (note that previously I used a fixed sampling scheme of five individuals for each of five ages). Next, a set of error terms is created based on the normal distribution with a mean of zero, and these are added to the true mean value for each age using the function (Appendix C.3.9). In some instances, it may be difficult to define the model and an alternate method of simulating data is necessary. One method that is very simple to implement is use of an observed data set as the true population and then generate samples by drawing at random with replacement from the population. The true value of the parameter(s) of interest is equal to the value obtained from the original data set. Such sampling with replacement can be done with a single call in S-PLUS (Appendix C.3.9). Mueller (1979) used this bootstrap approach in his examination of the estimation of Nei's genetic distance.

Of interest to evolutionary biologists is the question of the genetic difference between populations. Genetic variation can be measured by allelic frequency differences at a large number of loci but it would be desirable to compress these data into a single statistic that measures how far apart, genetically speaking, are the two populations. One such statistic is Nei's genetic distance, D,

$$D = -\ln\left(\frac{\sum_{i=1}^{n} \sum_{j=1}^{m_i} p_{1ij}p_{2ij}}{\sqrt{\sum_{i=1}^{n} \sum_{j=1}^{m_i} p_{1ij}^2 \sum_{i=1}^{n} \sum_{j=1}^{m} p_{2ij}^2}}\right) \tag{3.16}$$

where n is the number of loci, m_i is the number of alleles at the ith locus and p_{1ij}, p_{2ij} are the frequencies of the jth allele at ith locus in populations 1 and 2, respectively. An estimate of D, $\hat{D}$, is made by substituting the observed allelic frequencies (as D is the standard symbol I retain it here rather than using θ). No maximum likelihood estimator is available and the estimate is known to be biased when the number of loci sampled is small. Mueller (1979) investigated the utility of the delta and jackknife methods to estimate D and its SE. He solved the problem of generating realistic distributions of allelic frequencies by using an observed data set as the hypothetical population. From this set of data he drew n loci at random with replacement to generate a sample from the population. Because, the original data set is designated as the true population, the value of D estimated from the observed data set using Eq. (3.6) is the true value of the simulated population. Mueller selected three disparate data sets for study ($D = 0.0157$, 0.499, and 1.08) and from each drew 5, 15, or 30 loci.

As expected, the delta method produced biased estimates (Table 3.7), with the bias decreasing with the number of loci sampled (also as expected). The jackknife bias was considerably smaller than that obtained with the delta method and acceptably small when 15 or more loci were sampled (Table 3.7). The estimated confidence intervals are virtually identical for the two estimation methods and both are too small when the number of loci sampled was 5 or 15 and for one population for all three sample sizes ($D = 0.0157$, Figure 3.3). These results suggest that use of either the jackknife or delta method depends, in this instance, upon the true distribution of allelic frequencies. The reduction in bias of the jackknife gives this method the edge but one would be advised to use simulation

Table 3.7 *Bias (%) in the Jackknife ($\tilde{D}_J$) and Delta ($\hat{D}_D$) methods of estimating Nei's genetic distance. Simulations based on bootstrapping of three observed data sets (see text for details)*

True D	Number of loci	Bias in $\tilde{D}_J$	Bias in $\hat{D}_D$
0.0157	5	0.0072	7.95
0.0157	15	0.122	2.2
0.0157	30	0.135	1.26
0.499	5	17.4	23.6
0.499	15	0.066	5.7
0.499	30	0.17	2.7
1.08	5	86.4	143.0
1.08	15	3.35	9.32
1.08	30	0.33	3.32

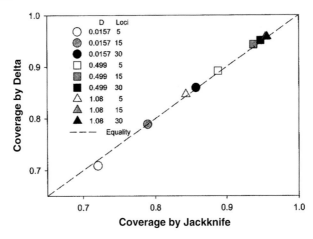

Figure 3.3 Proportion of cases in which true value of Nei's genetic distance fell within the estimated 95% confidence limits. Data from Mueller (1979).

to ensure the correct behavior of the estimator for a particular data set. The bootstrap Monte Carlo method described here can readily be used to generate the appropriate samples.

Summary

(1) The jackknife method proceeds by calculating the set of pseudovalues by sequential deletion of an observation (which might be a set of data as in the deletion of entire families in some genetical analyses) from the data set as given in Eq. (3.2).

(2) The jackknife estimate is the mean of the pseudovalues.

(3) The standard error (SE) of the estimate is estimated by the standard error of the pseudovalues.

(4) In some cases, a transformation of the estimate may improve the performance of the jackknife estimator.

(5) Hypothesis testing proceeds on the assumption that the pseudovalues are normally (or t) distributed.

(6) It cannot be assumed the jackknife will work in all circumstances and its reliability should be checked by simulation.

(7) Simulation of the data can be accomplished either by use of the specified statistical model or by bootstrapping the observed data set (Appendix C.3.9).

Further reading

Efron, B. (1982). *The Jackknife, the Bootstrap and Other Resampling Plans*. Society for Industrial and Applied Mathematics, Philadelphia.

Manly, B. F. J. (1997). *Randomization, Bootstrap and Monte Carlo Methods in Biology*. New York: Chapman and Hall.

Miller, R. G. (1974). The jackknife – a review. *Biometrika*, **61**, 1–15.

Exercises

(3.1) Generate 100 values from a normal distribution with mean zero and unit variance (i.e., $N(0,1)$). Jackknife the variance, test for normality of the pseudovalues (using the Shapiro–Wilkes test or other suitable test) and plot a histogram of the pseudovalues. (Hint: see Appendices C.3.1 and C.3.2.)

(3.2) Fit a linear regression to the data listed below using least squares regression and the jackknife. Test the hypothesis that the slope and intercept equal 0. Test the hypothesis that the slope equals 1 and the intercept equals 0.

x	1.63	4.25	3.17	6.46	0.84	0.83	2.03	9.78	4.39	2.72	9.68	7.88	0.21	9.08	9.04	5.59	3.73	7.98	3.85 8.18
y	2.79	3.72	4.09	5.89	0.75	−0.13	1.76	8.44	5.15	2.16	9.88	6.95	0.03	7.50	9.92	5.37	3.79	7.18	3.37 7.81

(3.3) Using the above data calculate the jackknifed correlation coefficient and test the hypothesis $\rho=0$. Do the analysis using no transformation and also using Fisher's *z*,

$$z = \frac{1}{2}\ln\left(\frac{1+r}{1-r}\right)$$

(3.4) Generate 1000 correlations using the following coding.

```
set.seed(1)                  # Set seed for random number generator
n <- 1000                    # Number of points
x <- runif(n,0,10)     # Construct X values evenly spaced from 1 to 10
error <- rnorm(n, mean=0, sd=1)   # Generate error term
y <- x + error               # Construct Y values
xy <- cbind(x,y)             # Data set to be examined
```

Calculate the pseudovalues for the untransformed correlation and for Fisher's z transformation. Examine the statistics, test for normality, and plot the data. Is the transformation useful?

(3.5) The table below shows the number of eggs laid by 20 female *Drosophila* of the indicated ages. Estimate the parameters of the function Eggs $= \theta_1(1 - e^{-\theta_2\text{Age}})e^{-\theta_3\text{Age}}$ using both MLE and the jackknife.

Ind	1	2	3	4	5	6	7	8	9	10	11	12	13	14	15	16	17	18	19	20
Age	1	3	2	4	1	1	2	5	3	2	5	4	1	5	5	3	2	4	2	5
Eggs	58	70	72	65	57	56	71	59	71	70	60	65	57	59	61	70	71	65	70	60

(3.6) Using the bootstrap method of simulating data, generate a sample of 10 observations from the data given in question 3.3, calculate the pseudovalues, test for normality and compare the results with the results from question 3.3.

List of symbols used in Chapter 3

ε	Error term
θ	Parameter to be estimated
$\hat{\theta}$	Estimate of θ
$\tilde{\theta}_{-i}$	Estimate of θ with the ith datum removed
$\tilde{\theta}$	Jackknife estimate ($=$ mean of pseudovalues)
σ	Standard deviation
C	Resource utilization index
D	Nei's genetic distance
G	Genetic variance–covariance matrix
K	Carrying capacity
MLE	Maximum likelihood estimate(s)
N	Population size or Total number of observations
$N(\mu,\sigma)$	Normal distribution with mean μ and standard deviation σ
P	Probability
S_i	ith pseudovalue
SE(.)	Standard error of term in parentheses
X	Data matrix
c	Constant
l_i	Survival to age i

m_i	Number of female offspring at age i
n	Number of observations or number of subgroups
p	Allelic frequency or proportion of resources used
r	Rate of increase of a population
t	Age
x	Observed value
$\bar{x}$	Mean value of x

4

The bootstrap

Introduction

In Chapter 3, I introduced the idea of using an observed distribution as a descriptor of a hypothetical distribution in order to test the efficacy of a statistical method. The bootstrap method takes a similar approach, in that it attempts to generate point estimates and confidence limits by taking random samples from the observed distribution. The underlying rationale behind the method is that the observed distribution is itself an adequate descriptor of the true distribution. The term bootstrap was given by Efron (1979) and derives from the saying "to pull oneself up by one's bootstraps," which refers to accomplishing something seemingly impossible by one's own efforts (supposedly from the book *Adventures of Baron Munchausen* by Rudolph Erich Raspe, though I have not been able to find the incident in my copy).

Suppose that our observed data consisted of a huge number of observations: in this case, it is clear that sampling from this distribution is equivalent to sampling from the original distribution. Herein lies the rub – if the sample is not huge then the observed distribution might be a poor descriptor. This is particularly true if the statistic to be estimated is very sensitive to outliers and the underlying distribution is skewed. The hope by many that a sample as small as 20 observations is an adequate representation of the underlying distribution is probably folly in the extreme (I have seen bootstrapping on samples as small as five, which is getting somewhat absurd). I cannot stress strongly enough that the bootstrap technique should not be used without theoretical or empirical verification that, in the particular circumstance proposed, it does actually work.

I shall follow the same route as in the previous chapter, presenting first an overview of the methods and then a series of examples illustrating the strengths and weaknesses of the approach, as well as its implementation.

Point estimation

Standard estimate

The bootstrap is not really a method for point estimation, as this is accomplished by the statistic that is itself to be bootstrapped and the bootstrap does not remove bias, as can the jackknife. However, if it fails in this regard, it is likely to fail in the more important issue of estimating confidence limits.

The basic steps in the bootstrap are as follows: suppose we have a set of n observations from which we estimate a parameter (or set of parameters) θ, the estimate being denoted as $\hat{\theta}$. To generate a bootstrap sample we randomly sample with replacement n observations from our original data set and calculate the estimate of θ, this bootstrap estimate being denoted θ_1^*. We repeat this procedure B times generating B bootstrap replicates $(\theta_1^*, \theta_2^*, \theta_3^*, \ldots, \theta_B^*)$. From this set of estimates we estimate the statistics of interest, generally the point estimate and its confidence limits. There are several alternatives for both.

The simplest bootstrap point estimate is the mean of the bootstrap replicates, θ^*,

$$\theta^* = \frac{1}{B} \sum_{i=1}^{B} \theta_i^* \tag{4.1}$$

Bias-adjusted bootstrap estimate

If $\hat{\theta}$ is a biased estimate of θ, then θ^* will itself be biased, because it estimates $\hat{\theta}$ rather than θ. The bias of $\hat{\theta}$ is defined as

$$\hat{\theta}_{\text{Bias}} = \hat{\theta} - \theta \tag{4.2}$$

which can be estimated by

$$\text{Est}\,(\hat{\theta}_{\text{Bias}}) = \theta^* - \hat{\theta} \tag{4.3}$$

Therefore, we can compute a bias-adjusted bootstrap estimate, θ_A^*, as follows: rearranging Eq. (4.2) gives $\theta = \hat{\theta} - \hat{\theta}_{\text{Bias}}$ and substituting the estimate of the bias (Eq. (4.3)) gives

$$\theta_A^* = 2\hat{\theta} - \theta^* \tag{4.4}$$

For other bias correction methods, see Efron and Tibshirani (1993) and Davison and Hinkley (1999).

Interval estimation

There are a number of ways to estimate confidence intervals, from the very simple SE method to the relatively complex accelerated bias-corrected percentile method. The different methods do not give the same interval, particularly if the distribution is highly skewed. A priori it is difficult to decide which might be the best method and one generally will have to resort to simulation. The number of replications for any single estimate typically places no strain on computer time, but verifying the behavior using simulation can definitely require very large amounts of computer time. Coding to calculate these different intervals is provided later in the examples section. To calculate a SE requires less replicates than the other methods discussed in this section (50–200) but in general 1000 replications are suggested (it is the default in the bootstrap routine of S-PLUS).

Method 1: standard error approach

If we can assume that the bootstrap estimator is normally distributed (or at least approximately so), we can estimate the 95% confidence interval as ± 2 SE of the estimate (to be precise ± 1.96SE, Figure 4.1). If B is small then the t value with $B-1$ degrees of freedom should be used instead of 1.96, but the bootstrap is so likely to be unreliable for such small samples that this is hardly worth

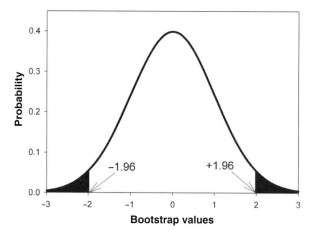

Figure 4.1 Standard error approach to estimate 95% bootstrap confidence interval. The distribution is assumed to be normal and the SE computed directly from the bootstrap distribution (bootstrap values are standardized to zero mean and unit variance).

considering. The SE of the estimate is estimated by the *standard deviation* of the bootstrap replicates

$$\mathrm{SE}(\hat{\theta}) = \sqrt{\frac{1}{B-1}\sum_{i=1}^{B}(\theta_i^* - \theta^*)^2}$$
(4.5)

Method 2: first percentile method

The distribution of bootstrap replicates is a descriptor of the distribution of the parameter(s) of interest. Therefore, rather than assuming normality, we can use the distribution itself to assign lower and upper confidence intervals (Figure 4.2). We rank the bootstrap replicates from lowest to highest; assuming no ties, we find the lower limit by moving down the column until we locate the

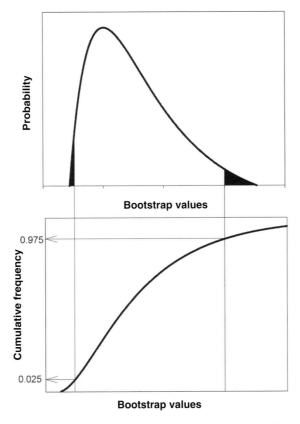

Figure 4.2 Percentile method of estimating 95% bootstrap confidence interval. The interval is selected to include 2.5% on the left and 2.5% on the right sides of the distribution (upper panel). To find these values the bootstrap values are ranked and the cumulative frequency curve constructed as shown in the lower panel.

bootstrap replicate value at which 2.5% of the replicates lie above the value. To locate the upper limit, we move down until 97.5% of the values lie above the bootstrap replicate value. If the distribution is symmetrical (e.g., normal, uniform), then the limits will be symmetric about the bootstrap estimate, otherwise they will be asymmetric.

Method 3: second percentile method

The first percentile method bootstraps the original data whereas in this method the attempt is to bootstrap the error distribution, the rationale being that any estimate is composed of the true value plus error (i.e., $\hat{\theta} = \theta + \varepsilon$, where ε is distributed in some fashion). The bootstrap distribution of errors is found by calculating the usual bootstrap replicate and then estimating the bootstrap replicate error as

$$\varepsilon_i^* = \theta_i^* - \hat{\theta} \tag{4.6}$$

The upper and lower values for the errors are found in the same manner as in the first percentile method and these are subtracted from the estimate

$$\hat{\theta} - \varepsilon_U < \theta < \hat{\theta} + \varepsilon_L \tag{4.7}$$

where ε_U is the upper limit for the error and ε_L is the lower limit (the "odd" arrangement in the above equation stems from the definition of the errors given in Eq. (4.6)).

Method 4: bias-corrected percentile method

The foregoing methods suffer from the same problem as the standard bootstrap estimator, namely that the initial sample is, to a greater or lesser degree, a biased sample of the distribution, i.e., these methods assume that $\theta^* - \hat{\theta}$ and $\hat{\theta} - \theta$ are distributed about zero. To correct for this bias, we assume that there exists some transformation, say $f()$, that normalizes the two quantities about a mean of $z_0\sigma$, where σ is the standard deviation of the distribution,

$$f(\theta^*) - f(\hat{\theta}) \sim N(z_0\sigma, \sigma) \quad \text{and} \quad f(\hat{\theta}) - f(\theta) \sim N(z_0\sigma, \sigma) \tag{4.8}$$

What is useful about this assumption is that we do not have to actually estimate the function, merely assume that it exists. We proceed using the following steps (Figure 4.3):

(1)   The value of z_0 is the value of z, the abscissa of the standard normal distribution, corresponding to the proportion of replicate bootstrap

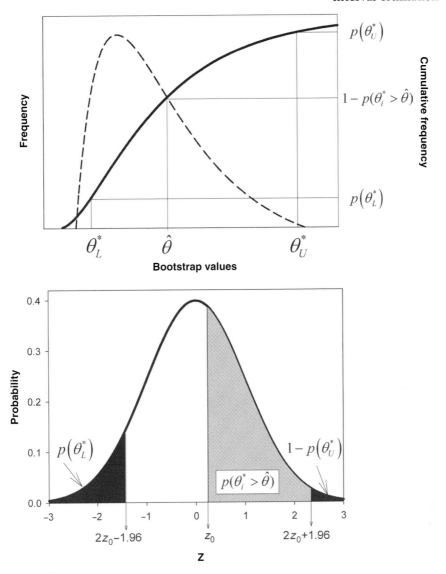

Figure 4.3 Bias-corrected percentile method of estimating 95% bootstrap confidence interval. The upper panel shows the distribution of bootstrap values (dotted line) and its cumulative distribution (solid line). The lower panel shows the standard normal distribution and the values to be estimated from it. The cross-hatched area and the solid area on the right tail is equal to the proportion of bootstrap replicates that are greater than the observed values, $p(\theta_i^* > \hat{\theta})$, and z_0 is the value on the abscissa corresponding to this proportion. The proportion corresponding to $p(z < 2z_0 - 1.96)$ is labeled $p(\theta_L^*)$. The lower 95% confidence limit for the bootstrap estimate, θ_L^*, is found by reading of the bootstrap value that corresponds to this proportion on the cumulative frequency curve (upper panel). The upper 95% confidence limit, θ_U^*, is found similarly.

estimates (θ_i^*) that exceed the observed estimate, $\hat{\theta}$. For example, if this proportion is 0.4013 then $z_0 = 0.25$.

(2) For the standard normal distribution, the upper and lower confidence limits are $-z_{\alpha/2}$ to $+z_{\alpha/2}$, where α is the type 1 error probability. Thus for the 95% confidence limit $+z_{\alpha/2} = 1.96$. For the transformed distribution, we can write the confidence region as

$$-z_{\alpha/2} < f(\hat{\theta}) - f(\theta) + z_0 < z_{\alpha/2} \qquad (4.9)$$

Rearranging this gives the confidence region for $f(\theta)$ as

$$f(\hat{\theta}) + z_0 - z_{\alpha/2} < f(\theta) < f(\hat{\theta}) + z_0 + z_{\alpha/2} \qquad (4.10)$$

Now here comes the elegant part: what we want to do is to find the bootstrap value that corresponds to the right and left sides of the inequality signs, i.e., for the upper confidence value we wish to find the value of θ^* for which the probability of observing $f(\theta^*)$ is less than $f(\hat{\theta}) + z_0 + z_{\alpha/2}$. Letting the upper confidence value for θ^* be θ_U^* and the probability be $p(\theta_U^*)$ we have

$$p(\theta_U^*) = \text{Prob}\left\{ f(\theta_U^*) < f(\hat{\theta}) + z_0 + z_{\alpha/2} \right\}$$

$$= \text{Prob}\left\{ f(\theta_U^*) - f(\hat{\theta}) + z_0 < z_0 + z_{\alpha/2} + z_0 \right\} \qquad (4.11)$$

$$= \text{Prob}\left\{ z < 2z_0 + z_{\alpha/2} \right\}$$

(3) The upper confidence bound for the bootstrap is the bootstrap value that corresponds to this probability in the cumulative frequency distribution of the bootstrap values (Figure 4.3).

(4) The lower confidence bound, θ_L^*, is found in a similar manner using

$$p(\theta_L^*) = \text{Prob}\left\{ z < 2z_0 - z_{\alpha/2} \right\} \qquad (4.12)$$

Method 5: accelerated bias-corrected percentile method

The above method makes the assumption that the standard deviation of $f(\hat{\theta})$ is a constant. However, standard deviations frequently increase with the value of the parameter and hence it might be more reasonable to assume that $f(\hat{\theta})$ is an increasing function of $f(\theta)$. The accelerated bias-corrected percentile method approaches this assumption by making the more limited assumption that $f(\hat{\theta})$ is a linear function of $f(\theta)$, namely, $\sigma = 1 + af(\theta)$. This addition leads to a more complex algorithm for computing the confidence bounds that involves the jackknife. The procedure is as follows:

(1) Calculate z_0 as in the bias-corrected percentile method.
(2) Estimate a by jackknifing the original data set and applying the equation

$$a = \frac{\sum_{i=1}^{n} (\tilde{\theta} - \tilde{\theta}_{-i})^3}{6\left[\sum_{i=1}^{n} (\tilde{\theta} - \tilde{\theta}_{-i})^2\right]^{1.5}} \tag{4.13}$$

(3) Estimate $p(\theta_U^*)$ and $p(\theta_L^*)$ using the equations

$$p(\theta_U^*) = \text{Prob}\left\{z < \frac{z_0 + z_{\alpha/2}}{1 - a(z_0 - z_{\alpha/2})} + z_0\right\}$$

$$p(\theta_L^*) = \text{Prob}\left\{z < \frac{z_0 - z_{\alpha/2}}{1 - a(z_0 - z_{\alpha/2})} + z_0\right\} \tag{4.14}$$

(4) Read off the relevant bootstrap values using the cumulative frequency function of bootstrap values. This method is termed "BCa" in S-PLUS.

Method 6: percentile-t-method

The previous percentile methods may be sensitive to skewed distributions and the distribution of bootstrap values may not, in this case, correctly mimic the true distribution. A proposed method to overcome this deficiency is the percentile-*t*-method, which transforms the bootstrap values into what is hoped to be a statistic that has *t* distribution. To use this method, it is necessary to be able to calculate a SE for the parameter being estimated. Assuming this to be true, the transformed bootstrap replicate, T_i^* is given by

$$T_i^* = \frac{\theta_i^* - \hat{\theta}}{\hat{\sigma}_{\theta_i^*}} \tag{4.15}$$

where $\hat{\sigma}_{\theta_i^*}$ is the estimated SE of θ_i^*. If there exists no analytical method for estimating this SE it can be estimated either by bootstrapping the bootstrap replicate or by the jackknife. Clearly, there is considerable computational effort required in this method! From the distribution of T_i^* values, we find the lower and upper values that satisfy the required confidence limits; for example, for the 95% confidence limits find the values of T_i^* that are 2.5 and 97.5% above the lowest value. Letting these values be T_L and T_U, we estimate the confidence interval of the parameter as

$$\theta_L^* = \hat{\theta} - T_L\hat{\sigma}_{\hat{\theta}}$$
$$\theta_U^* = \hat{\theta} - T_U\hat{\sigma}_{\hat{\theta}} \tag{4.16}$$

where $\hat{\sigma}_{\hat{\theta}}$ is an estimate of the SE of $\hat{\theta}$. Because of the very large computational effort required when the SE cannot be estimated except by bootstrapping the bootstraps, Tibshirani (1988) developed an alternate procedure known as the variance-stabilized bootstrap-t. Whereas the percentile-t-method may require 25000 bootstraps the variance-stabilized method can require as few as 3500 evaluations.

There are other methods of estimating confidence limits but, in general, the only way to decide which method is appropriate is to carry out a simulation analysis. Unfortunately, the performance of the different methods, as will be shown by the examples discussed later, can vary dramatically.

Hypothesis testing

The bootstrap is not used much for hypothesis testing and some caution should be exercised in determining the appropriate method. A simple test for $\theta = C$ is to see if C is contained within the estimated confidence limits. In general, assuming normality, we would construct the t-statistic $t = |\hat{\theta} - C|/\sigma_{\hat{\theta}}$, where $\sigma_{\hat{\theta}}$ is an estimate of the SE of $\hat{\theta}$. One might be tempted to then use the statistic $t = |\theta^* - C|/\text{SE}(\theta^*)$. However, this test does not take into account the variability that is inherent in the fact that the bootstrap estimate is an estimate from an estimate. A more reasonable approach is to compare the difference between the bootstrapped value and the estimate against the scaled difference between the estimate and the hypothesized value

$$t_i^* = \frac{|\theta_i^* - \hat{\theta}|}{\hat{\sigma}_{\theta_i^*}} \quad \text{vs.} \quad t = \frac{|\hat{\theta} - C|}{\hat{\sigma}_{\hat{\theta}}} \tag{4.17}$$

For each bootstrap replicate, the statistic t_i^* is calculated and compared with t. The probability under the null hypothesis (H_0: $\theta = C$) of obtaining the observed value, P, is then estimated as

$$P = \frac{n_{t_i^* > t}}{n} \tag{4.18}$$

where $n_{t_i^* > t}$ is the number of cases in which $t_i^* > t$. A simpler test is to compare only the numerators,

$$d^* = |\theta_i^* - \hat{\theta}| \quad \text{vs.} \quad d = |\hat{\theta} - C| \tag{4.19}$$

A potential problem with d^* is that it can be very sensitive to the variation in the sample, with outliers playing a perhaps overly important role. Of course, one can examine the distribution for potential problems.

The bootstrap can be applied to test hypotheses on multiple populations but a better approach is a randomization test, as discussed in the next chapter.

Examples of the use of the bootstrap

Bootstrapping the mean

Bootstrapping to obtain an estimate of the mean is useful both to illustrate the implementation of the method and to highlight some of the restrictions. Bootstrap procedures, like the jackknife, are becoming part of standard statistical packages, although some packages are better than others with respect to flexibility. The S-PLUS module is particularly good in this regard as it stores the bootstrap replicates and hence permits the estimation of types of estimates not given directly by the routine.

Appendix C.4.1 shows coding to generate 30 random normal values ($N(0,1)$), produce 1000 bootstrap values and then output the statistics computed by the routine "bootstrap." The distribution of bootstrap replicate values is, unsurprisingly, not significantly different from normal (Shapiro–Wilkes Normality Test, $W = 0.9987$, $P = 0.6771$, Figure 4.4). The output includes the bootstrap estimate, the SE, and percentiles estimated using Methods 2 (first percentile method, termed the "Empirical Percentiles" in the output) and 5 (accelerated bias-corrected percentile method, termed "BCa Confidence Limits" in the output).

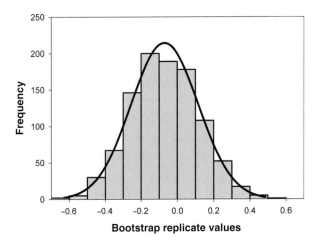

Figure 4.4 Distribution of 1000 bootstrap replicate estimates of the mean from an initial sample of 30 observations drawn from a normal distribution with mean zero and unit variance. Normal distribution superimposed using observed mean and variance.

Table 4.1 *Summary of coverage for three methods of bootstrap estimation (Method 1 = SE, Method 2 = EP, Method 5 = BCa) of the mean and variance from a normal distribution with mean zero and unit variance*

Parameter	N^a	Lower limit			Upper limit			Overall coverage		
		SE	EP	BCa	SE	EP	BCa	SE	EP	BCa
Mean	10	0.046	0.050	0.044	0.044	0.050	0.048	0.910	0.900	0.808
	30	0.032	0.040	0.038	0.020	0.022	0.022	0.948	0.938	0.940
Variance^b	10	0.002	0.002	0.034	0.234	0.232	0.164	0.764	0.766	0.802
	30	0.006	0.018	0.068	0.094	0.088	0.048	0.900	0.894	0.884

^{a}Sample size.
^{b}1000 bootstrap replicates, otherwise 250.

The limits using the SE (Method 1, ± 1.96 SE) are shifted downwards relative to either other estimate (-0.4352 to 0.2939 vs. -0.4257 to 0.2971 and -0.4179 to 0.3105).

To test the efficacy of these different methods, I generated 500 data sets of 10 and 30 observations each and computed the confidence limits for these using 250 bootstrap replicates (this is less than the generally recommended 1000 for the percentile methods but given the normality of the bootstrap in the present case the lower number should be sufficient. See Appendix C.4.2 for coding). Using either the uncorrected or bias-corrected estimate, a one-sample t-test showed no overall bias in the estimates ($P = 0.371$ and 0.399, respectively). For a sample size of 30, the SE method gave the overall confidence limit closest to the desired 95% (94.8%), but all three were acceptable (93.8 and 94% for the other two, Table 4.1). There appears to be a tendency for the lower limit to be somewhat optimistic (instead of excluding 2.5%, the three methods excluded 3.2, 4, and 3.8%).

A sample size of 30 is not an unreasonable size to mimic a symmetric distribution such as the normal distribution but what happens if the size is reduced? With a sample size of 10, the bootstrap performed very poorly, with the t-tests indicating bias ($P = 0.046$ and 0.042 for the uncorrected and bias-corrected estimate, respectively) and the confidence limits being excessively small (91, 90, and 88% for the three methods, Table 4.1). Increasing the number of bootstrap replicates to 1000 decreased the bias ($P = 0.1417$ and 0.1412 for the uncorrected and bias-corrected estimate, respectively), but did not improve the estimate of the confidence intervals (now 89, 88, and 89%). Similarly reduced confidence limits have been found when estimating the mean from a log-normal distribution, though a bimodal distribution tended to produce limits slightly too large (Table 4.2).

Take-home message: "the bootstrap cannot compensate for poor sampling!"

Table 4.2 *Overall coverage for various methods of bootstrap estimation of 95% confidence limits for several different statistical models. From Mooney and Duval (1993)*

Method	Mean of a log-normal	Mean of a bimodal	OLS regression[a]
Parametric[b]	0.920	0.970	0.937
Standard error (1)	0.809	0.957	0.924
First percentile (2)	0.916	0.961	0.928
Bias-corrected (4)	0.916	0.959	0.923
Percentile-*t* (6)	0.941	0.969	0.949

[a]Ordinary least squares regression with skewed error term.
[b]Confidence limits set as mean $\pm t_{0.025}$SE, where SE is the parametric SE and $t_{0.025} = 2.064$ (24 df). In all cases the sample size was 25.

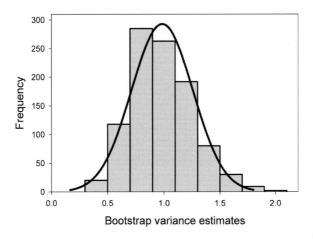

Figure 4.5 Distribution of 1000 bootstrap replicate estimates of the variance from an initial sample of 30 observations drawn from a normal distribution with mean zero and unit variance. Normal distribution superimposed using observed mean and variance.

Bootstrapping the variance

Whereas the mean is normally distributed, the distribution of variance estimates is skewed, as can be illustrated by bootstrapping the variance estimate in the previous example. As expected, the distribution of replicates is more skewed than for the mean, and is significantly different from normal ($W = 0.9827$, $P < 0.001$), though the deviation is actually not particularly pronounced (Figure 4.5). The observed estimate is 1.031 and the standard bootstrap estimate is 0.9851. The three estimated confidence limits are quite disparate: 0.452–1.518 (SE method), 0.520–1.571 (empirical percentile method), and 0.644–1.889

(BCa method). The results from the analysis of 500 replications using 1000 bootstraps to estimate confidence limits for each simulated data set indicates that the confidence limits are too small (Table 4.1), but there is little difference among the three methods. The problem of excessively small confidence limits is considerably exacerbated when the sample size is reduced to 10 (Table 4.1) and the method is quite unacceptable. This should come as no surprise really, as a sample size of 10 simply has very little information about the variance. The message to be reiterated throughout this chapter is that the bootstrap can only deal with the information contained in the sample and if the sample is small then in all likelihood the bootstrap will perform poorly. How "small" is "small" can, in general, only be decided by a simulation analysis. It is essential to test the efficacy of the bootstrap before using it!

More on sample size and confidence intervals: the Gini coefficient

In ecological research, much interest has been focused on size hierarchies. An index used to assess size inequality is the Gini coefficient of inequality defined as

$$\theta = \frac{\sum_{i=1}^{n} (2i - n - 1)x_i}{n^2 \mu} \tag{4.20}$$

where x_i is the size of the ith individual after all individuals have been ranked by size (smallest to largest) and μ is the population mean. An unbiased estimate of this coefficient is (Damgaard and Weiner 2000)

$$\hat{\theta} = \left(\frac{\sum_{i=1}^{n} (2i - n - 1)x_i}{n^2 \bar{x}} \right) \left(\frac{n}{n - 1} \right) \tag{4.21}$$

Obviously, the distribution of the data can vary enormously, and hence a general method of estimating the confidence limits is highly desirable. Dixon *et al.* (1987) tested the bootstrap approach using three simulated distributions: uniform, log-normal, and truncated normal (Appendix C.4.3 gives sample coding for a uniform distribution. Note that there is a typographical error in the equation given by Dixon *et al.*). For these distributions, it is possible to find analytical expressions for the expected value of the Gini coefficient (see note in Table 4.3) and hence evaluate the estimated confidence limits. Confidence limits were estimated using a variety of percentile methods and, as all gave more or less the same results, Dixon *et al.* present only the results for the empirical percentile method (Method 2). The results mirror those already observed for the mean and the variance, namely that the

Table 4.3 *Overall coverage using the percentile bootstrap method for the Gini coefficient from simulated populations [Adapted from Dixon* et al. *(1987)]*

		Sample size		
Distribution of data	Gini coefficient	20	100	250
Uniform $(1,19)^a$	0.30	0.918	0.954	0.943
Log-normal $(0,0.54)^b$	0.30	0.851	0.925	0.952
Truncated normal $(0,1)^c$	0.41	0.925	0.927	0.939
Log-normal $(0,1)^b$	0.52	0.782	0.879	—

[a](a,b) = minimum value, maximum value.
[b]Generate y from a normal distribution, $N(\mu,\sigma)$ and transform $x=\exp(y)$.
[c]Generate a random normal deviate, $N(\mu,\sigma)$ and accept if value greater than zero.
Analytical Population Gini coefficients:
[a]Uniform: $(b-a)/[3(b+a)]$.
[b]Log-normal: $2\Phi(\sigma/\sqrt{2})-1$, where $\Phi(x)$ is the cumulative normal.
[c]Truncated-normal: $4\Phi(0)+\sqrt{2}-3=\sqrt{2}-1$.

confidence limits are too small unless the sample size is large, in this case 250 (Table 4.3).

Yet more on sample size and the bootstrap: not all bootstraps are created equal

We have already found that not all bootstraps behave optimally and that sample size is a critical element (Tables 4.2 and 4.3). This pattern is particularly well illustrated by the analysis of Cordell and Carpenter (2000) on the estimation of a risk parameter. The mapping of loci associated with susceptibility to disease for which the actual mode of inheritance is unknown is commonly done using affected-sib-pair linkage methods. Three parameters are typically estimated by these methods: the probabilities that an affected-sib-pair share 0, 1, or 2 alleles identical by descent (which means that they were derived from the same mutational event) at a disease locus. Two methods of determining confidence limits based on asymptotic theory have been proposed, termed the profile likelihood method and the multivariate normal approximation. While such methods are valid for "large" samples (and what is meant by "large" is not clear) they may fail on "small" samples. Using simulation modeling, Cordell and Carpenter tested the efficacy of these two methods and compared them to nine possible bootstrap methods (Table 4.4). A total of 1000 simulations were run from which we expect 50 "significant" results by chance for any given method. Rearrangement of the goodness of fit formula for χ^2 gives the acceptable range

Table 4.4 *Percentage coverage for nine different bootstrap methods of estimating the probability that an affected-sib-pair share no alleles identical by descent. The results for the other two parameters (probability of one or two alleles shared) are qualitatively similar*

Method	Sample size				Reference[a]
	50	100	200	500	
Non-studentized pivotal (1)[b]	70.8	74.9	88.4	94.4	Efron (1981)
Bootstrap-*t* (6)	79.8	85.1	**95.4**	**96.3**	Efron (1981)
Variance stabilized bootstrap-*t*	70.8	85.1	**95.2**	**95.7**	Tibshirani (1988)
Percentile (2)	99.2	90.8	92.6	**94.9**	Efron and Tibshirani (1993)
Bias-corrected percentile (4)	78.5	90.4	92.8	**94.5**	Efron (1982)
Bias-corrected and accelerated (5)	76.7	93.1	**94.3**	**94.9**	Efron (1987)
Test-inversion bootstrap	97.4	96.5	**95.9**	96.5	Carpenter (1999)
Studentized test-inversion bootstrap	97.2	**96.0**	**95.5**	96.7	Carpenter (1999)
Bootstrap profile likelihood	97.8	93.6	**95.6**	**95.6**	Carpenter (1998)
Profile likelihood	98.2	**95.5**	92.1	**95.7**	Cordell and Olson (1997)
Multivariate normal approximation	99.5	90.7	92.8	**94.6**	Cordell and Olson (1997)

Taken from Cordell and Carpenter (2000). Bold font indicates values not significantly different from the expected 0.95.

[a]Paper describing the method.

[b]Numbers in parentheses give the number for the methods described in this chapter.

in coverage to be $0.95 \pm \sqrt{((3.84)(0.95)(0.05)/1000)} = 93.7\text{--}96.4\%$. At a sample size of 50, none of the methods give acceptable coverage, either being too small or too large (Table 4.4). The profile likelihood method and one bootstrap method are acceptable given a sample size of 100. Six bootstrap methods are acceptable when the sample size is increased to 200 and 500, though not the same six for the two sample sizes. No method is consistently acceptable but there is a tendency for the more complex bootstrap methods to perform better than the simpler ones. Without prior simulation, the poor performance of the methods at the lower sample sizes would not be known. To use such methods without verification runs the very real risk of using a method that in the particular circumstance produces wildly optimistic confidence limits (e.g., results for the first three bootstrap methods at a sample size of 50).

> *The bootstrap can produce confidence limits that are too large: an example from niche overlap*

In the previous chapter, we considered four measures of niche overlap and the ability of the jackknife to estimate the parameters and their confidence intervals. Mueller and Altenberg (1985), the authors of this study, also examined the utility of the bootstrap in a single case, that in which there are two resources, two species and two types of individuals within each species. Type I individuals of species 1 select the first resource with probability 0.80, whereas type II individuals of species 1 select this resource with probability 0.15. The probabilities for species 2 are reversed, so type I individuals select the resource with probability 0.15 and type II with probability 0.80. The proportion of type II individuals in the population is P_{II}. Because the possibility of two types of individuals in the population is not incorporated in the niche overlap indexes, Mueller and Altenberg termed P_{II} a contamination.

The four niche overlap measures, considered were

(1)   Coefficient of community (C_1): $C_1 = \sum_{i=1}^{2} \min(p_i, q_i)$
(2)   Morisita's index (C_2): $C_2 = 2 \sum_{i=1}^{2} p_i q_i / (\sum_{i=1}^{2} p_i^2 \sum_{i=1}^{2} q_i^2)$
(3)   Horn's   index   (C_3): $C_3 = (\sum_{i=1}^{2} (p_i + q_i) \log(p_i + q_i) - \sum_{i=1}^{2} p_i \log p_i - \sum_{i=1}^{2} q_i \log q_i)/2 \log 2$
(4)   Euclidean distance (C_4): $C_4 = 1 - \sqrt{\sum_{i=1}^{2} (p_i - q_i)^2 / 2}$.

The data were generated as follows: the first 10 individuals of each species were sampled, the probability of a type II individual sampled being P_{II}. A simple way to do this is to generate a uniform random number between 0 and 1; if this number is less than P_{II} the individual is designated as type II. The individual was then assigned 20 resource items dependent upon the resource utilization probabilities. Suppose, for example, a type I individual of species 1 was first selected; 20 random numbers are generated and the number of cases in which the random number is less than 0.80 totaled. This number represents the number of resource 1 items selected by this individual. The frequencies of resource 1 for the ith individuals of the two species can be written as two vectors $x_{i,1}, x_{i,2}, x_{i,3}, \ldots, x_{i,10}$ and $y_{i,1}, y_{i,2}, y_{i,3}, \ldots, y_{i,10}$, where x denotes species 1 and y denotes species 2. The original sample consists of two matrices with the rows representing individuals and the columns the resource items. A bootstrap replicate was formed by selecting 10 individuals (rows) with replacement from the two species matrices. For each individual, a sample of 20 resource items were chosen using the observed probability (e.g., for the ith individual of species 1 it would be x_i). A total of 1000 bootstrap replicates were so generated. The bias-adjusted bootstrap estimate was calculated and 95% confidence intervals calculated using the bias-corrected percentile method. To determine the actual confidence limits 1000 runs were made for three values of P_{II}.

The results of this simulation analysis are summarized in Table 4.5: in all cases the amount of bias was very small, even though in some cases it was

Table 4.5 *The percent bias, and overall coverage for the jackknife ($\tilde{C}_I$) and bootstrap (C_I^*) estimators of four indexes of niche overlap. As the results for the coefficient of community were identical to the Euclidean distance, only the former is shown. Adapted from Mueller and Altenberg (1985)*

Estimator	% bias $P_{II}=0$	Coverage	% bias $P_{II}=0.10$	Coverage	% bias $P_{II}=0.25$	Coverage
$\tilde{C}_1$	0.1	0.955	0.1	0.727	0.0^a	0.656
C_1^*	0.4	0.989	0.3	0.954	0.4	0.951
$\tilde{C}_2$	0.1	0.956	0.0^a	0.722	0.7*	0.644
C_2^*	0.5	0.985	0.2	0.949	0.3	0.954
$\tilde{C}_3$	0.0^a	0.952	0.6*	0.728	0.7*	0.659
C_3^*	0.1	0.985	0.0^a	0.950	0.1	0.955

*Bias significant.
^{a}0.0 = <0.05.

statistically significant. In the absence of any contamination ($P_{II}=0$), the jackknife gave correct confidence limits whereas the bootstrap produced confidence limits that were too large, enclosing approximately 99% rather than 95%. On the other hand, when there was contamination by a second type in the population, the jackknife consistently underestimated the confidence region but the bootstrap gave valid limits! The problem is how in a real circumstance does one determine the presence of contamination.

Bootstrapping when the unit is not the individual observation

In all the examples so far the data set has consisted of a single vector. In some cases, one might wish to bootstrap a data set in which the units to be bootstrapped themselves contain individual observations. An example is the estimation of quantitative genetic parameters. Recall that for the full sib design (e.g., *N* families with *n* individuals per family), the appropriate resampling unit is the family. The easiest method to do this is to use the blocking design described in Appendix C.3.6, replacing the "jackknife" routine with "`bootstrap (Data, H2.estimator, B=1000, trace=F)`." This example can also serve to illustrate the extreme caution one should use in adopting the bootstrap method. I generated bootstrap estimates for a small sample consisting of 16 families with four individuals per family, given a true heritability of 0.8 (see Appendix C.3.5 for coding to generate such data). To assess the accuracy of the bootstrap estimate and its estimated SE, I ran the simulation 100 times using 100 bootstraps per run to estimate the bootstrap estimate and SE (a suitable number for the SE

Table 4.6 *Comparison of heritability estimates using the standard (maximum likelihood) formula* $(\hat{h}^2)$, *the jackknife* $(\tilde{h}^2)$ *the standard bootstrap* (h^{2*}) *and the bias-adjusted bootstrap* (h_A^{2*}). *See text for details of simulation*

Statistic[a]	$\hat{h}^2$	$\tilde{h}^2$	h^{2*}	h_A^{2*}
mean	0.779	0.780	0.877	0.680
SE[b]	0.278	0.262	0.246	0.312
$\overline{Est(SE)}$[c]	0.267	0.285	0.213	0.213
LCL[d]	0.723	0.728	0.823	0.618
UCL[e]	0.834	0.832	0.926	0.742
Coverage	0.93	0.87	0.85	0.82

[a]Based on 100 runs of the simulation.
[b]Standard deviation of the 100 estimates.
[c]Mean of the 100 estimated SE using either the standard formula or bootstrap.
[d]Lower 95% confidence limit.
[e]Upper 95% confidence limit.

estimate but low if using the percentile methods). Both an estimate of heritability and its SE can be made using least squares methods (or restrictive maximum likelihood, which are identical given the balanced design used in the simulation) and these perform well, the mean estimate not being significantly biased (one-sample t-test, $t = 0.7585$, df $= 99$, $P = 0.4499$, Table 4.6). On the other hand, the bootstraps performed very poorly. Both the standard and the bias-adjusted estimates are significantly biased (one-sample t-tests, $t = 3.1433$, df $= 99$, $P = 0.0022$ for the standard bootstrap estimate; $t = 3.8338$, df $= 99$, $P = 0.0002$), with the standard being biased upwards and the bias-adjusted being biased downwards (Table 4.6. This is also indicated by the confidence limits on the 100 mean estimates)! Further, the estimated SE of the standard bootstrap are biased downwards, as has been observed in the previous examples for the bootstrap with small sample size. The bias in the bootstrap estimates leads to the estimated 95% intervals being smaller (85 and 82%).

The results for the jackknife are shown for comparison: in this case there is no bias in the estimate but the SE estimates tend to be underestimated (Table 4.6). The underestimate by the jackknife appear to be somewhat less severe than for the bootstrap, an observation frequently made when the two are compared: e.g., estimation of species number using presence and absence data (Mingoti and Meeden 1992) and species richness measures (reviewed in Hellman and Fowler 1999). For the estimation of population growth rate, discussed in the previous chapter, both the jackknife and bootstrap performed about equally well (Meyer *et al.* 1986).

Bootstrapping to estimate several parameters: linear regression

Thus far we have considered only single estimates, but in many cases we would wish to determine several parameters simultaneously. A simple example is linear regression, in which we wish to estimate both the intercept (θ_1) and the slope (θ_2). First, we shall investigate the situation in which the assumption of normal distribution of errors applies, i.e., $y = \theta_1 + \theta_2 x + \varepsilon$, where ε is a normal distribution with mean zero and standard deviation σ. Second, we shall examine the situation in which the error distribution has a mean of zero but is highly skewed. A convenient distribution to use is the gamma distribution, which has two parameters, called shape and rate (an alternate parameter is called "scale," and is simply the inverse of rate). "Shape" is a parameter that governs the skew (skewness$=2/\sqrt{\text{shape}}$), while "rate" is proportional to the inverse of the mean, which is equal to shape/rate. By varying the shape parameter, we can generate distributions that range from almost normal to very highly skewed (Figure 4.6). The one used in the present example was moderately skewed with shape equal to 2 (Appendix C.4.4, Figure 4.7).

The simulation was run with a sample size of 300 (Figure 4.7) and a sample size of 30 (Table 4.7). For the larger sample size, the least squares and jackknife estimates and SE are essentially identical. The bootstrap gives an intercept estimate (θ_1) that is twice as large in magnitude as obtained with the other two methods, but the true value of zero is still contained within the confidence limits. Overall, regardless of the error distribution there is no reason to use anything other than the least squares estimate, which is obviously considerably easier to calculate. The situation changes little when the sample size is drastically

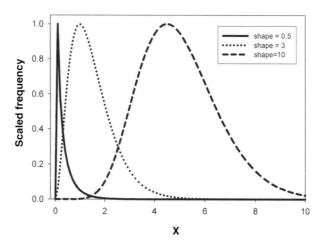

Figure 4.6 Three gamma distributions showing the effect of changing the shape parameter (rate parameter$=2$).

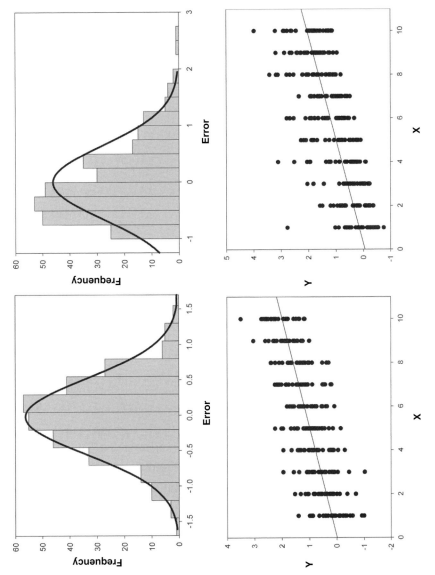

Figure 4.7 The distribution of the errors generated in the linear regression example ($n=300$) showing the fitted normal curves (upper panels) and the regressions (lower panels). Normal distribution of errors shown in left panels and gamma distributed errors shown in right panels.

Table 4.7 *Least squares (LS), jackknife (Jack.) and bootstrap (Boot.) estimates of linear regression coefficients (Equation, $y=\theta_1+\theta_2x+\varepsilon$). The true regression coefficients were $\theta_1=0$, $\theta_2=0.2$ and the error distribution (ε) was either normal with mean 0 and $\sigma = 0.5$, or a gamma distribution with shape and rate parameters equal to 2*

	θ_1		θ_2	
Method	Estimate	SE	Estimate	SE
Normal, Sample size = 300				
LS	-0.0093	0.0665	0.1990	0.0107
Jack.	-0.0093	0.0668	0.1990	0.0106
Boot.	-0.0180	0.0662	0.1990	0.0101
Gamma, Sample size = 300				
LS	-0.0535	0.0809	0.2095	0.0130
Jack.	-0.0535	0.0814	0.2095	0.0130
Boot.	-0.0538	0.0801	0.2095	0.0121
Normal, Sample size = 30				
LS	-0.0637	0.2037	0.2054	0.0328
Jack.	-0.0640	0.2216	0.2054	0.0338
Boot.	-0.0587	0.1935	0.2017	0.0317
Gamma, Sample size = 30				
LS	-0.0945	0.2462	0.1973	0.0397
Jack.	-0.0945	0.2670	0.1973	0.0413
Boot.	-0.0813	0.2548	0.1946	0.0445

reduced to 30, though the SE of the jackknife and bootstrap estimates are larger than those obtained with least squares. To see if the least squares confidence intervals are adequate, I made 1000 runs for both normal- and gamma-distributed errors (see Appendix C.4.5 for coding. Note that I have used the "by" routine rather than a loop. Thus, I first generate the entire data set (30 000 points) and then analyze by subsets of 30. This is more efficient than looping). For the normally distributed errors, the estimated 95% confidence limits enclosed the true values of the intercept (θ_1) and slope (θ_2) with probabilities 0.953 and 0.94, which is perfectly satisfactory. For the gamma distributed errors, the corresponding probabilities are 0.949 and 0.95, which is also perfectly satisfactory. Even with the extremely skewed gamma distribution shown in Figure 4.6 (shape=0.5), the probabilities are still 0.94 and 0.962!

Mooney and Duval (1993) carried out a similar analysis using a gamma distribution with a shape parameter of 3 and a sample size of 25 (Table 4.2, column labeled "OLS regression"). Instead of bootstrapping the values x and y, they use the bootstrapped residuals in the following manner. First, the least squares regression was fitted to the data, giving

$$\hat{y}_i = \hat{\theta}_1 + \hat{\theta}_2 x_i \qquad (4.22)$$

where $\hat{y}_i$ is the predicted value of y_i. Second, the residuals were calculated as

$$\hat{\varepsilon}_i = y_i - \hat{y}_i \qquad (4.23)$$

and the residuals calculated. Second, 25 bootstrap residuals were drawn from the set of observed residuals. These bootstrap residuals were added to the vector of estimated values of y

$$y_i^* = \hat{y}_i + \hat{\varepsilon}_i^* \qquad (4.24)$$

where $\hat{\varepsilon}_i^*$ is the ith bootstrap residual. Finally, the bootstrap estimates of the intercept and slope were obtained by regressing y^* on the x. Unfortunately, Mooney and Duval do not give the number of simulated runs made, nor whether the estimated coverages differ from the required 0.95. The results for least squares method (labeled "parametric" in Table 4.2) at 0.937 are very close to 0.95, but the first three bootstrap methods appear to underestimate the confidence region. The percentile-t method clearly gives appropriate coverage (0.949). The excellent coverage obtained using linear regression suggests that in only few cases of linear regression estimation will the bootstrap approach merit the effort required, particularly as one cannot be assured that the percentile-t method will generally work.

In summary, these results illustrate the considerable robustness of least squares regression to departures from normality. The use of either the jackknife or bootstrap approaches is not warranted in most cases of linear regression.

Bootstrapping to estimate several parameters: nonlinear regression

In the previous chapter, we explored the use of maximum likelihood and the jackknife to estimate parameter values and their associated SE of the two-parameter von Bertalanffy equation

$$l_t = \theta_1 (1 - e^{-\theta_2 t}) \qquad (4.25)$$

where l_t is the length at age t. As before, to generate suitable data, I simulated a population consisting of five age groups (1, 2, 3, 4, and 5) and drew from this population five individuals from each age group, the length of the ith individual in age group t, $l_{t,i}$, being generated by the expression $l_{t,i} = l_t + \varepsilon_{t,i}$, where

$\varepsilon_{t,i}$ is a random variable specific to this individual (see Appendix C.3.6 for coding). I set the true values at $\theta_1 = 100$ and $\theta_2 = 1$ and used three error distributions, all with zero mean: (1) A random normal with mean zero and standard deviation of 10 (this is the one shown in Box 3.7). This error distribution satisfies the assumption for the maximum likelihood method. (2) A uniform distribution from -5 to $+5$ (in S-PLUS, `Error <- runif(n, min=-5, max= 5)`, and (3) a random normal in which the standard deviation increased proportionally with age (in S-PLUS, `Error <- rnorm(n, 0, Age*2)`). The analysis showed that the MLE approach and jackknife were equivalent for the first two error distributions but that the jackknife performed better for the third type of error distribution. How well does the bootstrap perform?

The first issue to be decided is how to generate bootstrap values. One method is to select at random from the 25 individuals. However, this will result in a data set that is unbalanced relative to the original, because the bootstrapped data set will not generally contain five observations per age group (this is also an issue in the linear regression analysis discussed in the last section). An alternate approach in this circumstance is to bootstrap the residuals. For the present analysis, I have used the standard bootstrap, because it is very much easier to implement and there is no indication that bootstrapping the residuals is superior. Coding to bootstrap the von Bertalanffy function is given in Appendix C.4.6, and in Appendix C.4.7, I show one method of generating multiple samples to determine the actual coverage. One feature worth bringing the reader's attention to is the use of the "write" function. I found that the fitting routine (`nls`) would, on some interactions, fail to converge, which led to the program halting without saving any of the data. To avoid this, I had the program write the data directly to a text file at each cycle. This both saves the data as it is computed and reduces the memory demands of S-PLUS. Use of the option `"append=T"` means that multiple runs can be made without having to do any housekeeping on the file.

With a normal error distribution, the confidence limits on the two parameters using the SE approach (Method 1) were 0.96 and 0.94 (504 simulations), which is adequate. However, both the MLE and jackknife approaches also performed adequately for this type of error distribution and hence there would be little reason to use the computationally more intensive bootstrap. The situation that is more interesting is that in which the error variance increased with age, in which case the MLE approach didn't perform satisfactorily but the jackknife produced confidence limits reasonably close to 0.95 (0.944 and 0.963 for the first and second estimates, respectively); neither is significantly different from the required 0.95 ($\chi^2 = 0.7579$, df$=1$, $P=0.3840$; $\chi^2 = 3.5579$, df$=1$, $P=0.0593$). On the other hand, the bootstrap on the first parameter gave a coverage of 0.9321

(781 simulations), which is significantly different from the required 0.95 (χ^2=5.2457, df=1, P=0.0220), while the bootstrap on the second parameter gave a coverage of 0.9398, which is not significantly different from the required 0.95 (χ^2=1.7037, df=1, P=0.1918). It is possible that use of one of the percentile methods would produce a better fit to requirement. The fit is certainly very close and much better than the MLE method (which were 0.887 and 0.982, respectively, see Table 3.6). However, given the relative ease with which the jackknife can be fitted, there seems little reason to use the bootstrap, except as a check for consistency.

The bootstrap as a hypothesis testing method

Consider the problem of comparing the mean from two populations, presumed to differ only in this parameter. Suppose the observations $x_1, x_2, x_3, \ldots,$ $x_i, \ldots, x_n$ are distributed with probability $P(\theta)$, where θ is the mean of the distribution. The usual approach to this problem is to find a transformation that normalizes the distribution and then compare the means of the transformed variables or to compare the untransformed distributions using a non-parametric test such as the Mann–Whitney test. However, neither of these methods actually tests for a difference between the means. To illustrate, assume that the underlying distribution is log-normal, and that we have two samples, labeled x and y: in this case

$$\log x_i \sim N(\mu_x, \sigma^2), \quad \log y_i \sim N(\mu_2, \sigma^2) \tag{4.26}$$

The hypothesis that we wish to test is

$$H_0: \theta_x = \theta_y \tag{4.27}$$

But if we use a log transformation, which normalizes our data, the actual hypothesis under test is

$$H_0^L: \mu_x = \mu_y \tag{4.28}$$

Now θ is related to μ according to

$$\theta = e^{\mu + \frac{1}{2}\sigma^2} \tag{4.29}$$

Thus, a test of differences between μ_x and μ_y is not directly a test for differences between θ_x and θ_y. If the variances are the same (i.e., $\sigma_x^2 = \sigma_y^2$), then the tests are equivalent, but if they differ then it is possible to reject H_0 even if we cannot reject H_0^L. We can, of course, test for a difference between the variances, but it would be better to have a test that does not rely upon this intermediate step and, in any case, even if the variances are different, we would still wish to test for differences between means. Two such tests were proposed by Zhou et al. (1997), one being based on a likelihood approach and the other on the bootstrap. The likelihood test was termed the Z-score method and can be developed from the following considerations:

As discussed in Chapter 2, the ML estimators for μ_x and μ_y are, respectively

$$\hat{\mu}_x = \frac{1}{n_x} \sum_{i=1}^{n_x} \log x_i \quad \text{and} \quad \hat{\mu}_y = \frac{1}{n_y} \sum_{i=1}^{n_y} \log y_i \tag{4.30}$$

and the unbiased estimates of σ_x^2 and σ_y^2 are, respectively

$$\hat{\sigma}_x^2 = S_x^2 = \frac{1}{n_x - 1} \sum_{i=1}^{n_x} (\log x_i - \hat{\mu}_x^2)^2 \quad \text{and} \quad \hat{\sigma}_y^2 = S_y^2 = \frac{1}{n_y - 1} \sum_{i=1}^{n_y} (\log y_i - \hat{\mu}_y^2)^2 \tag{4.31}$$

The null hypothesis that we wish to test is $H_0: \theta_x = \theta_y$, and this is equivalent to testing $H_0: \log \theta_x = \log \theta_y$, or in terms of the log-normal distribution,

$$H_0: \mu_x + \frac{1}{2}\sigma_x^2 = \mu_y + \frac{1}{2}\sigma_y^2 \tag{4.32}$$

Zhou *et al.* (1997) proposed the Z-score statistic

$$Z = \frac{\left(\hat{\mu}_y + \frac{1}{2}S_y^2\right) - \left(\hat{\mu}_x + \frac{1}{2}S_x^2\right)}{\sqrt{\left(\frac{S_x^2}{n_x} + \frac{1}{2}\left[\frac{S_x^4}{n_x - 1}\right]\right) + \left(\frac{S_y^2}{n_y} + \frac{1}{2}\left[\frac{S_y^4}{n_y - 1}\right]\right)}} \tag{4.33}$$

Provided n_x and n_y are sufficiently large, under the null hypothesis, $H_0: \theta_x = \theta_y$, Z will be normally distributed with a mean of zero and unit standard deviation, i.e., $N(0,1)$.

The proposed bootstrap test is the t statistic method (Eq. (4.17)). We proceed by the following steps

(1) Estimate the observed t statistic by

$$t_{obs} = \frac{|\hat{\mu}_x - \hat{\mu}_y|}{\sqrt{\frac{S_x^2}{n_x} + \frac{S_y^2}{n_y}}} \tag{4.34}$$

(2) Estimate the grand mean, $\hat{\mu}_G$,

$$\hat{\mu}_G = \frac{1}{n_x + n_y} \left(\sum_{i=1}^{n_x} x_i + \sum_{i=1}^{n_y} y_i \right) \tag{4.35}$$

(3) Transform the two samples so that they have a common mean

$$X_i = x_i - \hat{\mu}_x + \hat{\mu}_G, \qquad Y_i = y_i - \hat{\mu}_y + \hat{\mu}_G, \tag{4.36}$$

(4) Sample with replacement n_x values of X_i and n_y values of Y_i.

(5) Compute the two bootstrap means X_1^*, Y_1^* and form the bootstrap t

$$t_1^* = \frac{|X_1^* - Y_2^*|}{\sqrt{\dfrac{S_{1,X}^{*2}}{n_X} + \dfrac{S_{1,Y}^{*2}}{n_Y}}} \tag{4.37}$$

where $S_{1,X}^{*2}$, $S_{1,Y}^{*2}$ are the estimated variances for the two bootstrap samples.

(6) Repeat steps 4 and 5 B times to obtain $t_1^*, t_2^*, t_3^*, \ldots, t_B^*$

(7) The estimated probability, P_{est}, under the null hypothesis of obtaining a value of t_i^* greater than t_{obs} is

$$P_{est} = \frac{\text{number of times } t_i^* > t_{obs}}{B} \tag{4.38}$$

To test the two methods Zhou *et al.* (1997) ran simulations with five different scenarios in which the data were drawn from two log-normal distributions with different means and variances on the transformed scale, but which had the same means on the original scale: for example in scenario 1, $\mu_x = 1.1, \sigma_x^2 = 0.4$ and $\mu_y = 1.2, \sigma_y^2 = 0.2$, giving means on the untransformed scale of $1.1+0.4/2=$ 1.3 and $1.2+0.2/2=1.3$. For each parameter combination, they used five sampling levels (25, 50, 100, 200, and 400). In addition to the bootstrap and Z-score methods Zhou *et al.* (1997) also tested for a difference using the standard t-test on the untransformed data (this test can be justified on the grounds of the central limit theorem from which it can be inferred that even though x and y may not be normally distributed their means (θ_x and θ_y) will be at least closer to normal).

Given that the means of the two distributions were the same, an appropriate statistical test will declare a significant difference with a probability of 0.05 (Type 1 error rate). All three tests typically declare too many significant differences (Figure 4.8). For the t-test and the bootstrap, there is a strong correlation between α and the parameter combination ($r^2=0.83$ and 0.79, respectively) but not between α and sample size ($r^2=0.05$ and 0.02, respectively). The Z-score method is weakly correlated with both the parameter combination and the sample size ($r^2=0.21$ in both cases). Most importantly, the Z-score method is consistently the best method and the estimated α never deviates much from 0.05 (Figure 4.8). On the other hand, both the t-test and the bootstrap can deviate

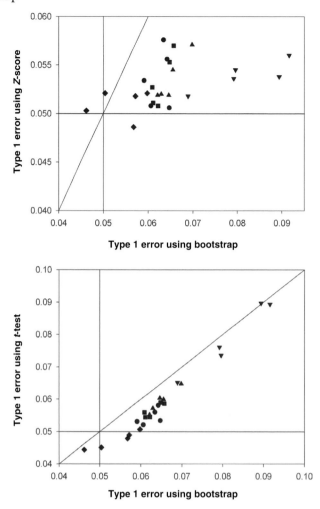

Figure 4.8 The estimates of type 1 error rate (based on 10 000 simulations) of the bootstrap and Z-score (top panel); bootstrap and *t*-test (bottom panel) methods when testing two log-normal distributions with the same mean. Five parameter configurations were used (see Table below) and five sample sizes for each design (25, 50, 100, 200, and 400). There was little effect of sample size (see text) and these are not distinguished in the plot.

Design	Symbol	μ_x	σ_x^2	μ_y	σ_y^2
1	●	1.1	0.4	1.2	0.2
2	■	1.05	0.5	1.2	0.2
3	▲	1.0	0.6	1.2	0.2
4	▼	2.5	1.5	3	0.5
5	◆	1.2	0.4	1.2	0.4

far too much to be acceptable. Interestingly, the bootstrap performs worse than the *t*-test!

If the data were not log-normally distributed, the performance of the Z-score method would decline, whereas that of the *t*-test or bootstrap might not, at least to the same degree. Whether the bootstrap would ever exceed the performance of the *t*-test remains to be demonstrated. The important point is that one should not presume the performance of the bootstrap but perform the necessary simulations to examine its behavior.

Bootstrapping phylogenies: the problem of deciding what a bootstrap actually measures

In complex measures, the definition of just what a bootstrap is actually measuring may not be entirely clear. This is well illustrated by the problem of phylogenetic tree construction. A major area of research in evolutionary biology is the analysis of phylogenetic relationships, that is how extant and extinct species are related by evolutionary descent. A simple example of a phylogenetic tree is shown in Figure 4.9. Four species are analyzed in this example, e.g., lemur (species A), human, chimpanzee, and gorilla. Species A, the lemur, is called an outgroup, defined as a species (or group of species) that is closely related to the species being studied and used to differentiate between shared derived and ancestral derived features. The object is to construct a tree that represents the evolutionary transitions among the species, the outgroup being used to designate ancestral characters. At the first evolutionary transition following the split from A we could have B (as shown) or C or D. For the present purposes we do not need to consider how the "best" tree is constructed (for a simple description see Futuyma, 1998) only that such a decision can be made. The question arises as to the confidence we can actually have in this tree. Felsenstein (1985) suggested the use of the bootstrap to derive confidence intervals for each node of the phylogeny. The method proposed by Felsenstein is illustrated in Figure 4.10.

The data set consists of a series of taxa (A,B,C,D) that form the rows of a matrix and a series of characters (1 through 10) that form the columns. The characters can be morphological traits, behavioral traits or, most often today, DNA sequence data. We could bootstrap these data by taking cells at random (with replacement) from the original data matrix, but Felsenstein argued that the appropriate bootstrap replicate is made by taking entire characters (i.e., columns). The argument rests upon the proposition that each character "evolved independently from the others according to a stochastic process that has among its parameters the topology and branch lengths of the underlying phylogeny" (Felsenstein 1985, p. 784).

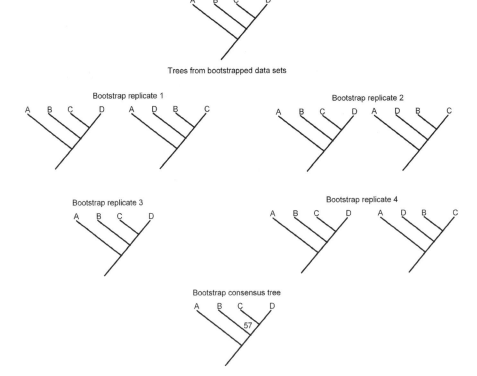

Figure 4.9 A simple, illustrative example of the use of the bootstrap to analyze phylogenetic relationships. The top tree shows the "best" tree obtained from the observed data. Organism "A" is the outgroup, which means that it is basal to all other species and is not included in the bootstrap analysis. The first, second, and fourth bootstrap replicates each produced two equally-likely trees, whereas the third replicate produced only one. The bootstrap consensus tree was produced using a 50% majority rule, which means that only clades that occurred in at least 50% of all trees are considered. The bootstrap "support" (57) for the transition between B and C–D is equal to the number of trees with this transition (4) divided by the total number of trees (7) multiplied by 100 (to give a percentage). Figure redrawn from Soltis and Soltis (2003).

In Figure 4.10, four possible bootstrap replicates are shown: note that, because of sampling with replacement, characters can appear more than once in a replicate or even not at all. From each matrix the "best" phylogenetic tree is constructed, as shown in Figure 4.9. It is possible, as shown in Figure 4.9, that several trees may be "equally-likely." From the set of replicate phylogenies, a final bootstrap tree is produced most typically using a rule such as the 50% majority

OriginalData Set

Taxa	Characters									
	1	2	3	4	5	6	7	**8**	9	*10*
A	C	G	A	A	C	C	A	**C**	T	*T*
B	C	G	A	A	C	C	G	**G**	T	*T*
C	G	G	T	A	C	C	G	**G**	A	*T*
D	G	G	T	A	G	C	G	**C**	A	*T*

Four Bootstrap Data Sets

Bootstrap replicate #1

Characters									
1	2	3	4	5	6	7	**8**	9	10
8	*10*	7	4	1	*10*	2	**8**	5	3
C	T	A	A	C	T	G	**C**	C	A
G	T	G	A	C	T	G	**G**	C	A
G	T	G	A	G	T	G	**G**	C	T
C	T	G	A	G	T	C	**C**	G	T

Bootstrap replicate #2

Taxa	Characters									
	1	2	3	4	5	6	7	**8**	9	10
SC =	1	**8**	*10*	4	2	9	2	**8**	5	6
A	C	C	T	T	A	G	T	**G**	C	C
B	C	C	G	A	A	T	G	**G**	C	C
C	G	G	A	A	G	A	A	**G**	G	C
D	G	G	C	T	A	C	C	**C**	G	C

Bootstrap replicate #3

Characters									
1	2	3	4	5	6	7	8	9	10
8	2	5	7	1	6	9	4	4	*10*
A	G	C	C	C	C	T	A	A	T
A	G	C	C	G	C	T	A	A	T
T	G	G	G	G	C	A	A	A	T
T	C	G	G	G	C	A	A	A	T

Bootstrap replicate #4

Taxa	Characters									
	1	2	3	4	5	6	7	8	9	10
SC =	7	**8**	5	**8**	9	6	4	*10*	1	5
A	A	C	C	C	T	C	A	T	C	C
B	G	G	C	G	T	C	A	T	C	C
C	G	G	C	G	A	C	A	T	G	G
D	G	C	G	C	C	A	C	A	T	G

Figure 4.10 A simple example of constructing bootstrap replicates from a matrix of character values for different taxa. Bootstrapping is done across characters (here shown as DNA bases), which means bootstrapping the columns. Thus, in bootstrap replicate #1 column 8 (bold font) from the original sample occurs in columns 1 and, by chance, 8, while column 10 (italic font) occurs in columns 2 and 6. "SC=" designates the original column number. Modified from Soltis and Soltis (2003).

rule. In the example, the bootstrap consensus tree has a split that divides B from the clade C–D. This division occurs in 57% of the bootstrap phylogenies and is used as a measure of the confidence we have in this node of the phylogeny. The use of the bootstrap to analyze phylogenies has proved outstandingly popular, as can be judged by the observation that Felsenstein's paper promoting the procedure has received more than 7000 citations. But what does this bootstrap "support" actually represent?

Felsenstein (1985, p. 786) concluded that "Bootstrapping provides us with a confidence interval within which is contained *not the true phylogeny* (my italics), but the phylogeny that would be estimated on repeated sampling of many characters from the underlying pool of characters. As such it may be misleading if the method used to infer phylogenies is inconsistent." Suppose, for example, we used traits that are analogous rather than homologous (i.e., traits due to convergent evolution, such as the wings of insects and birds, rather than due to common descent, such as the wings of birds and bats), then it is conceivable to get a tree that has very high bootstrap support but is simply wrong. The use of many characters may avoid this problem. Molecular data can indeed provide a large data base but the problem could still arise if we are using the molecular data for a single gene: in this case we may be examining the evolutionary history (which could involve convergent evolution and hence analogy) of the gene and not necessarily that of the species. These issues are certainly well known, and I raise them here to illustrate the important point that the bootstrap cannot solve the problem of an incorrect biological assumptions or poor data. Felsenstein suggested that if a grouping occurs in 95% of the bootstraps then it can be taken to be significant: note again that he is not saying that this is the "true" phylogeny only that it occurs more frequently than expected by chance given the observed distribution of characters.

The issue of what the bootstrap proportion (e.g., 0.57 for B and C–D split in Figure 4.9) actually means was taken up by Hillis and Bull (1993, p. 183) who noted that it is necessary to distinguish three concepts: "repeatability," "accuracy," and "precision." The precision of a bootstrap "is the degree to which bootstrap proportions based on a finite set of pseudovalues are expected to match the values that would be obtained from an infinite set of pseudovalues," whereas accuracy is "the probability that a specified group is contained in the true phylogeny," and repeatability is "the probability that a specified group will be found in an analysis of an independent sample of characters." Felsenstein's interpretation of the bootstrap proportions is that they are measures of repeatability (Hillis and Bull 1993, p. 183). Precision is easily taken care of, since it requires simply a large enough sample. Accuracy and repeatability are different matters entirely.

Hillis and Bull (1993, p. 187) made the observation that "Although boot-strapping was introduced as a measure of repeatability, bootstrap results commonly are interpreted as a measure of accuracy (e.g., in a framework of hypothesis testing)." Using simulations, Hillis and Bull examined the relation-ship between the bootstrap support and the actual probability of obtaining the correct phylogeny. The answer depended on the type of phylogeny; for some phylogenies the bootstrap support underestimated the accuracy, with a bootstrap support of 70% corresponding to a probability of 95% that the clade is real, whereas in other types of phylogenies the bootstrap can overestimate the accuracy. Hillis and Bull (1993, p. 192) concluded that "The strong positive relationship between high bootstrap proportions and phylogenetic accuracy does indicate a use for bootstrapping. However, bootstrap results should not be interpreted directly as estimates of either repeatability or accuracy under most conditions. They are poor estimates of repeatability and are usually very conservative estimates of accuracy...The values cannot be directly compared among studies."

Felsenstein and Kishino (1993) responded to the challenge of determining why Bull and Hillis obtained the results that they did and showed that the problem lay not in the bootstrapping procedure per se but rather the equating of the bootstrap proportion with accuracy. In summary, Felsenstein and Kishino (1993, p. 199) suggested that "when systematists see the group of interest occur a fraction P of the time among the bootstrap samples, that they should regard $1-P$ as a conservative assessment of the probability of getting that much evidence favoring the group if it is not present...For example, when the group is seen to have $P=0.85$, that much evidence favoring the group would be expected less than 15% of the time if the group were not on the true tree." Efron *et al.* (1996) further examined the issue and produced a better estimate of the error, though requiring the more elaborate and extensive bootstrap methodology of the accelerated bias-corrected percentile method (Method 5).

Summary

(1) A bootstrap replicate consists of taking from a sample of size n, the same number of observations, by random selection with replacement. The parameter(s) of interest are computed in the same manner as with the original sample. The process is repeated a large number of times (typically 200 for the estimation of SE and 1000 for the estimation of confidence intervals) and the set of bootstrap estimates used to estimate the mean and confidence interval of the parameter(s) of interest.

(2) The bootstrap estimate, θ^*, is the mean of the bootstrap replicates. This estimate is biased and the simplest bias-adjusted estimate, θ_A^* is $\theta_A^* = 2\hat{\theta} - \theta^*$.

(3) The SE of the estimate is estimated by the standard deviation of the bootstrap replicates.

(4) Confidence intervals can be estimated by a variety of approaches but there is no way, in general, of deciding which method, if any, is appropriate, except by simulation.

(5) Small samples typically, but not invariably, result in bootstrap confidence intervals that are smaller than required. What is considered "small" can only be decided by simulation analysis.

(6) The bootstrap can be used for hypothesis testing but care should be exercised in setting up the test statistic.

Further reading

Davison, A. C. and Hinkley, D. V. (1999). *Bootstrap Methods and their Applications.* Cambridge: Cambridge University Press.

Efron, B. (1982). *The Jackknife, the Bootstrap and Other Resampling Plans.* Society for Industrial and Applied Mathematics, Philadelphia.

Efron, B. and Tibshirani, R. J. (1993). *An Introduction to the Bootstrap.* New York: Chapman and Hall.

Manly, B. F. J. (1997). *Randomization, Bootstrap and Monte Carlo Methods in Biology.* New York: Chapman and Hall.

Mooney, C. Z. and Duval, R. D. (1993). *Bootstrapping: A nonparametric approach to statistical inference.* Newbury Park: Sage Publications.

Exercises

(4.1) Generate 100 values from a normal distribution with mean zero and unit variance (i.e., $N(0,1)$). Bootstrap, using the default of 1000 bootstrap replicates, the median, output the summary statistics from bootstrap, test for normality of the replicates (using the Shapiro–Wilkes test or other suitable test) and plot a histogram of the bootstrap replicates.

(4.2) Fit a linear regression to the data listed below using least squares regression and the bootstrap. Using the unstandardized differences, test the hypothesis that the slope and intercept equal to 0. Test the hypothesis that the slope equals 1 and the intercept equals 0. Compare the results to the parametric test.

---

x	1.63 4.25 3.17 6.46 0.84 0.83 2.03 9.78 4.39 2.72 9.68 7.88 0.21 9.08 9.04 5.59 3.73 7.98 3.85 8.18
y	2.79 3.72 4.09 5.89 0.75 −0.13 1.76 8.44 5.15 2.16 9.88 6.95 0.03 7.50 9.92 5.37 3.79 7.18 3.37 7.81

(4.3) Using the above data calculate the bootstrap correlation coefficient and test the hypothesis $\rho=0.96$. Use Fisher's z transformation, $z = 0.5\ln\left(\frac{1+r}{1-r}\right)$. Compare the results with the parametric test.

(4.4) Use the coding below to generate a sample of 20 correlated points in which the error distribution is a gamma distribution with shape and rate parameters equal to 2 and a mean of 0. Test the hypothesis $\rho=0$ using the bootstrap and parametric tests (see question 3). Which result do you think is the more reliable?

```
                                   # Set seed for random number generator
set.seed(0)
n         <- 20                    # Number of points
                          # Construct normal distribution of x values
x         <- rnorm(n,0,1)
shape     <- 2                     # Set shape parameter
rate      <- shape                 # Set rate parameter
                          # Calculate mean of gamma distribution
mu        <- shape/rate
                               # Generate error term with mean zero
error     <- rgamma(n,shape,rate)-mu
y         <- 0.5*x + error         # Construct y values
Corr.df   <- data.frame(cbind(x,y))   # Put in a single dataframe
```

(4.5) Generate 10000 data sets using the same protocol as in question 4.4. For each set calculate r and use the data to determine the probability of obtaining a value of r that is greater in magnitude than zero. Use coding in question 4.4 to generate ten data sets and analyze these using both the parametric and bootstrap tests. Which appears to be the better test in this circumstance?

(4.6) The table below shows the number of eggs laid by 20 female *Drosophila* of the indicated ages. Estimate the parameters of the function Eggs $= \theta_1(1 - e^{-\theta_2 \text{Age}})e^{-\theta_3 \text{Age}}$ using the bootstrap. Compare your results to those of the MLE and the jackknife, previously calculated (Exercise 3.4).

Ind	1	2	3	4	5	6	7	8	9	10	11	12	13	14	15	16	17	18	19	20
Age	1	3	2	4	1	1	2	5	3	2	5	4	1	5	5	3	2	4	2	5
Eggs	58	70	72	65	57	56	71	59	71	70	60	65	57	59	61	70	71	65	70	60

(4.7) Generate 10 data points from a normal distribution with mean 4 and unit variance. Compute the coefficient of variation, CV. Use the bootstrap to estimate the standard deviation. Check the answer by generating 10000 data sets, estimating the CV and computing the standard deviation of the resultant data set. Compare the bootstrap results with the simulation results.

(4.8) The following data were gathered on the size of a particular species within a community:

$$x : 2, 10, 10, 11, 14, 15, 16, 18, 18, 18$$

Calculate the Gini coefficient for these data and use the simplified method of hypothesis testing (Eq. (4.19)) to compare Gini coefficients ranging from 0.2 to 0.4 with the observed. Compare your results with the confidence limits from the bootstrap.

List of symbols used in Chapter 4

Symbols may be subscripted

E	Error term
ε_i^*	Estimate of the ith bootstrap error
θ	Parameter to be estimated
$\hat{\theta}$	Estimate of θ
$\hat{\theta}_{\text{Bias}}$	Bias in estimate
θ_i^*	Estimate from the ith bootstrap replicate
θ^*	Bootstrap estimate
$\tilde{\theta}_{-i}$	Estimate of θ with the ith datum removed
$\tilde{\theta}$	Jackknife estimate
μ	Mean

σ	Standard deviation
$\hat{\sigma}_{\theta_i^*}$	Estimate of SE of tth bootstrap replicate
B	Number of bootstrap replicates
C	Resource utilization index, or constant
Z	Z-score statistic
MLE	Maximum Likelihood Estimation
$N(\mu, \sigma)$	Normal distribution with mean μ and standard deviation σ
P	Probability
S_x^2	Estimate of variance of x
T_i^*	Value of ith transformed bootstrap replicate
SE(.)	SE of term in parentheses
a	constant in accelerated bias-corrected percentile method
d	Absolute difference between parameter estimate and hypothesized value
d_i^*	Absolute difference between parameter estimate and ith bootstrap estimate
l_t	Length at age t
n	Number of cases
p, q	In niche overlap indexes, proportion of resources used
t	Student's t
x	Observed value
$\bar{x}$	Mean value of x
y	Observed value
z	Abscissa of the standard normal distribution

5

Randomization and Monte Carlo methods

Introduction

Monte Carlo methods are used extensively in this book to generate models with which to illustrate or test particular statistical models or approaches. A **Monte Carlo model** is one in which there are one or more random components, e.g., the model $y=x+\varepsilon$, where ε is some randomly distributed variable. Kendall and Buckland (1982) define the Monte Carlo method as a method that denotes "the solution of mathematical problems arising in a stochastic context by sampling experiments ... the solution of any mathematical problem by sampling methods: the procedure is to construct an artificial stochastic model of the mathematical process and then to perform sampling experiments on it." In this chapter, I restrict my attention to the use of a Monte Carlo model to test a given statistical hypothesis. Randomization and the bootstrap can be considered as particular forms of the Monte Carlo method but their extensive and increasing use promote them to statistical methods in their own right. Monte Carlo models tend to be "tailor-made" for the particular problem under study, whereas the bootstrap and randomization methods can be more readily generalized, as illustrated in the last chapter by the routines now available on many computer software packages. Therefore, I shall first discuss randomization and then Monte Carlo techniques in general.

Randomization is first and foremost a technique for hypothesis testing, though it is possible to use it to construct confidence limits. Because of the few constraints it places upon the data, randomization is an extremely useful method. However, as with all statistical methods, it is not a panacea for bad data and its limitations must be recognized. The general principle underlying randomization is that if the data under study come from a common population then the observed statistic will be no different from that obtained if the data were assigned at random to the groups making up the data set. By comparing the observed statistic with that from many randomized data sets, we can estimate the probability of obtaining a statistic as deviant as that obtained from the original data. A limitation of the method should immediately strike the reader: if the observed data come from several statistical populations that differ in more than one parameter (e.g., both mean and variance) then the randomization procedure cannot work. This is a problem that is not peculiar to randomization: the standard *t*-test, for example, assumes that two populations differ in no more than the mean. The general approach to testing by randomization, and the construction of confidence intervals, can be most readily understood by considering a specific example, namely the difference between two means. Following this discussion, I shall present a variety of examples to illustrate the general utility of randomization and the factors that need to be considered in setting up the test.

Randomization – general considerations for hypothesis testing

An illustration of the approach: testing the difference between two means

Consider the problem of testing for a difference between two means. Randomization can specifically address the issue of a difference between two means in the case of both normal and non-normal distributions, subject to the restriction that a difference in means is the only difference between the two distributions. The parametric *t*-test compares two means under the assumption of a normal distribution of means, whereas the non-parametric Mann–Whitney test is a test of differences in central tendency, which for a highly skewed distribution will be the median, not a test for differences in the mean.

In comparing two means, the null hypothesis is

$$H_0: \mu_1 = \mu_2 \quad \text{or} \quad H_0: \mu_1 - \mu_2 = 0 \tag{5.1}$$

Thus, under the null hypothesis it should make no difference how we distribute the observations between the two groups. The randomization test is constructed as follows, *assuming that the two distributions differ only, if at all, in their means*:

(1) Compute a statistic that measures the difference between the two means. Several possible candidates are available:

    (a) the absolute difference between the two means
    (b) the *t*-statistic
    (c) the difference between the residuals of the two groups, $r_{i,k} = x_{i,k} - \bar{x}_k$, where $r_{i,k}$ is the *i*th residual from the *k*th ($k = 1, 2$)

group, $x_{i,k}$ is the ith observation in the kth group, and $\bar{x}_k$ is the mean of the kth group.

(2) Distribute the observations at random between the two groups. Suppose, for example, the original two groups are (1.2, 3.2, 4.4, 4.2) and (2.2, 5.5, 3.0, 3.1); a randomized data set might be (3.2, 5.5, 2.2, 4.2) and (1.2, 4.4, 3.0, 3.1). Note that this procedure differs from the bootstrap in that sampling is done without replacement (hence the alternate name for a randomization test is a **permutation test**).

(3) Repeat step 2 a large number of times, say N.

(4) Count up the number of cases, say n, in which the statistic from the randomized data set is larger than that from the original data.

(5) Estimate the probability, P, under the null hypothesis of obtaining a deviation as large or larger than observed as

$$P = \frac{n+1}{N+1} \tag{5.2}$$

The "extra" 1 is required because the observed value is itself a sample and hence should be counted, though clearly if N is very large then the extra 1 is not critical.

(6) If $P < 0.05$, the observed difference is declared to be unlikely under the null hypothesis and the null hypothesis is rejected in favor of the alternate that the means are different.

S-PLUS coding to do a randomization test for two means is shown in Appendix C.5.1. To make up the original sample, 20 data points are generated from a normal distribution, $N(0,1)$, and randomly assigned to two groups. Given that these data really did come from the same distribution, we would expect that a test for a difference between two means to give a significant result only 5% of the time. The test statistic used in the present example is the absolute difference between the two means, but the other possible statistics can be readily programmed into the routine that calculates the statistic. Though, as will be seen in later examples, the estimated probability can depend upon the test statistic chosen, in the present case simulations have shown that all three statistics have the same performance (Manly 1997, pp. 105–7). In 5000 randomizations more than 50% of the data sets generated absolute differences greater than observed: thus we have no reason to reject the null hypothesis (see output in Appendix C.5.1). It is possible in S-PLUS to use the bootstrap routine to do a randomization test (Appendix C.5.2): it is generally faster to use this routine whenever possible.

How many samples are required?

In general, we are interested in distinguishing a probability as small as 0.05 from a larger value. An approximate and simple way of estimating the relevant sample size required is to consider the standard error (SE) on the estimated probability: from the binomial we have

$$\mathrm{SE}(\hat{P}) = \sqrt{\frac{\hat{P}(1-\hat{P})}{N}} \tag{5.3}$$

where, for simplicity, I have absorbed the "extra" 1 into N. We require that $\hat{P} + 2\mathrm{SE}(\hat{P}) < 0.05$: Figure 5.1 shows such values for a range of N and $\hat{P}$. One thousand randomizations will generally be sufficient to determine significance unless P is close to 0.05, in which case one might have to increase the number of randomizations to as many as 10000. Of course, if the actual P is substantially larger than 0.05, many fewer randomizations will be required. In principle, one could set up the analysis in a sequential manner, first estimating P based

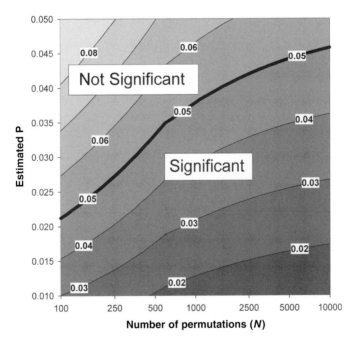

Figure 5.1 Contour plot showing the approximate upper confidence value as a function of the number of permutations and the estimated probability. Contours show the value of $\hat{P} + 2\mathrm{SE}(\hat{P})$. Values lying above the 0.05 contour line are "not significant" given the stated number of permutations.

on say 100 randomizations and then estimating the required number using a rearrangement of Eq. (5.3)

$$N = \frac{4\hat{P}(1 - \hat{P})}{(0.05 - \hat{P})^2} \qquad (5.4)$$

If $\hat{P}$ is much greater than 0.05 the above equation is superfluous, although if $\hat{P}$ is reasonably close to 0.05 (say 0.1) it would be worthwhile to increase the number of randomizations to 500 to ensure that the estimate is stable. Appendix C.5.3 shows the coding to calculate the required number using the previous data set but changing the expected mean of the second group by increasing the values in the second group by 1. The estimated P using 100 randomizations is 0.0297, which is not significantly less than 0.05, as indicated by the required N of 280. The probability from the parametric two-sample t-test is 0.039. In this case, increasing the sample size to 500 is necessary. One can simply do 400 more randomizations, changing the starting seed and then combining the result with the initial 100 (not necessary in present case, because the computational time is insignificant). In fact, with $N = 500$, $\hat{P} = 0.051\,896\,21$ which certainly requires further randomizations to verify the correct P. Using $N = 5000$ gives $\hat{P} = 0.046\,790\,64$ and a required N in excess of 17000! Further increasing N to 10000 gives $\hat{P} = 0.043\,595\,64$, which using Eq. (5.3) gives a SE of the estimate as 0.00204. Thus, the larger number of randomizations indicates that the P value is significant but marginal (it should be noted that one cannot simply keep increasing the sample size until, by chance, a significant result is found). In general, unless each randomization is very time consuming, I recommend commencing with 1000 randomizations.

Randomization – interval estimation

Method 1: estimating the standard error from the randomization procedure

The primary function of randomization is hypothesis testing but it is useful on many occasions to provide, in addition to a probability, an estimate of the standard error, or confidence interval. Manly (1991) suggested a protocol for estimating the confidence interval that requires a large number of extra randomizations. The object is to find the two constants, *Low* and *High*, that when subtracted from one of the groups produces a probability of 0.025 (the lower and upper values of the 95% confidence interval). This is done by trial and error, which means running a separate set of randomizations for each "guess." To illustrate this method Manly used data on the mandible lengths of male and female golden jackals (here shown in ranked order),

Males: 107,110,111,112,113,114,114,116,117,120
Females: 105,106,107,107,108,110,110,111,111,111

A standard t-test of these data indicates a highly significant difference (t=3.48, df=18, P=0.0026 for a two-tailed test but P=0.0013 for a one-tailed test based on the hypothesis that males will be larger than females) and a 95% confidence region for the difference of 1.91–7.69. Based on 10000 randomizations, the estimated probability for a two-tailed test is 0.0038, confirming the results of the t-test.

With a calculation that can be accomplished very fast, as is the case for the two-sample t-test, the confidence limits can be quickly found by iteration: the upper panel in Figure 5.2 shows a search using a large increment and only 1000 randomizations per value and the lower panel shows the results of a refined search using 10000 randomizations per value (see Appendix C.5.4 for coding, producing the data used in the lower panel). From the lower panel in Figure 5.2, the upper confidence value can be estimated as 7.72, but the lower confidence value presents some difficulty, because the curve is not "smooth." The reason for the lack of smoothness is in part due to the fact that different randomizations will produce slightly different values and, in the present case, a consequence of the original data containing a relatively large number of the same values (e.g., 111 in the female data), which results in randomized data sets that have the same average difference (this is demonstrated more in detail later). Taking the midpoint of the intersection of the *Low* curve with the 0.025 line gives a value of 1.9. Thus, the 95% confidence region from the randomization procedure is 1.90–7.72, which compares favorably to the parametric estimate of 1.91–7.69 (using his randomization method, Manly obtained a region 1.92–7.72, which is consistent with the results presented here).

Garthwaite (1996) modified the procedure to reduce the number of randomizations but the number of randomizations is still large and the algorithm is rather complex. Here, I present three approximate but simple methods of estimating the SE using only the original set of randomizations. For this purpose I shall use, in addition to the jackal data, two other data sets. The first set is that analyzed by Garthwaite on the effect of malaria on the stamina of the lizard *Sceloporis occidentalis*, stamina being measured as the distances 15 infected and 15 uninfected lizards could run in two minutes:

Infected: 16.4,29.4,37.0,23.0,24.1,25.0,16.4,29.0,36.7,28.7,30.2,21.8,37.0,20.3,28.3
Uninfected: 22.2,34.8,42.0,32.9,26.4,31.0,32.9,38.0,18.4,27.5,45.5,34.0,46.0,24.5,28.7

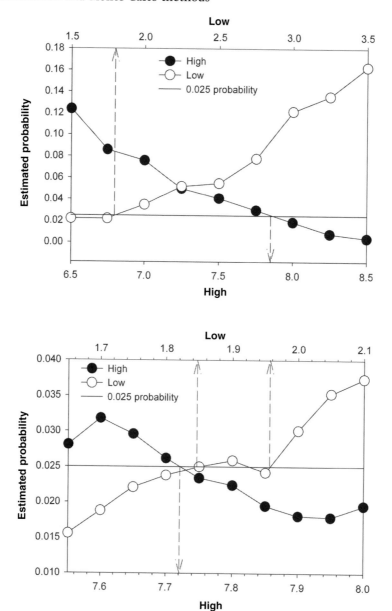

Figure 5.2 Illustration of Manly's method of estimating the lower (Low) and upper (High) confidence limits using randomization. The upper panel shows the results over a wide range and 1000 permutations per value. The lower panel shows a refined search using 10000 permutations per value. Dotted lines show estimated values for the confidence interval.

Table 5.1 *Estimates of the SE and 95% confidence limits (LC=lower, UC=upper) for the difference in stamina between infected and uninfected* Sceloporus occidentalis

Method	SE	LC[a]	UC
Lizard data			
Parametric	2.85	−0.23	10.95
Garthwaite	2.80[b]	−0.30	10.69
Normal approximation	2.73	−0.23	10.95
Average percentile	2.72	−0.21	10.93
Percentile	2.72[b]	−0.29	10.87
Gamma-distributed data			
True value[a]	0.18	0.16	0.86
Parametric	0.22	0.08	0.93
Normal approximation	0.20	0.09	0.92
Average percentile	0.21	0.07	0.94
Percentile	0.21[b]	0.07	0.94
Jackal data			
Parametric	1.38	1.91	7.69
Manly	1.38[b]	1.90	7.72
Normal approximation	1.45	1.76	7.84
Average percentile	1.62	1.40	8.20
Percentile	1.62[b]	1.40	8.20

[a]The sign is arbitrary and depends on which mean is subtracted.
[b]"Average" SE estimated by $(UC−LC)/([2][1.96])$.

Analysis using a two-tailed t-test reveals a marginally non-significant difference between the two groups ($t=1.9658$, df$=28$, $P=0.0593$). The estimated probability from 10 000 randomizations is 0.0563, which is very close to that obtained with the parametric test. The method of Garthwaite produces a slightly smaller confidence interval and a SE than the parametric method (Table 5.1).

The second data set, consisting of two samples, each of size 20, was generated from two gamma distributions, both with shape parameter=rate parameter=3, but with one incremented by 0.5 (Figure 5.3). Thus, the two samples drawn from these distributions differ only in their means. To determine the true distribution of the mean differences between two samples, I generated 10 000 samples from which the SE could be directly estimated. Despite the high skew in the parent populations (upper panel, Figure 5.3) the distribution of mean differences is remarkably normal. A t-test on the single sample indicates a significant

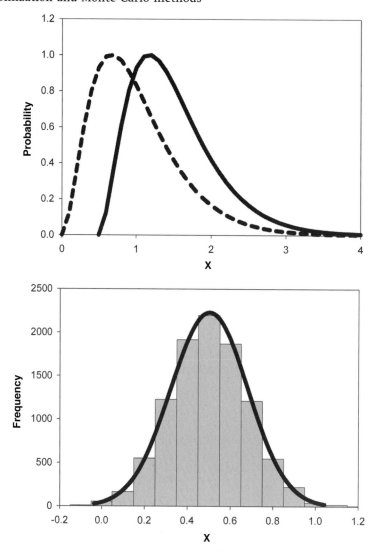

Figure 5.3 Probability density plots of two gamma distributions (upper panel) differing only in their mean value (scale parameter=rate parameter=3, with 0.5 added to rightmost distribution). Lower panel shows the distribution of 10 000 simulated data sets comprising the mean difference between 20 samples drawn from each of the two gamma distributions. Solid curve shows fitted normal curve.

difference between the groups (mean difference=0.504, t=2.4172, df=38, P=0.0205) and 10 000 randomizations give an almost identical probability of 0.0186. The estimated SE from the parametric analysis is slightly larger than the correct value (0.22 vs. 0.18, Table 5.1) and the confidence region is thus larger than the correct width (Table 5.1).

Method 2: normal approximation

A very simple procedure is to assume that the parameter of interest, θ, is normally distributed so that θ/SE is normally distributed with unit variance. Under this scenario, we can estimate the SE by

$$\text{Est(SE)} = \frac{\hat{\theta}}{x} \qquad (5.5)$$

where x is the value of the abscissa on the normal curve giving a probability of $\hat{P}/2$ (Roff and Bradford 1996). If the original sample is relatively small, the appropriate value of x using the t distribution can be used. Coding for this and the next two procedures is shown in Appendix C.5.5.

The necessary assumption of this method, that the data be normally distributed, can be readily examined using the randomized values. The estimated SE for the lizard data is smaller than that obtained by Garthwaite but the confidence region is virtually identical to the parametric estimate and shifted relative to Garthwaite's estimate (Table 5.1). For the gamma-distributed data the normal approximation is almost the same as estimated by the parametric method, and slightly larger than the true value (Table 5.1). This method appears to overestimate the SE for the jackal data and produce an enlarged confidence interval (Table 5.1).

Method 3: average percentile method

This method is similar to Method 2 in that it estimates the SE but differs from Method 2 in that it avoids the normality assumption by using the randomized distribution directly. We wish to find a value $C_{0.95}$ such that when added or subtracted from the observed difference generates a probability of 0.05. To locate this value, we sort the set of 10 000 *absolute* values of the randomized replicate differences into ascending order and find the value that is at the 95% point in the sorted list. We find the value of t corresponding to 0.025 with the appropriate degrees of freedom (e.g., 28 for the lizard data) and estimate the SE as

$$\text{Est (SE)} = \frac{C_{0.95}}{t_{0.025,\text{df}}} \qquad (5.6)$$

Assuming that the distribution is symmetrical, the lower and upper confidence limits (LC and UC, respectively) are estimated as

$$\text{LC} = \hat{\theta} - C_{0.95}, \quad \text{UC} = \hat{\theta} + C_{0.95} \qquad (5.7)$$

For both the lizard data and the gamma-distributed data, the estimates of SE and confidence limits are virtually identical to those obtained with Method 2 but for the Jackal data, the SE estimate appears to be overestimated and the confidence range too broad (Table 5.1).

Method 4: percentile method

This method follows the same rationale as Method 3 except that the upper and lower limits are found separately. To locate these values, we sort the set of 10 000 values of the randomized replicate differences into ascending order and find the values that are at the 2.5% and at the 97.5% points in the sorted list. The confidence limits are then

$$LC = C_{0.025} + \hat{\theta}, \quad UC = C_{0.975} + \hat{\theta} \tag{5.8}$$

The SE can be approximated as $(UC-LC)/(2t_{0.025,df})$. The results using this method are more or less the same as Method 3 (Table 5.1).

In summary, for the three data sets examined here the normal approximation appears to be the most satisfactory of the three approximate methods. If very accurate estimates are required then the values obtained from this approach can be used as starting points in the more computationally intensive method suggested by Manly.

Examples illustrating randomization tests

Single-factor (one-way) analysis of variance

Analysis of variance is remarkably robust to the assumption of normality (Sahai and Ageel 2000, pp. 85–6) but if normality cannot be established then the possibility that the results are incorrect cannot be excluded. Randomization can be used to verify or replace the results of the parametric analysis. *Remember, however, that randomization makes the assumption that the component distributions differ in no more than their means.* If there is departure from equal variances then the randomization test may be in error. This issue is taken up in the next section: here I shall assume that the distributions differ only in the mean and could be non-normal.

Although one might use the mean squares as the test statistic rather than the F value, there are two reasons why this is not recommended. First, use of the F-statistic permits an immediate comparison with the parametric analysis (i.e., differences between the two analyses cannot be attributed to differences in the statistic used) and, second, simulation suggests that the F-statistic generally produces a more powerful test than the mean squares (Gonzalez and Manly 1998).

Randomization can be done on the original data or the residuals: again, simulations suggest that the use of the original data is sufficient.

To illustrate the randomization method of one-way ANOVA, I shall use an empirical data set comprising the monthly consumption (milligrams of dry biomass) of ants by eastern horned lizards (data from Manly 1997):

Month	Observations
June:	13, 105, 242
July:	2, 8, 20, 59, 245
August:	40, 50, 52, 82, 88, 233, 488, 515, 600, 1889
September:	0, 5, 6, 21, 18, 44

Analysis of variance of these data indicates no significant effect of month on ant consumption ($F_{3,20}=1.64$, $P=0.21$). However, a Kruskall–Wallis test does indicate a significant effect ($\chi^2=11.0$, df=3, $P=0.012$). Inspection of the residuals suggests considerable lack of normality (Figure 5.4), that is confirmed using Lilliefor's test ($P <0.0001$). Given the small number of observations, the lack of normality is disturbing (with large sample sizes even small deviations from normality, that do not affect the ANOVA test, can be detected) and casts doubt on the ANOVA.

Inspection of the data also suggest that there may be heteroscedasticity (variances are: June, 13 279; July, 10 416; August, 319 724; September, 258). Ignoring this for the time being, we shall test for variation among groups using a randomization test with F as the test statistic and randomization of the original data. Two different methods of coding for this test are presented in Appendices C.5.6 and C.5.7. Appendix C.5.6 gives a very general method that is set in a framework that could be transported to another language such as FORTRAN. The data set consists of two columns, amount eaten and month. Randomization of the observations, but not the months (which would be redundant) is done using a loop and at each iteration the F statistic is calculated and stored. Appendix C.5.7 shows a somewhat different approach in that the observations are randomized and stored N times (the number of randomizations), leading to a data file consisting of three columns: an index number giving the randomization number (note that the first entry is the original data), the month data (replicated without randomization), and the observations. The F-statistics are then calculated using the "by" routine of S-PLUS. This method is slightly faster than the first method: for 999 replications the first method took 64 s, whereas the second took 55 s.

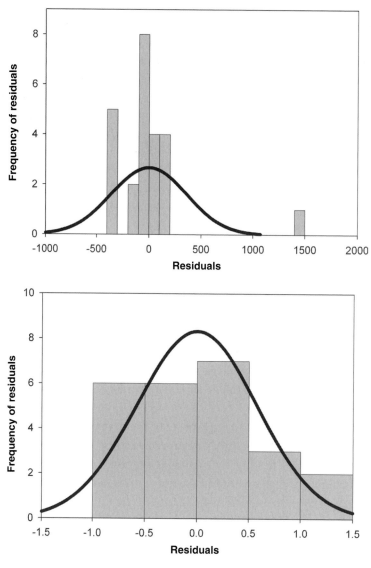

Figure 5.4 Frequency distributions (histograms and fitted normal curves) for the residuals from the fitted ANOVA model to the monthly consumption of ants by horned lizards. Top panel shows residuals using untransformed data. Bottom panel shows residuals using $\log(x+1)$ transformation.

The randomization gives a probability of 0.2 (the slight differences in the outputs of the two methods of coding is due to differences in the randomization procedure. In the first method, randomization is repeated on the previously randomized data set, whereas in the second it is always done on the original data. This makes no conceptual difference), which is pleasingly close to that obtained from the ANOVA. On the other hand, the lack of a difference between groups could be due to a confounding influence of heteroscedasticity: inspection of the data certainly suggests that the variances differ substantially (in order of months the variances are: 13 279, 10 416, 319 724, 258). The obvious approach at this point is to seek a transformation that would reduce or eliminate the differences among the variances. A $\log(x+1)$ transformation appears to do the trick, the variances of the transformed observations being, respectively, 0.407, 0.545, 0.325, 0.339. The ANOVA now indicates highly significant variation among groups ($F_{3,20}=6.08$, $P=0.004$), and the residuals are not significantly different from normal (Figure 5.4, Lilliefors test, $P=0.429$). From 4999 randomizations of the log transformed data the estimated probability is 0.0026, which, as before, is pleasingly close to that obtained from the ANOVA. In this example, there really is no need to resort to a randomization test, only to transform the data. The important message is that randomization itself has assumptions and these must be respected. We now turn to data in which we know that the only differences are due to differences in the mean.

I generated three groups of size 10 from an exponential distribution ($p(x)= e^{(-x/\mu)}/\mu$, $x>0$), all three with an initial mean, μ, of one (in S-PLUS, x < rexp(10, rate=1)). To the last group I added a constant, thereby changing only its location. In the first example, I added 0.8: an analysis of variance indicated no significant difference among the groups ($F_{2,27}=2.73$, $P=0.083$), whereas the randomization test gave $P=0.035$. For the second example, I added 1.0 to the third sample: the ANOVA indicated a marginal value of $P=0.0442$, whereas the randomization test gave a highly significant value of $P=0.012$. In these two cases, the randomization test is to be preferred.

In Chapter 2, I introduced the concept of the threshold trait in genetics (Figure 2.4) and the estimation of the heritability of liability from a selection experiment. Heritability of liability may also be estimated from full-sib (offspring with the same mother and father) families. The liability cannot itself be measured except in as much as it manifests itself in the dichotomous phenotype. The problem is then to measure the heritability of the underlying liability from the presence/absence appearance in the siblings. The heritability of any quantitative trait can be estimated from a one-way analysis of variance on full-sib data as twice the intra-class correlation coefficient (assuming no non-additive effects),

$$t = \frac{MS_{AF} - MS_{AP}}{MS_{AF} + (k-1)MS_{AP}} \tag{5.9}$$

where MS_{AF} is the mean square among families (among groups), MS_{AP} is the mean square among progeny (within groups), and k is the adjusted number per family if the number per family varies ($k = (T - \sum n_i/T)/(N - 1)$, T is the total number of individuals, n_i is the number in the ith family and N is the number of families). With a threshold trait, the problem is to estimate the heritability of the underlying trait from the relative proportion of the two phenotypic classes among full-sibs. We proceed in two steps: first we compute the intraclass correlation using the 0, 1 data of the manifested phenotypes, and second convert this to the heritability of the trait on the underlying liability scale by the formula

$$h^2 = 2t\frac{p(1 - p)}{z^2} \tag{5.10}$$

where p is the proportion of a designated morph in the population (which morph is chosen is entirely arbitrary. The proportion p is estimated as the average proportion across all families, $\sum p_i/N$), and z is the ordinate on the standardized normal curve corresponding to a probability p. For further discussion of threshold traits, see Roff (1997).

Because a dichotomous trait is necessarily coded as 0, 1, it is clearly not normally distributed. Thus, a randomization approach can be an important supplement in the testing of the significance of the heritability estimate. If only a single cage is employed per full-sib family, it is impossible to separate effects due to familial causes from effects due to a common environment. Therefore, at least two cages per family should be used and a nested ANOVA employed to separate cage effects from family effects. In an analysis of wing dimorphism (=a long-winged [macropterous] morph capable of flight and a short-winged [micropterous] flightless morph) in the cricket *Gryllus firmus* I used three methods (Roff, unpublished):

(1)   a nested ANOVA using all individuals categorized as 0 (=macropterous) or 1 (micropterous).

(2)   a one-way ANOVA using the mean proportion per cage (2 estimates per family, both the raw proportions and arcsine square-root transformed values were used: the results did not differ).

(3)   a randomization test conducted as follows. First, the heritability was computed by pooling the two cages per family. Next, cages were paired at random and the heritability computed for this sample: 999 such randomized heritabilities were computed. The probability of obtaining a value of h^2 as large or larger than that observed was estimated by the proportion of heritabilities from the randomized set plus the observed h^2 that were as large as or larger than the observed value. A somewhat better method would have been to compute the heritability using the nested ANOVA for each set rather than pooling the cages.

Table 5.2 *Probability values from three methods of statistical analysis for an effect of family on the distribution of macroptery (=long-winged, as opposed to the alternate morph, microptery, which is short-winged) in G. firmus (A=one-way ANOVA, R=randomization, NA=nested ANOVA)*

Env.[a]	Line[b]	Sex	p[c]	A	R	NA
15/25	L1	F	0.92	0.350	0.578	0.820
15/25	L1	M	0.77	<0.001	0.004	<0.001
15/25	C1	F	0.74	<0.001	0.006	<0.001
15/25	C1	M	0.53	0.002	0.002	0.001
15/25	L2	F	0.91	0.013	0.016	0.002
15/25	L2	M	0.79	0.030	0.017	0.006
15/25	C2	F	0.60	<0.001	0.001	<0.001
15/25	C2	M	0.30	<0.001	0.001	<0.001
17/30	S1	F	0.08	0.002	0.003	0.002
17/30	S1	M	0.03	0.622	0.430	0.618
17/30	C1	F	0.50	0.001	0.001	<0.001
17/30	C1	M	0.27	<0.001	0.001	<0.001
17/30	S2	F	0.04	<0.001	0.039	0.113
17/30	S2	M	0.02	<0.001	0.009	0.008
17/30	C2	F	0.53	<0.001	0.001	<0.001
17/30	C2	M	0.38	<0.001	0.001	<0.001

[a]Env=Environment, where A/B, A=photoperiod (number of hours/day of light), B=temperature (°C).
[b]L=line selected for increased proportion macropterous, S=line selected for decreased proportion macroptery, C=control line. The number refers to replicate (2 per line).
[c]Proportion macropterous.

Sixteen different comparisons were made (proportions differ between the sexes, two rearing environments, and 8 lines, Table 5.2). The three methods of statistically testing for differences attributable to family (nested ANOVA, one-way ANOVA, randomization) give very similar results, despite the fact that in some cases the proportions are close to 0 or 1, producing a highly skewed distribution. In 13 of the 16 tests, all three methods indicate highly significant variation among families. All three tests indicate no significant effect due to family in L1 female offspring or S1 male offspring. Phenotypic variation is low (proportion macropterous, $p=0.92$ and 0.03, respectively) in both cases, and hence the lack of significance probably reflects the low power of the tests under these conditions. In one case, S2 females, the nested ANOVA indicates no significant variation among families ($P=0.113$), while the other two tests indicate significant variation ($P<0.001$ and $P=0.039$ for the ANOVA and randomization methods, respectively). As with the previous cases, phenotypic variability is very low (proportion macropterous, $p=0.04$). These results suggest that the analysis of variance is very robust to fairly extreme skew. Nevertheless, a randomization test is a useful additional test.

Analysis of variance and the question of how to randomize

The above nested randomization test raises an important point with respect to the level of randomization. For a nested design, we could either randomize all individuals or randomize the nesting variable (cages in this example). The null hypothesis is that there are neither effects of cage nor family. However, there could easily be both family and cage effects, or any combination of the two. My interest was in determining the presence of family effects primarily: therefore, I considered it more conservative to randomize over cages rather than individuals, because the latter procedure might confound the two effects.

The issue of how to randomize is particularly acute for complex analyses of variance designs, the simplest being the two-way ANOVA. There are two primary ways in which the data could be randomized: (1) randomize observations across all cells, or (2) randomize only across rows or columns. The argument here is that the testing of interaction is not really possible with randomization, because randomization as carried out in the first method confounds effects. From a simulation study, Gonzalez and Manly (1998) concluded that for small data sets randomizing across all cells and using the *F*-statistic is generally appropriate. To illustrate this, I ran the following set of simulations. Twenty data points were generated, distributed across two treatments, with two levels in each treatment: individual observations were generated using the formula $x_{ijk}=T1_i+T2_j+(T1_i)$ $(T2_k)+\varepsilon$, where x_{ijk} is the kth observation in the ith level of treatment T1 and the jth level of treatment T2, and ε is an exponentially distributed error term (therefore, non-normal) with mean 1. The levels within each treatment were set at 1 and 2: thus an observation at level 1 in treatment T1 and level 2 in treatment T2 was determined as $x_{12k}=1+2+(1)(2)+\varepsilon=5+\varepsilon$. Various combinations of effects were simulated by setting one or more of the treatments to zero (e.g., $x_{12k}=1+2+(0)(1)(2)+\varepsilon=3+\varepsilon$ simulates the situation in which there is no interaction). The results of this study show that the randomization tests are virtually identical to the ANOVA results (Table 5.3), even though the error variance was far from normal. These results indicate, as in the case of the heritability of threshold traits, that analysis of variance is remarkably robust to the assumption of normality: in fact, despite my attempts, I could not create a circumstance in which the analysis of variance failed and produced different results from the randomization test. This does not mean that the randomization test is not required in this case: given that the assumptions of ANOVA are not fulfilled, one cannot be assured that the results will be correct.

Table 5.3 *Results of a simulation examining the probabilities from a two-way analysis of variance (P) and a randomization test (R). One hundred randomizations were performed for each test*

Model	P(T1)		P(T2)		P(T1*T2)	
	P	R	P	R	P	R
T1+T2+T1*T2+ε	<0.01	<0.01	0.76	0.83	<0.01	<0.01
T1+T2+0*T1*T2+ε	0.31	0.44	<0.01	0.01	0.35	0.50
	<0.01[a]	<0.01	<0.01	0.02	0.80	0.84
T1+0*T2+T1*T2+ε	<0.01[a]	<0.01	0.01	0.01	0.76	0.83
	<0.01	<0.01	<0.01	<0.01	0.12	0.09
	<0.01	<0.01	0.17	0.18	0.10	0.14
0*T1+T2+T1*T2+ε	0.05	0.07	<0.01	<0.01	0.76	0.80
T1+0*T2+0*T1*T2+ε	0.31	0.29	0.35	0.38	0.44	0.50

[a]Different random number seeds: 2, 20, 10.

Anderson and ter Braak (2003) undertook a detailed study of randomization tests for multi-factorial analyses of variance. They define two types of randomization tests, exact randomization tests and approximate randomization tests. An exact randomization test for any term in an ANOVA model satisfies two conditions: (1) the units that are randomized are identified by the denominator mean-square of the F-ratio appropriate to the test used by the conventional ANOVA and (2) randomizations are restricted to occur within the levels of terms of either smaller order or of the same order as the term being tested. Main effects are first order effects, two-way interactions are second order effects etc. These two conditions can be illustrated by reference to three types of ANOVA: one-way ANOVA, nested ANOVA, and a two-way mixed-model ANOVA (Figure 5.5).

In a one-way ANOVA, there is only a single factor (A) and hence only a single main effect and a single F-test, $F_A = MS_A/MS_R$, where MS refers to "mean square" and R is the residual term. The exchangeable units are thus the individual samples and there are no restrictions on the redistribution of units. In a two-factor, nested ANOVA the fixed effect, here factor A, is tested using the F-statistic, $F_A = MS_A/MS_B$ and thus the exchangeable units are the components of factor B (e.g., the cages in the heritability experiment described above). These units can be exchanged as shown in Figure 5.5. The nested effect, here factor B, is tested by $F_{B(A)} = MS_B/MS_R$. The exchangeable units are thus the basic sampling elements (e.g., the individuals within the cages in the liability experiment) but the problem is that there is no way to redistribute these units without also potentially affecting the effect of factor A. Therefore, no exact randomization test is possible with factor B. In the case of a two-way mixed model ANOVA

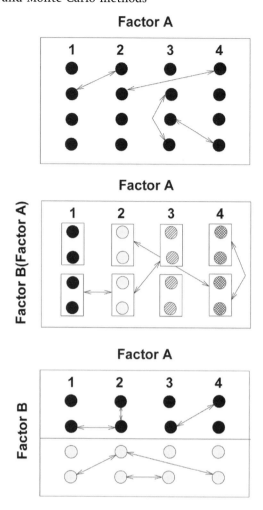

Figure 5.5 Diagrammatic illustration of how exact permutation tests can be constructed for one-way ANOVA (top panel), nested ANOVA (middle panel) and two-way mixed-model ANOVA. The arrows indicate the types of permissible permutations. Modified from Anderson and ter Braak (2003).

the fixed effect, here identified as factor A, is tested using $F_A = MS_A/MS_{A \times B}$, the random effect, B, is tested with $F_B = MS_B/MS_R$ and the interaction with $F_{A \times B} = MS_{A \times B}/MS_R$. Now, as with the nested ANOVA, it is not possible to redistribute the sampling elements necessary to test B, or the interaction, without also potentially altering the effect due to A and thus no exact randomization test for B or the interaction is possible. If randomization of sampling elements is restricted to within the levels of B (Figure 5.5) then an exact randomization test for A is possible.

Even if an exact randomization test is possible, there may be too few observations to make such a test have reasonable power. In this case, and in those cases in which exact tests are not possible, an approximate test can be carried out by doing a randomization test by permuting all observations. Where exact and approximate tests are possible, it is advisable to do both and compare the results. For example, in the case of the nested ANOVA do an exact test for factor A followed by an approximate test of both A and B: if the results of the two tests disagree then one would suspect that the approximate test is confounded. Anderson and ter Braak (2003) ran a simulation for the nested ANOVA case with four different error distributions: normal N(0,1), uniform (1,10), log-normal and a cubed exponential. For each combination of sample size and error distribution 1000 simulations were run and the number of "significant" tests recorded. A correct test should produce a significant result with a probability of 0.05: with 1000 runs the confidence region is 0.036–0.064 (i.e., 36 and 64 "significant" results). The exact randomization test for factor A never produced probabilities outside the confidence region, whereas the regular ANOVA and the approximate test based on randomization across both levels did produce several probabilities when the error distribution was highly skewed (Figure 5.6). Anderson and ter Braak also tested the use of the residuals but in this case the results were worse than with the raw data. Nevertheless, based on their simulations of other types of ANOVAs, they recommend using residuals rather than the raw data and provide tables showing how such residuals should be calculated for different types of two-way and three-way ANOVA designs.

The question of balance

A particularly serious concern in the analysis of variance is the issue of unbalanced designs, for in this case the significance tests lack theoretical justification (Shaw and Mitchell-Olds 1993). Several solutions to this have been proposed but none is general, whereas randomization does present a general approach. To illustrate a typical case in which a data set might be unbalanced, Shaw and Mitchell-Olds (1993) examine hypothetical data from an experiment on the effects of conspecifics on the height of a set of plants. They consider two treatments (e.g., no conspecifics or a fixed number of conspecifics) and also the potential effect of initial plant height, recorded as a two-group categorical variable: thus the appropriate statistical test is a two-way ANOVA. Even though the researchers may have begun the experiment with equal numbers per cell there is likely to be a loss of individuals, leading to unequal representation per cell, as shown in Table 5.4. Different answers are obtained, depending on the type of sums of squares used (Table 5.5). With type I sums of squares the answer depends upon the order in which variables are entered into the model: in the

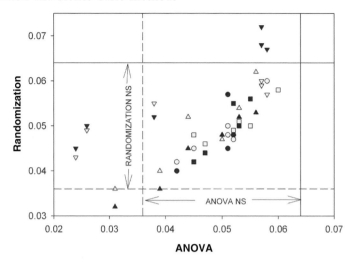

Figure 5.6 Plot of type I error rates, based on 1000 simulations, for three statistical methods of testing for the main effect in a nested ANOVA. Results exact (grey) and approximate (black) randomization tests are plotted against the normal-theory F-test. Data taken from Anderson and ter Braak (2003). The axes show the proportion of cases deemed "significant" by the corresponding test. Simulations producing values to the left or below the dotted lines produce too few "significant" values, whereas those to the right or above the solid lines produce too many "significant" values. Symbols designate error distributions; ● normal $N(0,1)$, ■ uniform $(1, 10)$, ▲ log-normal, and ▼ $\exp(1)^3$.

Table 5.4 *Example hypothetical set of unbalanced data in a two-way analysis of variance. Data from Shaw and Mitchell-Olds (1993)*

	Factor B (No. of conspecifics)	
Factor A (Initial size class)	0	1
1	50	57
	57	71
		85
2	91	105
	94	120
	102	
	110	

present example, factor A is significant whichever order is used but factor *B* is marginally non-significant if entered second and highly non-significant if entered first (Table 5.5). In contrast, order does not matter for type III sums of squares: use of this method gives the same answer as type I sums of squares

Table 5.5 *Analysis of the data shown in Table 5.4, using either type I or type III sums of squares. In the case of type I sums of squares the order of inclusion matters and hence two ANOVA tables are presented for this analysis. Also shown is the probability,* P_{rand}*, estimated from 1000 randomizations (Appendix)*

Source	df	SS	F	P	P_{rand}
Type I SS with factor A entered first					
Factor A	1	4291.2	40.17	0.0004	0.002
Factor B	1	590.2	5.52	0.051	0.049
A × B	1	11.4	0.11	0.753	0.731
Error	7	747.7			
Type I SS with factor B entered first					
Factor B	1	35.3	0.33	0.583	0.550
Factor A	1	4846.0	45.37	0.0003	0.001
A × B	1	11.4	0.11	0.753	0.749
Error	7	747.7			
Type III SS					
Factor A	1	4807.9	45.01	0.0003	0.001
Factor B	1	597.2	5.59	0.050	0.061
A × B	1	11.4	0.107	0.753	0.770
Error	7	747.7			

with factor A entered first. The general view is that type III sums of squares are preferable, but there is still doubt as to the validity of the estimated probabilities.

To do a randomization test on the above hypothetical data, we should retain the same structure in that each cell should contain the same number of observations as in the original data: this ensures that any effect due to the imbalance is maintained throughout the randomization process. For a two-way ANOVA, the data are coded in three columns as shown in Appendix C.5.8 and the randomization is accomplished by only randomizing the data column (the coding shown in C.5.8 might not appear to be the most efficient but use of "ssType=3" seems to disrupt other coding. Run time is still only about three minutes for 1000 iterations). The intriguing results of the randomization tests is that they generate almost exactly the same probabilities as obtained from the ANOVA test (Table 5.5). From this we can infer that the ANOVA is actually quite robust, at least in this instance, to the imbalance in the data. On the other hand, randomization gives us no help in deciding which is the best sums of squares to use (in a completely balanced design the answer does not depend on the sums of squares used – an ideal statistical test would have the same property for unbalanced data).

Testing for homogeneity of variances

Whereas analysis of variance is quite robust to the assumption of normality, theory suggests that it will be very sensitive to inequality of variances, particularly if sample sizes vary among cells (Sahai and Ageel 2000, pp. 86–8). It is thus important to determine homogeneity of variances. A test that appears to work well and is relatively robust to non-normality is Levene's test (Conover *et al.* 1981; Manly 1997). Levene's test consists of transforming within cells using the transformation

$$x_{ij}^* = \left| x_{ij} - \bar{x}_i \right| \qquad \text{or} \qquad x_{ij}^* = \left| x_{ij} - M_i \right| \tag{5.11}$$

where x_{ij} is the jth observation in the ith cell, $\bar{x}_i$ is the mean of the ith cell and M_i is the median of the ith cell. Coding to apply Levene's test using the cell means is given in Appendix C.5.9: the file resulting from this analysis can be subjected to a randomization test by modification of the coding in Appendix C.5.6 or C.5.7. Applying Levene's test and its randomized version to the data on ant consumption by horned lizards indicates a marginal effect by both tests (Appendix C.5.9), showing how difficult it is to identify even gross difference in variances as being significant.

More on homogeneity: the χ^2 test

A frequently used statistical routine is the χ^2 contingency test. However, the use of this distribution is strictly valid only with reasonably large samples: Cochran (1954) suggested that a reasonable rule of thumb is that no expected frequency should be less than 1.0 and that less than 20% of the expected frequencies should be less than 5.0. When data do not conform to these criteria the usual solution is to group data, most often by considering only the relative frequency of the most common type. Such a procedure necessarily loses information and should be avoided if at all possible.

To illustrate the problem and its solution we shall examine data on mitochondrial DNA variation among different populations of the American shad (*Alosa sapidissima*). From 14 separate rivers, Bentzen *et al.* (1988) identified 10 different mitochondrial DNA genotypes (Table 5.6).

The total number of fish sampled was 244, many genotypes being at very low frequency. Of the 140 cells, 92 (66) have expected values less than 1.0 and only 13 (9.3) have expected values greater than 5.0. Thus, following Cochran's rules, to use the χ^2 test it is necessary to combine cells. Only the most frequent genotype is represented at reasonably high frequency in all samples, and hence,

Table 5.6 *Distribution of mitochondrial genotypes of shad sampled from 14 separate rivers. Data from Bentzen* et al. *(1988)*

River	Mitochondrial genotypes									
1	13	15	1	0	0	0	0	0	0	0
2	8	0	2	5	2	1	0	0	0	0
3	8	0	0	2	0	1	2	0	0	0
4	11	4	0	1	1	0	0	0	0	0
5	9	1	0	1	7	0	0	1	1	0
6	12	2	3	0	2	0	2	0	0	1
7	11	1	0	0	5	0	1	1	0	0
8	17	0	0	0	3	0	0	1	0	0
9	10	0	0	0	1	0	0	0	0	0
10	12	1	1	2	0	0	1	1	1	0
11	6	0	3	0	1	0	0	0	0	0
12	12	0	0	2	0	0	0	0	3	0
13	16	0	0	4	0	0	0	0	1	0
14	7	0	0	0	0	0	0	0	0	0

following the usual procedure, we would combine all the rare genotypes. Now no cells have an expected value less than 1.0, and less than 20 (17.8) have expected values less than 5.0. The calculated χ^2 for the combined data set is 22.96, which just exceeds the critical value (22.36) at the 5% level. The estimated value of χ^2 for the uncombined data is 236.5, which is highly significant ($P < 0.001$) based on the χ^2 with 117 degrees of freedom. However, this result is suspected because of the very low frequencies within many cells.

Rather than combining cells and thus losing information, we can use randomization to test if the observed χ^2 value is significantly larger than expected under the null hypothesis of homogeneity among the rivers. A description of the algorithm used is illustrated in Figure 5.7 and the necessary coding is given in Appendix C.5.10. None of the χ^2 values obtained from the 999 randomizations of the data set exceed the observed χ^2, the largest value obtained being only 175. Thus, we conclude that, far from being marginally significant, the heterogeneity among samples is highly significant ($P < 0.001$). The cumulative frequency distribution of χ^2 values from the randomized set is an almost perfect match to the predicted cumulative frequency distribution of χ^2 values (Figure 5.8).

Figure 5.8 also shows the results of an analysis of a second data set, that by Avise *et al.* (1987) on mtDNA variation in Atlantic and Gulf Coast populations of the hardheaded catfish (*Ariusfelis*). Because of extremely low sample sizes, Avise *et al.* (1987) combined different genotypes into two clusters and combined

In doing the randomization we must keep the row and column totals constant. This can be achieved by the following algorithm. Let n_{ij} be the number of observations in the ith row of the jth column, and let the total number of observations be N. We create a matrix, **M** with two columns and N rows, and into each column we enter the row by column coordinates for every observation, with, for example, the first column containing the row coordinates and the second column the column coordinates. For example suppose our initial data matrix, is 2×2 consisting of two genotypes (G1, G2) and two sites (S1, S2) with entries:

Site	Genotype		Row totals
	G1	G2	
S1	1	2	3
S2	2	0	2
Column totals	3	2	5

The required 5×2 **M** matrix (with label headings superimposed for clarity) would be

$$[r \quad c]$$
$$\begin{bmatrix} 1 & 1 \\ 2 & 1 \\ 2 & 1 \\ 1 & 2 \\ 1 & 1 \end{bmatrix}$$

There are five entries, each entry denoting a single cell entry. Thus the total number in cell 2,1 (row, column) is equal to the number of entries "2,1" in **M**. To create a randomized matrix we randomize one of the columns of the **M** matrix and then count up entries to construct the new matrix. Thus, for example, suppose we randomize the "column" entries of the **M** matrix to get the sequence 1, 2, 1, 2, 1. The "new" **M** matrix and data matrix would be

$$[r \quad c]$$
$$\begin{bmatrix} 1 & 1 \\ 2 & 2 \\ 2 & 1 \\ 1 & 2 \\ 1 & 1 \end{bmatrix} \Rightarrow \begin{bmatrix} 2 & 1 \\ 1 & 1 \end{bmatrix}$$

Coding to do the above procedure is given in Appendix C.5. Note that in the coding I use two vectors rather than a single **M** matrix.

Figure 5.7 Description of algorithm to randomize cell entries in χ^2 contingency test.

different geographic sites into Gulf or Atlantic categories (i.e., two "genotypes" and two "sites"). Even in this radical lumping, one of the four cells contained only three fish and two of the cells contained expected values less than 5. The probability obtained using the randomization method is 0.26 and from the tabulated values $P=0.16$. Because of the low sample sizes and restricted

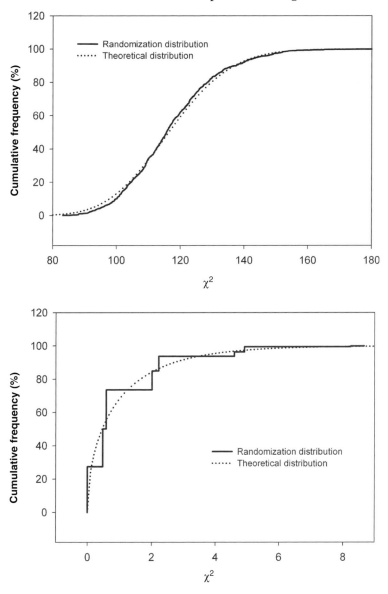

Figure 5.8 Cumulative frequency distributions of χ^2 from the randomized data set and the theoretical χ^2 distribution. Upper panel shows data from mtDNA in shad (data from Bentzen *et al.* 1988) and lower panel shows data on mtDNA variation in the hardheaded catfish (Data from Avise *et al.* 1987).

number of cells, the cumulative distribution of χ^2 is distinctly stepped (Figure 5.8), but it closely corresponds to the cumulative χ^2 distribution.

Roff and Bentzen (1989) analyzed four other data sets in addition to the two above, and in all four cases found that the probability obtained from the standard χ^2 test qualitatively matched the probability obtained from the randomization. This should not be taken as sufficient grounds to rely upon the standard test, because its behavior is uncertain at low sample sizes. It does suggest that a preliminary analysis with the standard test should give a good indication of the results of the randomization test.

Linear and multiple regression: dealing with non-normal errors

Regression is known to be very robust to failure of the assumptions of a normal distribution of errors, but randomization testing makes fewer assumptions about the error distribution and in this sense can provide a useful check. The obvious method of randomization is to randomly pair the X and Y values. Other methods of randomization, such as randomization of residuals seem to give the same answer in most cases.

Manly (1997) suggested that the use of randomization in regression can be justified on one of three grounds: (1) the X, Y pairs are separate independent random samples from a population in which all possible pairings are equally likely, (2) the data come from an experiment in which the X values are randomly assigned to experimental units and the Y values have the same distribution for all values of X, and (3) if it is assumed that if X and Y are independent then all possible pairing can occur. The crux of these three justifications is that under the null hypothesis all possible pairings are equally likely.

Randomization of linear regression coefficients is readily accomplished (see Appendix C.5.11, note that one can either use the coefficients as the test statistics or the t values; in the coding shown, the coefficients are used) but its utility is likely to be rather small except in those cases in which errors are grossly non-normal or extreme. Randomization testing can also be applied to testing multiple regression, in which case the dependent observations are randomized with respect to the multiple dependent variables. An alternative to randomizing the observations is to randomize the residuals, as suggested by ter Braak (1992). In this approach, the model is fitted to the data, the residuals are extracted, and then these used in place of the dependent variable in the randomization process. Simulations suggest that this method does not improve the performance of the randomization test and may in some cases reduce its power.

To demonstrate that the usual parametric test and randomization test can give different answers (Manly 1997), following a model suggested by Kennedy and Cade (1996), produced twenty "observations" on independent variables X_1, X_2, and the dependent variable Y using the following algorithm:

(1)   19 values of X_1 were drawn from a uniform distribution between 0 and 3.
(2)   The 20th data point of X_1 was set at the extreme value of 33.

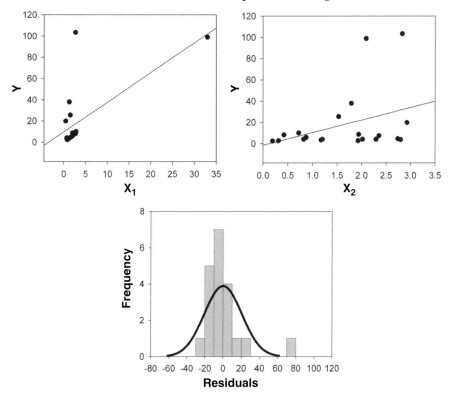

Figure 5.9   The upper panel shows the plots of Y vs. X_1 and X_2, using the simulated data of Manly (1997). Solid lines show fitted regression lines. Lower panel shows the distribution of the residuals from the fitted multiple regression equation and the superimposed best fit normal distribution.

(3)   20 values of X_2 were drawn from a uniform distribution between 0 and 3.

(4) Y values were calculated from $Y = 3X_1 + \varepsilon^3$, where ε was a random error taken from an exponential distribution.

A plot of Y on each variable separately suggests that there is something amiss with the error structure, one point lying far above the fitted regression line of Y on X_1 (Figure 5.9). Rather surprisingly, given that we know how the data were generated, the error variance about X_2 appears to increase with X_2. A multiple regression analysis using least squares indicates that the overall model is highly significant and that there is a highly significant effect due to X_1 but not due to X_2 (Table 5.7). A stepwise regression analysis retains both variables in the model. However, an analysis of the residuals from the fitted model shows a highly significant deviation from normality (both Lillifor's test and Shapiro–Wilkes test give $P < 0.0001$). Given the relatively small number of observations, the failure to

Table 5.7 *Results of analysis of the multiple regression model described in the text and the analysis of the results of applying three methods of analysis to 5000 simulated data sets*

| | Probability | | |
Method	θ_1	θ_2	Full model
Analysis of single data set[a]			
Least squares analysis	0.0015	0.1054	0.0017
Randomization of residuals	0.0510	0.0984	0.0510
Randomization of Y	0.0274	0.01152	0.0074
Analysis of 5000 simulated data sets with $\theta_1=\theta_2=0$[b]			
Least squares analysis	0.052	**0.024**[c]	0.054
Randomization of residuals	0.071	0.041	0.056
Randomization of Y	0.051	0.053	0.050

[a]5000 randomizations.
[b]100 randomizations per analysis.
[c]Numbers in bold indicate values significantly different from 0.05 for simulation.

conform to the normal distribution should be taken as strong evidence that the data do not conform to the assumptions of least squares regression.

Applying the randomization test to the above data produces a qualitatively identical result, except that the significance levels are reduced (Table 5.7). Randomization of the residuals indicates no significant effects! To examine which of these methods of analysis is most appropriate, Manly ran the following simulation: he constructed twenty sets of observations by taking the twenty error values $(E_1, E_2, \ldots, E_{20})$ and creating Y using $Y_i=\theta_1 X_{1i}+\theta_2 X_{2i}+E_j$, where E_j is an error value drawn at random (without replacement) from the error set. When θ_1 and θ_2 both equal zero a correct test will detect a significant effect with a probability of 0.05. Only the randomization of the observations produced the correct number of "spurious" significant probabilities (Table 5.7). When one or both of the slopes differed from zero, the least squares analysis typically gave more significant tests when the null hypothesis was not true (i.e., more power, Figure 5.10). These analyses suggest that, in the present case, when the null hypothesis is true, randomization is the preferred method, but it is less powerful than the least squares method. Which method is to be preferred depends upon which type of error is most important.

Correlation and regression: dealing with the problem of non-independence

Regression is quite robust to the distribution of residuals but the entire principle of hypothesis testing breaks down when the dependent and independent variables are not independent. A very common case in which this

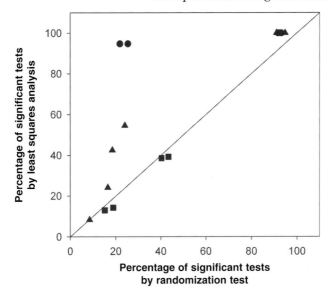

Figure 5.10 Summary of power analysis of simulation model of the multiple regression model $Y_i = \theta_1 X_{1i} + \theta_2 X_{2i} + E_j$ (see text for further explanation); •: tests on θ_1 in models in which $\theta_1 > 0$; ■: tests on θ_2 in models in which $\theta_2 > 0$; and ▲: tests on the overall multiple regression model in which at least one slope was greater than zero.

occurs is when a researcher, for example, asks if some variable, such as fecundity (*y*) varies as a function of per unit mass, i.e., does fecundity per gram of salmon vary with salmon size (*x*)? This violates the basic assumption of linear regression, because it is equivalent to regressing the reciprocal of the independent variable on itself, i.e., *y/x* vs. *x*. One way to avoid this problem is to regress fecundity on size (*y* vs. *x*, or some suitable transformation) and then examine the resulting equation with respect to variation on a per unit mass basis. An example discussed in Roff (1997, p. 136) is egg size and body size in birds: for passerines (song birds) egg mass, *y*, is related to body mass, *x*, according to the equation $y = 0.258x^{0.73}$. This equation tells us that egg mass increases with body mass but it is perhaps not immediately clear how the relative egg size varies. The correlation between the two variables (on a log scale to linearize the relationship) is 0.96, which is sufficiently high that algebraic manipulation of the regression equation is reasonable. Dividing both sides by body mass gives relative egg mass $= 0.257x^{-0.27}$, which immediately shows that relative egg mass is declining with body mass. Passerine body weights range from about 4g to 1200g, giving a variation in relative egg mass from 18 to 4% (for non-passerines the range is even larger, with a humming bird producing an egg that is 29% of its body size but an ostrich producing an egg that is only 2% of its body size). To test if the relationship between relative egg size and body size is significant (which it clearly is in this case), we cannot employ the usual linear regression or correlation analysis. A very simple solution is to use randomization, where we randomize that component of the dependent variable that is not contained within the dependent variable (e.g., in the present example, we randomize absolute egg and body size and then compute relative egg size for each randomized data set). By this process the "built-in" relationship between the two variables is taken into account.

A more complex version of the above problem arose in a study of song sharing in the passerine, the American redstart (Shackell *et al.* 1988). Males of this species divide their song repertoire according to functional situations. Prior to and during the arrival of females, each male sings one particular song repetitively in preference over the others. The details of this song vary considerably among males, particularly with respect to the last three phones of the song. The question Shackell *et al.* (1988) wished to address was "do neighboring birds tend to sing the same song?" They measured 15 components of the last three phones (Figure 5.11) and to reduce the problem of non-independence among measures within the same phone they used the first principal component of each phone. The null hypothesis is that there is no correlation between neighbors in their song components. A simple test of this hypothesis is to calculate the correlation between the songs of all the birds and their neighbors. The problem is that there is a lack of independence between the two variables. Consider bird x with neighbor y: bird x appears first as an independent variable and bird y as a dependent variable, but then bird x also appears as a dependent variable, because it is neighbor to bird y (Figure 5.11)!

The expected value of the regression coefficient is, in fact, negative. This can be demonstrated as follows. The distribution of birds on the x-axis is obviously the ascending ranking of birds. Consider now the first bird; all of its neighbors have larger values. Similarly the neighbors' values of the last bird on the x-axis must be less than his. For the second bird on the x-axis only 1 value at most, that of the first bird, can be less than his. Similarly, for the second-to-last bird only, the last bird can have a larger value. Thus, there must be a bias towards higher values at the low end on the x-axis and low values at the high end.

This problem of lack of independence was solved by randomization, namely, estimating, via simulation, the probability of observing by chance alone a correlation coefficient as large as or larger than that obtained from the data. Shackell *et al.* constructed 5000 randomized sets from each data set (1982 and 1984). From these, the null distribution of the correlation coefficient was generated for each year (note that because of the use of principal components the correlation coefficient is equal to the regression coefficient). Each randomized data set has to satisfy two requirements. First, if bird i is neighbor to bird j, bird j must be

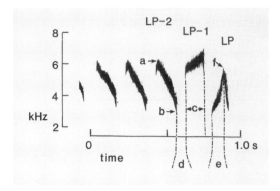

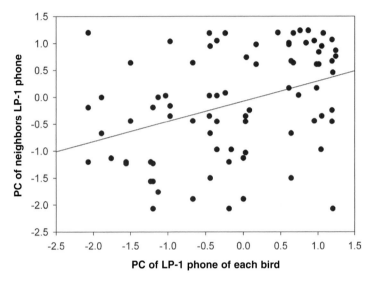

Figure 5.11 Spectrograms of phone variables used in analysis of redstart song. Phones are designated as last (LP), next-to-last (LP-1) and second-to-last (LP-2). Measured variables: (a) maximum frequency, (b) minimum frequency, (c) duration of phone, (d) interval between phones, (e) proportion of total duration spent in upward sweep in frequency, and (f) number of inflection points. Lower panel shows the distribution of LP-1 in 1982, with the least squares regression line.

neighbor to bird i. Second, the frequency distribution of the number of neighbors/bird must remain the same (for a description of the appropriate algorithm see Shackell *et al.* 1988). For the 1982 data set, the randomization procedure indicated that the expected correlation was approximately -0.03. Although the expected value of the correlation is negative as predicted, the bias is small.

For both data sets (1982, 1984) the randomization procedure indicated a significant correlation for LP-1 but not the other two components (Table 5.8). Further analysis showed that the significant correlation arose from the single

Table 5.8 *Correlations between neighbors using the first principal component of LP-2, LP-1 and LP. Taken from Shackell et al. (1988)*

Principal component	1982 correlation between neighbors	P	1984 correlation between neighbors	P
LP-2	0.0082	0.425	0.2637	0.034
LP-1	0.3749	0.005	0.2665	0.031
LP	0.2722	0.028	0.0782	0.238

variable labeled "f" in Figure 5.11. Thus, Shackell *et al.* were able to conclude that indeed the songs on neighboring redstarts do resemble each other more than expected by chance. This particular analysis illustrates the great utility of the randomization approach, for in this case the use of the parametric statistics is clearly not valid.

Comparing distance matrices: the Mantel test

Ecologists frequently deal with spatially distributed data, such as density per quadrat, as illustrated in Figure 5.12. Visual examination of these data suggests that there is a pattern to the distribution, with high counts tending to be clustered. One way to examine the data is to compare the two matrices comprising the spatial distance between points and the difference in measured values (Z in Figure 5.12). Denoting these as $\mathbf{M}_x$ and $\mathbf{M}_y$, respectively, we have

$$\mathbf{M}_x = \begin{bmatrix} 0 & x_{12} & \cdots & x_{1n} \\ x_{21} & 0 & \cdots & x_{2n} \\ \cdots & \cdots & \cdots & \cdots \\ x_{n1} & x_{n2} & \cdots & 0 \end{bmatrix} \quad \mathbf{M}_y = \begin{bmatrix} 0 & y_{12} & \cdots & y_{1n} \\ y_{21} & 0 & \cdots & y_{2n} \\ \cdots & \cdots & \cdots & \cdots \\ y_{n1} & y_{n2} & \cdots & 0 \end{bmatrix}$$

Note that $x_{ij}=x_{ji}$ and that the diagonal elements are obviously zero. To compare the above two matrices we could use the **Pearson product moment correlation**, pairing up corresponding elements across the two matrices and omitting points either above or below the diagonal, as these are simply duplicate values (lower plot Figure 5.11). Practically, to carry out such a test we first reorder the matrices into two vectors, composed of the above- or below-diagonal elements of the two matrices:

$$\mathbf{V}_x = \begin{bmatrix} x_{12} & x_{13} & \cdots & x_{23} & x_{24} & \cdots & x_{(n-1)(n-2)} & x_{(n-1)(n-1)} \end{bmatrix}$$
$$\mathbf{V}_y = \begin{bmatrix} y_{12} & y_{13} & \cdots & y_{23} & y_{24} & \cdots & y_{(n-1)(n-2)} & y_{(n-1)(n-1)} \end{bmatrix}$$

Taking the results of the parametric analysis at face value, the correlation is highly significant ($r=0.160$, $n=435$, $P<0.01$), but, because the data clearly fail to

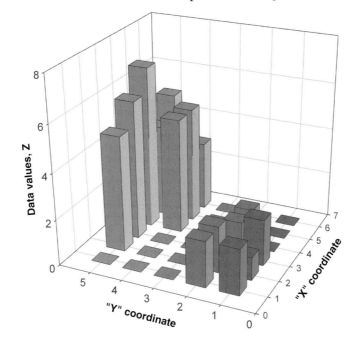

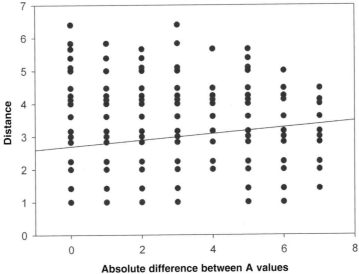

Absolute difference between A values

Figure 5.12 The upper plot shows the hypothetical spatial distribution of data, such as counts per quadrat. There are a total of 30 paired points, making a 30×30 matrix of distances and difference measures (in this case simply the absolute difference). These two matrices, called "Distance" and "Difference," can be calculated using the coding shown in Appendix C.5.12. Lower plot shows plot of the absolute differences against distance. Solid line shows fitted regression.

satisfy the assumptions of the test, such a result cannot be so taken. The solution, called the Mantel test, is to use randomization to generate many estimates of r and calculate the proportion of cases in which $|r_{observed}| > |r_{random}|$, where $r_{observed}$ is the correlation from the observed data set and r_{random} is the correlation from a randomized data set. The Mantel test is easily accomplished by first converting the matrices into the vectors, $\mathbf{V}_x$ and $\mathbf{V}_y$, and then treating the data in the same manner as in the regression problem (Appendix C.5.13). For the hypothetical data, the Mantel test indicates a highly significant correlation between the two measures ($P=0.004$). Dietz (1983) examined several other measures of association and found that Spearman's rho or a variant of Kendall's tau, both non-parametric measures of association, had higher power for highly skewed distributions than the Pearson product-moment correlation and was very similar when the data were not skewed. Spearman's rho is readily incorporated by replacing each vector with its ranks and then proceeding as before.

Comparing matrices: other tests

In some cases we may be interested in comparing matrices that have the same units: for example, the matrix of phenotypic or genetic variances and covariances of two different populations or species. The Mantel test has been used for such a purpose but it has several disadvantages. First, it is a test of correlation between matrices but not of equality of matrices: matrices may be correlated and equal, correlated and proportional, correlated but not proportional or uncorrelated (Figure 5.13). Second, the Mantel test does not take into account the variability that may be inherent in the matrix entries; for example, there can be considerable variance in the estimates of the genetic variances and covariances.

We can extend the randomization procedure to take into account sampling variation. Consider the question of comparing two phenotypic variance–covariance matrices (hereafter, referred to simply as covariances). The null hypothesis is that the two sets of covariances come from the same statistical population. Note that we do not assume that they have the same means. However, for the randomization procedure, it is necessary to transform the values such that the corresponding trait means are equal. Under the null hypothesis, the individuals could have come with equal probability from either population: therefore, a randomized data set is created by randomly assigning individuals to the populations and then calculating the new matrices (if the trait means had not been equalized prior to the randomization then the covariances would differ by virtue of different means). In this case, the individual traits are retained as a unit, the individual being the unit of randomization. If we were

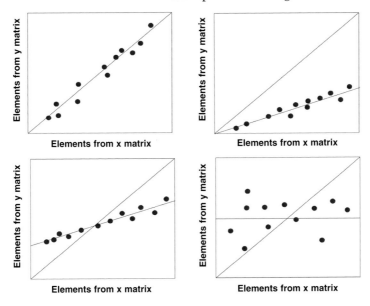

Figure 5.13 Diagrammatic illustration of four possible ways in which two matrices can covary.

comparing genetic covariance matrices the unit of randomization would be the families. It is important to keep in mind what is being tested.

After construction of the two randomized matrices the test statistic (plausible candidates are discussed below) is determined, and compared with the observed value. The entire process is repeated, say, 4999 times and the probability of obtaining the observed result under the null hypothesis estimated from the proportion of times the statistic from the randomized data sets exceeds that from the observed data set.

There are a number of alternate test statistics, each being sensitive to particular characteristics of matrix structure:

(1)   Mantel test: is one candidate and its limitations have already been discussed.

(2)   Maximum likelihood (Anderson 1958; Shaw 1991): this test proceeds in three stages, (a) calculate the elements of the two matrices separately using maximum likelihood: let the log-likelihoods of these be LL_1 and LL_2, and the combined log-likelihood be $LL_{1,2}$, (b) calculate the elements of the matrix under the null hypothesis that the data come from the same population: let the log-likelihood so calculated be designated LL_0, (c) the log-likelihood ratio for comparing the two hypotheses is $2(LL_{1,2} - LL_0)$ and is tested against the χ^2 distribution with one degree of

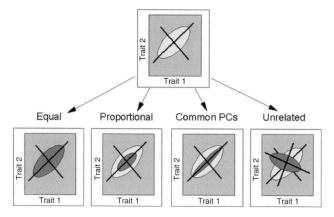

Figure 5.14 Schematic view of the Flury hierarchy for two covariance matrices consisting of two traits. The ellipses represent the covariance structure with the axis orientation denoting the principal components and the spread of the ellipses along each axis the eigenvalues. The analysis proceeds from right to left (equality to unrelated structure). With more than two matrices another possible finding is for some but not all PCs to be common.

freedom, (d) the foregoing test is sensitive to deviations from multivariate normality and hence a randomization test is a useful addition. This can be done by randomizing the data as described above and recalculating the log-likelihoods and using $2(LL_{1,2}-LL_0)$ as the test statistic.

(3) Flury hierarchy (Flury 1988; Phillips and Arnold 1999): this extends the maximum likelihood approach by comparing the structure of the matrices using the principal components (Figure 5.14). As with the previous approach, the analysis can proceed using maximum likelihood or by randomization.

(4) T method (Willis *et al.* 1991; Roff *et al.* 1999): A very simple test in which we compare the sum of the absolute differences between the matrix elements, i.e.,

$$T = \sum_{i=1}^{C} \left| \hat{\theta}_{i1} - \hat{\theta}_{i2} \right| \tag{5.12}$$

where, as above, the matrix has been written in vector format with $\hat{\theta}_{ij}$ the estimate of $\hat{\theta}_i$ of the *j*th matrix (*j*=1, 2), and *C* is the number of distinct elements in each matrix ($=0.5n[n+1]$, where *n* is the number of rows and columns). An alternative would be to use the squared difference, which would be consistent with a least squares approach (I have not found the results to differ).

(5) Jackknife-followed-by-MANOVA (Roff 2002; Roff *et al.* 2004): This has been discussed in Chapter 3. Randomization can be used either at the start, thereby producing a randomization of the entire process, or by randomizing the pseudovalues between the two sets, thereby producing a randomization test of the MANOVA component.

(6) Reduced major axis regression (Roff 2000): consider the model

$$\theta_{i1} = A + B\theta_{i2} \tag{5.13}$$

which assumes a linear relationship between the elements of the two matrices. There is no a priori reason why the elements should be so related but empirically this is a frequent observation. While the above model suggests a linear regression approach, the specification of dependent and independent variables is arbitrary, and simple linear regression is excluded because there will be variance in both dependent and independent variables. If, as will generally be the case for the comparison of covariances, this variation is approximately the same along both axes then we could use **reduced major axis (RMA) regression**. Under the null hypothesis $A=0$ and $B=1$, and, therefore, deviations are assessed by comparing $|A_{obs}|$ with $|A_{random}|$, whereas deviations of B are assessed by comparing $|B_{obs}-1|$ with $|B_{random}-1|$. A non-significant A but significant B indicates that the matrices are proportional and the proportionality constant is significantly different from 1. If both A and B are significant the elements of the two matrices are linearly related but not proportional.

Testing for density-dependence

An issue that has long plagued ecologists is the detection of density-dependence in populations. First, what is meant by density-dependence? Let the population sizes at two consecutive time periods be N_t and N_{t+1}. The rate of increase in the population is given by N_{t+1}/N_t. Ratios are always rather difficult to work with and so we take logs, giving the rate of increase (on a log scale) to be $\ln(N_{t+1})-\ln(N_t)$, which for simplicity I shall denote as $d_t=x_{t+1}-x_t$. Density dependence is defined to occur when d_t is a function of x_t. A simple model of density-dependence is

$$x_{t+1} = r + \theta x_t + \varepsilon \tag{5.14}$$

where r is termed a drift parameter and accounts for long-term density-independent changes in population size, θ is a coefficient describing negative density-dependence ($\theta<1$) and ε is an independent random variable with mean zero. Pollard and Lakhani (1987) suggested the following randomization method for testing the null hypothesis $\theta \geq 1$: (1) For the observed data set calculate the correlation between d and x, say r_{obs}. (2) Randomize the vector of d and recalculate the correlation to produce a randomized value of the correlation, say r_{rand}. (3) From the set of randomized correlations estimate, the probability of obtaining a value of r_{obs} as small or smaller than observed. This test is simply a one-tailed randomization test of the correlation coefficient. A number of other tests have been proposed but simulations have shown that this test performs best (Holyoak 1993). The above test can be modified to take into account nonlinear relationships between population size and growth rate, a popular model being the log-linear functional relationship $x_{t+1} = r + \theta N_t + \varepsilon$ (Saitoh *et al.* 1999).

Monte Carlo methods: two illustrative examples

In the first example, we observe two individuals of a plant species that are some distance d apart (Figure 5.15). The area that could be occupied by a plant of this species is shown as the surrounding circle. A question we might wish to answer is, "are the two individuals further apart (i.e., show interference) than expected by chance?" To answer this question we could proceed as follows: (1) pick two points within the circle at random and determine the distance between them, (2) repeat this procedure a large number of times (say N) and record the number of times the distance between the randomly placed points (say n) exceeds that observed, (3) estimate the probability of obtaining a distance as great or greater than that observed as $(n+1)/(N+1)$. For an example of Monte Carlo methods applied to spatial patterning see, for example, Couteron *et al.* (2003).

In the second example, we measure the body size of two animal species that we suspect might show character displacement (Figure 5.15). The question we wish to address is, "Are the two animals further apart in morpho-space (here a single body size measure) than might be expected by chance?" To answer this question we proceed in the same manner as in the previous example, in this case selecting points within some fixed interval. Studies addressing the size-ratio question are considered in greater detail below.

The above two examples are very simple but highlight the components of this approach, namely, (1) we have a set of observed measurements, (2) we have a theoretical model that can be used to generate the set of measurements under the null hypothesis, (3) as with randomization, we estimate the required probability by comparing the observed data with the randomly generated data. Monte Carlo models are generally tailored to a specific hypothesis and, thus far,

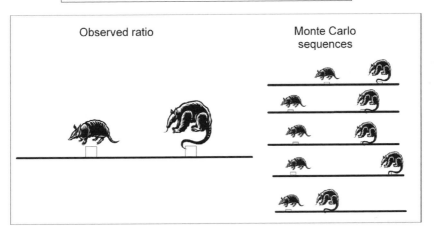

Figure 5.15 Two hypothetical examples in which Monte Carlo methods could be employed. In the upper figure, we wish to test the hypothesis that the distance between the two plants (left-side) is greater than expected by chance. In the lower figure, we wish to test the hypothesis that the ratio in size between the two co-occurring species (left-side) is greater than expected by chance. In both cases, we generate pairs of points within the allowable space (right-side) and use the distribution of distances/ratios so generated to estimate the probability of obtaining a distance/ratio as large or larger than observed.

have tended to use a variety of techniques for assessing significance. There is, however, a very general approach first introduced by Besag and Clifford (1989) that can remedy this situation, though there remain cases in which alternate approaches will be easier. I shall begin by considering such a case, namely the question of validating models in step-wise regression. Following this, I introduce the **Generalized Monte Carlo test**. Thereafter, to illustrate the diversity of questions that can be addressed and approaches hitherto used, I consider some Monte Carlo models that have been used in ecological studies.

Validating step-wise regression

In an analysis of the Canadian lynx cycle Arditi (1989) obtained three "contending" regression models for lynx abundance:

Model 1: $Y=0.43X_1 - 0.65X_2+0.61X_3$ $R^2=0.64$

Model 2: $Y=0.43X_1 - 0.67X_2+0.43X_3+0.37X_4$ $R^2=0.75$

Model 3: $Y=0.34X_1 - 0.47X_2+0.46X_3+0.38X_5 - 0.33X_6$ $R^2=0.79$

where $X_1,\ldots,X_6$ are weather factors (e.g., average temperature during a particular month, lagged to take into account lags in population response). These three equations were the result of stepwise examination of a data set in which there were a possible 120 (24 months × lags of 0–4 years) predictor variables! These three models could be evaluated using the cross-validation technique described in the next chapter. However, whereas this approach would allow one to decide among the three models it would not tell one if the particular subset regression is statistically significant, because in this case such an enormous number of equations were examined that it is possible that the final models are a consequence of a type I error. To evaluate this possibility, Arditi constructed a Monte Carlo model in which (a) the weather data were retained as observed, thereby preserving their structural relationships, and (b) constructing a set of lynx abundances by randomly sampling from a log-normal distribution with the constraint that the sequence of abundances were serially correlated with the same value as observed in the real data. For each randomly constructed data set, Arditi used stepwise regression to select the best 3, 4, and 5 variable models and compared their R^2 with the observed. For the analyses using all 120 predictor variables the observed value of R^2 was not significantly greater than expected by chance ($P > 0.2$ in all three models, Table 5.9). Arditi also considered analyses in which the number of predictor variables was reduced: in all cases the same stepwise regressions were obtained using the observed data set and for the smallest set of predictor variables the observed R^2 was significantly larger than expected ($P < 0.03$ in all three models, Table 5.9). These results point out the necessity of carefully selecting the predictor variables: Arditi notes that lags longer than 2 years make no sense on biological grounds, which makes one wonder why they were selected in the first place as possible predictor variables.

Table 5.9 *Results of the Monte Carlo test of lynx data described in text. Shown is the proportion of cases in which, for a given combination, the R^2 from the Monte Carlo model equaled or exceeded the observed R^2. Estimates based on 2000 simulations per combination. Modified from Arditi (1989)*

Predictor variables			Model		
Months	Lags	Total	1	2	3
24	0–4	120	0.393	0.252	0.258
24	0–2	72	0.213	0.108	0.108
12	0–2	36	0.067	0.027	0.021
8	0–2	24	0.026	0.007	0.004

As Arditi (1989, p. 33) himself notes, "The practice of 'blindly' offering all sorts of variables, under the assumption that the procedure will automatically select the relevant ones, must be avoided."

Generalized Monte Carlo significance tests

To illustrate the generalized Monte Carlo approach, I shall consider the question of whether a set of body size ratios are larger than expected by chance, and thus (perhaps) indicative of competitive displacement. The specific situation to be analyzed is shown in Figure 5.16: there are four species in the community under study and a possible pool of 20 species that the researcher deems could be members of the community. To obtain an index of the community-wide ratio the species are ranked according to size and the three pairwise ratios calculated: 13/7, 39/13, and 100/39. Because of the large differences in size and because ratios generally do not have well-behaved distributions, the researcher uses the mean of the log of the ratios, which is equivalent to using the geometric mean, $GM = \text{antilog}\left((1/3)\left[\sum_{i=2}^{4} \log(x_i) - \sum_{i=1}^{3} \log(x_i)\right]\right)$. This measure has an interesting property: it depends only upon the two extreme sizes, $GM = \text{antilog}\left((1/3)[\log(x_3) - \log(x_1)]\right)$! The fact that the intervening size ratios only play a role in so far as they determine the divisor (1/3 in this case) is unsettling and perhaps argues against this measure of size ratio. So, the sometimes poor distributional properties of ratios notwithstanding, the researcher decides to use the arithmetic mean instead, which we shall denote as $\hat{\theta}_{\text{Obs}}$ (for the present purpose the importance lies not in the particular statistic chosen, only that one is chosen). The probability of obtaining a mean ratio as large or larger than that observed, P, could be obtained by generating N communities of four species selected at random, without replacement from the species pool and for each calculating the mean ratio for each such community, say $\hat{\theta}_i$, for the ith random

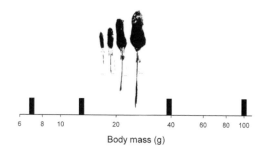

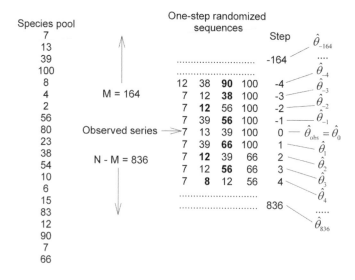

Figure 5.16 A simple illustration of the generalized Monte Carlo method applied to body size ratios. The upper plot shows four hypothetical desert rodents (based on Figure 1 of Bowers and Brown 1982). The observed series of body masses is 17, 13, 39, 100 and the possible hypothetical species pool from which these four species could have come is shown on the bottom left (20 species ranging from 2 through 100 g). The total number of stepwise random sequences is 1001. A random number is drawn which places the observed sequence at position 165, which for simplicity is here denoted as position 0. Two series of sequences, denoted the Monte Carlo sequences, beginning from the observed sequence (=Monte Carlo sequence #0) are generated by stepwise replacement of one species selected at random from the Monte Carlo sequence with a randomly selected species from the species pool, subject to the restriction that the same species cannot appear twice in any single Monte Carlo sequence. In one "direction" 164 steps are made whereas in the other "direction" 836 steps are made. At each step, the relevant statistic is calculated and the required probability estimated by comparing the observed value with the values from the Monte Carlo sequences.

community: then, as usual, P is estimated as $P = [n(\hat{\theta}_i \geq \hat{\theta}_{Obs}) + 1/(N + 1)]$, where $n(\hat{\theta}_i \geq \hat{\theta}_{Obs})$ is the number of times $\hat{\theta}_i \geq \hat{\theta}_{Obs}$ and N is the total number of random communities.

In the present case, the randomization method is easy because it is operationally easy to sample without replacement from the species pool. However, there are cases in which this may not be easily accomplished. One such circumstance is shown in Figure 5.17: there are six islands (A, B, C, D, E, F) and on each island there occurs a maximum of four species from some assumed "guild," taxon or group of species that the researcher suspects might be so similar that competitive interactions could structure the islands-wide community, i.e., the sets of species occurring on the islands is not a random

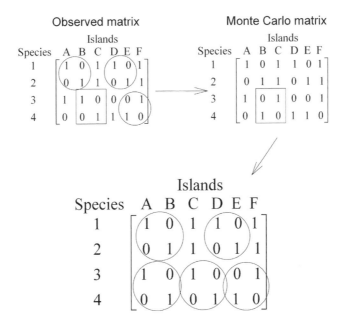

Figure 5.17 A hypothetical matrix describing the distribution of four species on six islands. The four unit patterns within the circles or square of the observed matrix denote those components of the matrix that can be permuted while keeping marginal totals constant. A single step Monte Carlo change is made by selecting one of these components at random (e.g., the one in the square) and permuting the entries as shown. This produces the Monte Carlo matrix which now has six possible changes. This process is repeated in a "forward" and "backward" direction to produce a series of communities with which the hypothesis that the observed community is randomly constructed can be tested.

assembly of species. Each random community must satisfy the requirement that the marginal totals remain the same (i.e., the number of species per island and the number of occurrences of each species remain constant), which places a severe constraint on the number of permissible combinations. It can be shown that all possible combinations can be formed by permuting the four unit pattern highlighted in Figure 5.17. Thus, a random community could be formed by randomly selecting and permuting a large number of such four unit patterns (each randomization may itself create more possible randomization patterns). While such a process would create random communities it would also be very time consuming. An easier approach is given by the generalized Monte Carlo method which requires only single step changes for each community. To illustrate this approach we shall first return to the size ratio problem.

Besag and Clifford (1989) noted that a particular statistic (such as the observed mean size ratio), say $\hat{\theta}_{Obs}$, can be viewed as one in a series of single step Monte Carlo changes. Let the number of steps in the series be $N+1$ (1001 in Figure 5.16): we first assign $\hat{\theta}_{Obs}$ to a position within this series by selecting a random number between 1 and N, say M (=164 in Figure 5.16). Next, we make a single change at random to the data and recompute the required statistic, say $\hat{\theta}_1$: in the present case, this would mean swapping species between the species pool and the observed sequence (no duplication of species is allowed and thus sampling is without replacement). This single step process is repeated $N-M$ times to give a "forward" sequence $(\hat{\theta}_1, \hat{\theta}_2, \hat{\theta}_3, \ldots, \hat{\theta}_{N-M})$. Similarly, the single step process is done M times giving a "backward" sequence $(\hat{\theta}_{-1}, \hat{\theta}_{-2}, \hat{\theta}_{-3}, \ldots, \hat{\theta}_{-M}$: "forward" and "backward" are used for convenience and refer to two sequences commencing for the specified number of steps from the observed sequences). We now estimate the probability that the observed statistic is significantly larger (or smaller) than expected in the same manner as for the randomization procedure. This method is known as the **serial method** (Besag and Clifford 1991). An alternate method is called the **parallel method**; because it is typically more computer intensive and less efficient than the serial method (Manly 1993), it will not be considered here. To ensure that the results are not due to the particular sequence generated, the **serial generalized Monte Carlo** test should be repeated by generating a series of random "starting" positions. The set of probabilities so obtained can be used to estimate the mean and SE of the probability.

The above approach is particularly useful for cases, such as the community structure case shown in Figure 5.17. In this particular case, there are problems with identifying species exhibiting interactions that structure the community: this issue is taken up in the next section.

Community assembly rules

Perhaps one of the largest, and certainly one of the most contentious, areas of ecological study in which Monte Carlo models have been used to test statistical hypotheses is that of community structure. A variety of questions have been addressed, which can be arranged roughly into the three categories of niche overlap, species co-occurrence and the size structure of communities. To illustrate the disparity in approaches, I shall consider the latter two (for papers dealing with niche overlap see, for example, Joern and Lawlor 1980; Lawlor 1980; Cole 1981; Kochmer and Handel 1986; Tokeshi 1986; Vitt *et al.* 2000). It is not my intent to enter the debate that continues on the merits of one particular null model over another but to present an overview of the types of models used to highlight the care with which null models and their tests must be constructed (for general discussions refer to Strong 1982; Harvey *et al.* 1983; Strong *et al.* 1984; Connor and Simberloff 1986; Losos *et al.* 1989; Pleasants 1990).

Patterns of species co-occurrences

The general approach has been to reallocate species (individuals) to islands (patches) keeping the total number per island and/or number of species occurrences constant (Table 5.10). Gotelli (2000) provides a general analysis of the consequences of different sets of assumptions. Several different indexes and tests have been proposed: the best is to compare the observed index against the randomized value using the same method as for randomization. Manly (1995) has advocated keeping both row and column totals constant as this reflects inherent constraints on the species dispersal capabilities and the island characteristics. As a test statistic he proposed

$$S = \sum_{i=1}^{R} \sum_{j=1}^{R} \frac{(O_{ij} - E_{ij})^2}{R^2} \tag{5.15}$$

where O_{ij} is the observed number of times that species i co-occurs with species j, E_{ij} is the expected number based on the null model and R is the total number of species. Because, by definition, $O_{ii} - E_{ii} = 0$, it is not necessary to specify $i \neq j$ in the above equation. The probability of obtaining a value of S as large, or larger, than that observed is estimated from the set of S values obtained from the generalized Monte Carlo method. Interactions between particular species pairs can be tested to find those particular pairs that deviate more than expected by chance: to account for multiple comparisons these individual tests must be Bonferroni-corrected.

Table 5.10 *Examples of tests of the null model that species assemblages are random. Some studies may have used several different measures but only one is selected here for illustration*

Organism		Measure	Test	Reference
Island birds and bats	Marginals kept constant	Frequency of pairs, triplets not found	χ^2	Connor and Simberloff 1979
Birds on the Bismark archipelago	Probability of species occurrence in each cell	Species pairs	χ^2	Gilpin and Diamond 1982; Diamond and Gilpin 1982
Bird species in Galapagos archipelago	Number of species per island not fixed but number of species occurrences fixed	Species pairs	t-test of Similarity index[a]	Alatalo 1982
Pollinator communities	Species drawn at random from a species pool	Overlap *vs.* no overlap	Binomial	Armbruster 1986
Vanuatu birds	Marginals kept constant	Species pairs	Similarity index tested by randomization data	Wilson 1987
Fish in stream pools	Using total number collected, individuals (not species) assigned with either marginals constant or pool totals only constant	Frequency occurrence matrices	χ^2 test for homogeneity among pools. Values from randomized sets compared to observed.	Capone and Kushlan 1991
Fish in lakes	Marginals kept constant	Frequency of observed and "expected" co-occurrences	Kolmogorov–Smirnov test	Jackson *et al.* 1992
Desert perennial community	Number of occurrences in study patches kept constant	Various indexes	Calculate P from comparison with values from random communities	Silvertown and Wilson 1994
Stylidium (plant) communities	Randomly selected from species/morph pool	Number of overlaps in pollination niche	P from randomization	Armbruster *et al.* 1994

[a]Similarity index=(Observed number of islands shared-expected number of islands shared)/Standard deviation of expected number of islands shared.

Size structure and community organization

The general conceptual model used is that of a linear array of body size or some morphological structure (Figure 5.16). The null model is that the position on the size axis is random. To construct the null model, the general procedure is to (1) construct a pool of candidate species, (2) draw from the pool the requisite number of species, (3) calculate the required statistic, (4) repeat 2 and 3 a large number of times to generate the sampling distribution of the statistic, and (5) compare the observed value of the statistic with the sampling distribution to obtain a probability of observing the event. The actual statistic used to compare observed and predicted values varies widely and there appears to be no consensus on the best method (Table 5.11). The disparity in approaches can be readily appreciated by a brief overview of the methodologies of several studies:

Body size in desert rodents (Bowers and Brown 1982)

The list of candidate species was restricted to those which occurred at a frequency greater than 5% in a sample. Many samples were drawn from a particular desert (three deserts were studied) and the overall frequency of species calculated. The expected frequency of observed pairs of species was then calculated as the product of the observed relative abundances. This differs from other analyses in that species are not being allocated to the sample from a candidate list. The Monte Carlo aspect of the analysis comes in the testing procedure. Species pairs were assigned to one of two categories, >1.5 ratio or <1.5 ratio: thus we have a 2×2 matrix with the columns being body size ratio and the rows the presence or absence of co-occurrence of every possible specie pairs. Testing for heterogeneity in the association was done using χ^2. The problem with this test in this circumstance is that there is a lack of independence because any given species appears multiple times. Bowers and Brown used simulation to determine the distribution of χ^2 values under the null hypothesis of random association (the reader is referred to the article for the specifics of the simulation).

Island birds (Strong 1979)

Synthetic island avifaunas were constructed by randomly selecting species from a candidate list of mainland species, taking no account of relative abundances. Species were ranked and adjacent size ratios calculated. The observed size ratios were compared with the simulated data by three tests, all of which essentially compare the expected numbers above and below the predicted mean (Table 5.11). Hendrickson (1981) repeated the test using the median. Brandl and Topp (1985) used the same approach to analyze size variation in carabid beetles of Central Europe.

Table 5.11 A brief survey of studies employing Monte Carlo techniques to test the hypothesis of character-displacement in communities

Organism	Morphology	Measure	Test	Reference
Bumblebees	Proboscis	Mean pairwise ratio (larger/smaller)	t-test	Ranta 1982a
Bumblebees	Proboscis	Average overlap[a]	P[b], separate tests combined using Fisher's method	Hanski 1982
Beetles (Pterostichus spp.)	Body size	Pairwise ratios: (1) mean, (2) minimum, (3) standard deviation	Proportion of observed above or below median from random community	Brandl and Topp 1985
Water beetles	Body size	Mean size difference	Results are "clear" without test	Ranta 1982b
West Indies birds	Body size	Size differences and ratios: (1) mean, (2) minimum, (3) standard deviation	Proportion of observed above or below median from random community	Case et al. 1983
Desert rodents	Body size	Ratios divided into two groups (>1.5, <1.5)	χ^2 2×2 contingency test	Bowers and Brown 1982
Pond snails	PCs of shell components	2 Euclidean distances (mean, farthest neighbor)	t-tests	Dillon 1981
Desert lizards	Body components	5 Euclidean distances (nearest neighbor, SD of, farthest neighbor, SD of, mean)	No statistical tests	Ricklefs et al. 1981
Island birds	Body measures	Ranked pairwise ratios of single traits	Binomial using mean, Wilcoxon signed-rank tests, χ^2	Strong et al. 1979
Island birds[c]	Body measures	Ratios of single traits	Binomial using median, one-sample Kolmogorov-Smirnov test	Hendrickson 1981
Bird-eating hawks	Body measures	Ratios of pairs, triplets, quartets and quintets	One-sample Kolmogorov-Smirnov test	Schoener 1984

[a]Average overlap calculated as $\sum_{i=1}^{s-1} \sum_{j=i+1}^{s} \max(2 - d_{ij}, 0)/[s(s-1)/2]$, s is the number of species and d_{ij} is the difference in proboscis length between species pair i and j.

[b]Probability estimated by number of times statistic from random community was less than the observed statistic.

[c]Same fauna as analyzed by Strong (1979). See Strong and Simberloff (1981) for a rebuttal.

Island birds (Case *et al.* 1983)

Synthetic avifaunas for the West Indies were formed by randomly assigning the observed number of species to each island from a candidate list that took into account frequency of occurrence in the West Indies. Thus, for example, if a species occurred on *k* islands it would be represented *k* times in the candidate list. The analysis was also done using a list from the mainland. Both size differences and ratios between adjacent pairs were used, with the test being a comparison of the expected and observed numbers above and below the median.

Bird-eating hawks (Schoener 1984)

Avifaunas of species pairs, trios, and quartets were constructed using the list of all bird-eating hawks (47 species). Schoener then computed all possible size ratios: for pairs there is one possible set, whereas for trios there are two (adjacent pairs and pairs with an intervening species) and for quartets there are three (pairs separated by 0, 1, and 2 intervening species). Schoener compared the difference in cumulative frequency between observed and predicted under four different scenarios ("uncorrected," "corrected" for geographic range, "corrected" for species occurrence in several sets, "corrected" for ratios between the same species).

Proboscis length in North European bumblebee communities (Ranta 1982a)

The species pool was made up of all North European bumblebee species and species selected either at random or only from the list of "abundant" species. Proboscis lengths were ranked and adjacent size ratios computed. The observed distribution of ratios was compared to the simulated data using a *t*-test.

Proboscis length of bumblebees in Europe and North America
(Hanski 1982)

Hanski used the same approach as Ranta to construct the random communities but used an index of average overlap (see footnote to Table 5.11). He estimated the probability of obtaining an index as small or smaller than the observed index using the distribution from the simulated data set. He then combined the probabilities from different communities using Fisher's method to obtain an overall probability.

Summary

(1) Randomization, or permutation, is primarily a hypothesis testing method, though it can be used to generate confidence intervals.

(2)   A randomization test consists of first calculating a statistic (or set of statistics), θ_{obs} from the observed set of data. Under the null model of no difference the data are then randomly reallocated to the groups and the statistic(s) recalculated, θ_{rand}. This process is repeated to produce a large number of values of θ_{rand}. The probability of obtaining a value of θ_{obs} as large or larger under the null hypothesis is then estimated as $P=(n+1)/(N+1)$, where n is the number of θ_{rand} greater than or equal to θ_{obs} and N is the number of randomizations.

(3)   The number of randomizations required depends upon how close P is to 0.05. Typically one should commence with 1000 randomizations, increasing this number if the estimated confidence limits about P include 0.05. The required number can be roughly estimated from $N = 4\hat{P}(1 - \hat{P})/(0.05 - \hat{P})^2$.

(4)   The results of randomization tests frequently match those of their parametric equivalent. While this indicates that parametric tests are remarkably robust to a deviation from their underlying assumptions, it should not be taken as indicating that such tests can be substituted for randomization tests. Without theoretical justification or detailed simulation analysis, we cannot know under what circumstances the parametric tests will fail. Therefore, whereas the parametric tests provide an approximate first test, they should be accompanied by the appropriate randomization procedure.

(5)   Monte Carlo methods are similar to randomization and bootstrap methods except that randomized data sets are constructed using a specified null model. Typically these models are designed for a very specific test. The generalized Monte Carlo significance test permits a general approach to Monte Carlo testing and can considerably reduce the number of computations required.

Further reading

Crowley, P.H. (1992). Resampling methods for computation-intensive data analysis in ecology and evolution. *Annual Review of Ecology and Systematics*, **23**, 405–48.

Edgington, E.S. (1987). *Randomization Tests*. New York: Marcel Dekker, Inc.

Manly, B.F.J. (1997). *Randomization, Bootstrap and Monte Carlo Methods in Biology*. New York: Chapman and Hall.

Potvin, C. and Roff, D. (1996). Permutation tests in ecology: A statistical panacea? *Bulletin of the Ecological Society of America*, **77**, 359.

Exercises

(5.1) Using the data shown below do a *t*-test to compare the two means. Compare the results with a randomization test using first *t* as the test statistic and then the mean difference.

x	−0.79	0.79	−0.89	0.11	1.37	1.42	1.17	−0.53	0.92	−0.58
y	−0.88	−0.17	−1.16	−1.23	2.14	0.86	1.36	−1.46	0.74	−2.15

(5.2) Using the data below, test for a linear regression between *y* and *x* using least squares regression and randomization. Test the hypothesis that the slope and intercept equal 0. Test the hypothesis that the slope equals 1. Compare the results to the parametric test.

x 1.63 4.25 3.17 6.46 0.84 0.83 2.03 9.78 4.39 2.72 9.68 7.88 0.21 9.08 9.04 5.59 3.73 7.98 3.85 8.18

y 2.79 3.72 4.09 5.89 0.75 −0.13 1.76 8.44 5.15 2.16 9.88 6.95 0.03 7.50 9.92 5.37 3.79 7.18 3.37 7.81

Note: To use randomization to test the slope against some specified value, say β_0, we observe that under the null hypothesis, $\beta=\beta_0$ and hence $z=y-\beta_0 x$ will be independent of *x*. Therefore, we regress *z* on *x* and test for a significant regression.

(5.3) A common method of testing for a difference between two variances is the variance ratio test. Letting the two estimated variances be s_1^2 and s_2^2, where $s_1^2 > s_2^2$ an *F*-statistic is calculated as $F = s_1^2/s_2^2$. Use this test on the data below and also compare the variances using randomization by (a) using *F* as the test statistic and (b) the difference between the two variances as the test statistic.

x −0.06 −1.51 1.78 0.91 0.05 0.53 0.92 1.75 0.73 0.57 0.17 0.31 0.66 0.01 0.16

y 1.86   0.44 0.59 0.18 −0.59 −1.16 1.01 −1.49 1.62 1.89 0.10 −0.44 −0.06 1.75 1.74

(5.4) The data below show a time series of population counts over a 30 year time period

1, 1, 6, 1, 3, 11, 15, 10, 9, 18, 21, 33, 40, 45, 44, 48, 44, 39,

40, 36, 46, 48, 50, 58, 60, 73, 83, 94, 99, 102

Use the method of Pollard and Lakhani (1987) to test for positive density-dependence.

(5.5) A hypothesis frequently encountered in the literature is that survival decreases with density. In particular this hypothesis has been advanced for the survival of chicks in relation to clutch size. This has led some researchers to plot survival vs. clutch size. However, survival is calculated as number of surviving chicks/clutch size and hence there is a clear lack of independence between the two variables. Use randomization to test the hypothesis that survival decreases with clutch size in the data given below. Compare your results with the parametric analysis.

Clutch size: 1, 1, 2, 2, 2, 3, 4, 4, 4, 5, 5, 6
N surviving: 1, 1, 1, 1, 1, 2, 2, 2, 3, 1, 2, 3

(5.6) Is there evidence of an effect of factor A, B or their interaction on X?

Factor A   1, 1, 1, 1, 1, 2, 2, 2, 2, 2, 2
Factor B   1, 1, 2, 2, 2, 1, 1, 1, 1, 2, 2
X 13, 9, 9, 8, 9, 16, 16, 14, 11, 7, 13

(5.7) Repeat the analysis of the data given in question (5.6) but this time randomize the treatment combinations rather than X. Does it make a difference?

(5.8) The amphipod *Gammarus minutus* occurs in streams both within and outside of caves. These populations are genetically distinct. The data below show measurements of three measures of eye structure, the number of ommatidea, the width and length of the eye. A reasonable hypothesis is that the eyes of individuals from the population within the cave will be different (smaller but here we consider a two-tailed test) from those outside the cave. To test this the researcher uses a multivariate analysis of variance, obtaining the output:

```
***Multivariate Analysis of Variance Model ***
Short Output:
Call:
mANOVA (formula=cbind (OMMATIDI, EYE. L, EYE. W)~ HABITAT,
data=GammarusData, na.action=na.exclude)
Terms:
HABITAT Residuals
Deg. Of Freedom   1     33
1 out of 3 effects not estimable
Estimate effects may be unbalanced
Analysis of Variance Table:
Df Pillai  Trace approx.  F num df   den df  p-value
HABITAT 1  0.85125         59.1333    3  31     0
Residuals 33
```

Because of the message "Estimated effects may be unbalanced", the output is questionable. Therefore, a randomization analysis is necessary to confirm the mANOVA results. Using the approach shown in Appendix C.5.8 conduct the analysis.

OMMATIDI	EYE.W	EYE.L	HABITAT
5	71	48	CAVE
4	70	47	CAVE
4	71	50	CAVE
8	91	62	CAVE
4	62	41	CAVE
3	82	51	CAVE
2	78	40	CAVE
3	74	49	CAVE
6	100	61	CAVE
5	92	60	CAVE
6	82	50	CAVE
4	52	50	CAVE
4	66	43	CAVE
3	82	51	CAVE
6	92	63	CAVE
6	89	51	CAVE
6	72	51	CAVE
5	74	51	CAVE
7	92	56	CAVE
4	78	56	CAVE
16	150	80	RESURGENCE
18	150	80	RESURGENCE
17	141	79	RESURGENCE
12	118	73	RESURGENCE
16	133	78	RESURGENCE
16	145	79	RESURGENCE
26	198	91	RESURGENCE
16	140	80	RESURGENCE
11	121	64	RESURGENCE
21	183	100	RESURGENCE
14	130	75	RESURGENCE
16	138	80	RESURGENCE
18	143	80	RESURGENCE
22	148	80	RESURGENCE
20	151	78	RESURGENCE

List of symbols used in Chapter 5

Symbols may be subscripted

ε	Error term
θ	Parameter
μ	Mean
A	Intercept of major axis regression
B	Slope of major axis regression
C	(1) Amount to be added/subtracted from estimate to obtain upper/lower confidence value. (2) Number of distinct elements in a matrix
E	Expected number of species co-occurrences
GM	Geometric mean
LC	Lower confidence value
M	Median or matrix
MS	Mean square
N	(1) Number of permutations. (2) Population size
O	Observed number of species co-occurrences
P	Probability
R	(1) Total number of species. (2) Probability from randomization test
S	Manly's test statistic for species co-occurrence
SE	Standard error
T	Symbol for T method of matrix comparison
$T1, T2$	Coefficients in ANOVA simulation
UC	Upper confidence value
V	Vector
X, Y	Observation
k	Group designator
n	Number of cases in which statistic from randomized data exceeds observed
r	(1) Residual. (2) Drift parameter in density-dependence analysis
x, y	Observed values

6

Regression methods

Introduction

Regression is probably one of the most powerful tools in the data analysis package of the biologist, particularly when considered within the very broad framework of general linear models. Nevertheless, there are a number of problems with the approach that can be resolved by the use of computer-intensive methods. The most difficult problem, and that which is the focus of the present chapter, is the problem of determining which variables to include in a regression and how to include them. For example, should a predictor variable, X, be entered simply as X or would a better fit be obtained using a polynomial form such as X^2, or even a more general function, which we might not have any a priori reason to formulate? With a single predictor the problem is not very acute, because one can plot the data and visually inspect the pattern of covariation with the response variable, Y. But suppose the pattern is clearly non-linear and none of the usual transformation methods (e.g., log, square-root, arcsine, etc.) linearizes the data: the computer intensive methods outlined in this chapter can be used to both describe the pattern of covariation and to test its fit relative to other models. With multiple predictors the situation can be very problematic if the predictors are complex functions or there are non-linear interactions between predictors.

In this chapter, I shall consider four approaches: first, the use of cross-validation to decide among several competing models when different stepwise methods of including or excluding predictors are used. Secondly, I examine several methods, called local smoothing functions, of describing a response as a function of one or two predictors for data sets in which no functional form can be assigned on biological grounds. Thirdly, I introduce generalized additive models as a solution to dealing with the situation in which the predictor variables may be unknown functions. Finally, I consider the use of tree models to tease apart complex interactions where linear regression methods are unlikely to be appropriate. The range of models considered in this chapter by no means exhausts the computer-intensive approaches to regression problems but they provide an overview of the variety of methods available and should alert the reader to the advantages of taking alternate approaches to regression.

Cross-validation and stepwise regression

The problem of multiple, contending models

Consider the results of the following hypothetical experiment designed to investigate the effects of inbreeding (the breeding between close relatives) and morphology on the reproductive capacity of female sand crickets. The response variable is ovary weight, which is an excellent index of fecundity, and the predictor variables are the inbreeding coefficient (*F*, which varies between 0 and 1), head width (*HW*) and wing morph (long-winged or short-winged). The two morphological traits are included because previous experiments have shown that both of these affect fecundity. Using a multiple regression approach, we can specify the **full** or **saturated model** as

$$
\begin{aligned}
y_i = {}& \theta_0 + \theta_1 F + \theta_2 Morph + \theta_3 HW \\
& + \theta_4 (F)(Morph) + \theta_5(F)(HW) + \theta_6(Morph)(HW) \\
& + \theta_7(F)(Morph)(HW)
\end{aligned} \tag{6.1}
$$

The above model contains seven parameters or variables, rather than the three measured variables (*F*, *Morph*, *HW*). This model is highly significant (Table 6.1) but individual analyses of the terms indicates that not all contribute significantly to a reduction in residual variance.

However, the individual tests do suggest that all three predictor variables are statistically important. On the other hand, all variables are highly significantly correlated, although in no case is the correlation very large (the high level of significance comes from the very large sample size). The question one would like to answer is "Does a model that has fewer parameters account for the relationship between ovary weight and the predictor variables equally as well?" Three general approaches are typically taken to address this question. All three approaches involve the addition and/or elimination of model components and are frequently referred to as a group as **stepwise regression**, although the three types are specifically called **backward deletion, forward selection**, and **stepwise regression**. Backward deletion proceeds by first putting all terms in the model and then sequentially deleting those terms that do not contribute significantly to the regression. Forward selection works in the opposite direction, beginning with

Table 6.1 *Analysis of multiple regression equations with ovary weight as the response variable and predictor variables, F (inbreeding coefficient), Morph (0 or 1), and HW (head width). Hypothetical data for the sand cricket. Left side shows fit for the saturated model. Right side shows results of different stepwise procedures (* = variable retained) using the default parameters in SYSTAT (SY) and S-PLUS (S+)*

Effect	$\hat{\theta}_i$	$|t|$	P	Backwards SY	Backwards S+	Forwards SY	Forwards S+	Both S+
F	−13.292	2.949	0.0033	*	*	*	*	*
Morph	−1.061	2.233	0.0260	*	*	*	*	*
HW	−0.001	0.774	0.4395		*		*	*
(F)(Morph)	13.518	2.994	0.0029	*	*			*
(F)(HW)	0.029	2.958	0.0032	*	*			*
(Morph)(HW)	0.002	2.576	0.0103	*	*	*	*	*
(F)(Morph)(HW)	−0.030	3.038	0.0025	*	*			*
R^2 for each of the stepwise models =				0.41	0.41	0.39	0.39	0.41

Analysis of variance for regression

Source	df	Mean-square	F	P
Regression	7	0.3652	45.51	<0.00001
Residual	463	0.0080		

Pairwise correlations
(correlations below diagonal, P values above)

	Ovary wt	F	Morph	HW
Ovary wt		<0.0001	<0.0001	<0.0001
F	−0.449		<0.0001	<0.0001
Morph	0.216	0.180		0.0014
HW	0.443	−0.420	−0.147	

a model with the single predictor variable that accounts for most of the variance. Addition of components to the model proceeds by the sequential addition of a variable within the remaining set and stops when the addition no longer satisfies an input criterion. In both these methods, a variable that is eliminated from (backwards deletion) or added to (forward selection) the model is not considered in future tests. In stepwise regression, the inclusion or deletion of any of the parameters can occur at each step. The procedure commences with an initial model, such as the saturated model, but then sequentially adds or deletes terms until some criterion is satisfied. The decision to include or exclude a term in any of the three methods is based on some measure of the additional decrease in residual variance accounted for by the relevant variable. For example, SYSTAT uses the probability value, with the default value set at 0.15, whereas S-PLUS uses the change in Akaike's information criterion (AIC),

$$AIC = -2LL_{max} + 2k \qquad (6.2)$$

where LL_{max} is the maximum log-likelihood and k is the number of parameters. For a regression model, the log-likelihood (Chapter 2) is

$$LL_{max} = constant - \frac{N}{2}\ln\sigma^2 - \frac{\sum_{i=1}^{N}(\hat{y}_i - y_i)^2}{2\sigma^2} \qquad (6.3)$$

where σ^2 is the error variance, $\hat{y}_i$ is the predicted value of y_i (sum of squared differences = residual sums of squares) and N is the number of observations. If, as is usually the case, σ^2 is unknown we use its estimate $\hat{\sigma}^2 = \sum_{i=1}^{N}(\hat{y}_i - y_i)^2/N$ and AIC is thus

$$AIC = constant + N\ln\left(\frac{\sum_{i=1}^{N}(\hat{y}_{i,k} - y_i)^2}{N}\right) + 2k \qquad (6.4)$$

where $\hat{y}_{i,k}$ is the predicted value for the model with k parameters. The change in AIC with the addition of a parameter is

$$\Delta AIC = N\ln\left(\frac{\sum_{i=1}^{N}(\hat{y}_{i,k+1} - y_i)^2}{N}\right) - N\ln\left(\frac{\sum_{i=1}^{N}(\hat{y}_{i,k} - y_i)^2}{N}\right) + 2 \qquad (6.5)$$

The change in the value of AIC used to stop the stepping process is arbitrary and, as with the SYSTAT approach, the default value in S-PLUS is liberal, the reasoning being that it is better to include marginal effects rather than omit potentially contributing variables.

The results of applying these three types of stepwise regression to the cricket data is shown in Table 6.2; what is immediately apparent is that there is neither agreement between statistical packages nor among the three stepwise regression methods. Forwards addition produces the simplest model whereas the backwards and stepwise methods include all but one (SYSTAT) or all (S-PLUS) of the parameters. The two "simple" models obtained from forwards addition account for 39% of the variance whereas the other models all account for 41% of the variance. Which model should one choose? In general, what we are interested in is the ability of the model to predict values not used in the construction of the model, otherwise the model has no generality. One solution is to compare the performance of the models using cross-validation.

Cross-validation

In this method, a portion of the data is set aside and the remainder is used to estimate the regression. This regression equation is then used to predict the values of the data points set aside: fit is judged by the correlation between the predicted and observed values in the "new" data set. The data set used to fit the model is known as the **training set** and that used to test it is called the **testing set**. There are three approaches to cross-validation:

(1) **Holdout method**: the data set is split into two parts as described above.

(2) ***K*-fold cross validation**: the data set is split into k subsets of approximately equal size and the cross-validation performed k times using each subset once as the testing set. This procedure can be accomplished as follows, using, as an example, 10-fold cross-validation: first generate a sequence of integers from 1 to 10. Repeat this sequence until all data rows have been accounted for (thus the last row could be indexed by an integer other than 10). If the data were originally sorted in some manner, the sequence of integers should be randomized to ensure that each subset is a random sample from the data. Iterate across the integer values deleting the data with the selected integer index from the data: the deleted data is the testing set. Fit the model to the remaining data (the training set) and use the function so estimated to predict the values for the testing set. Calculate the correlation between predicted and observed data within the testing set.

Instead of dividing the data set into k parts one can also take out a kth portion at random from the data set to use as the testing set, and then repeat this procedure k times, or even more.

(3) **Leave-one-out cross-validation**: this is the extreme case of K-fold cross-validation in which a single observation is omitted and thus the whole process repeated N times (the number of cases in the data set).

There is no real guide as to which method is the best, except that the holdout method is probably the worst because it provides only a single example of cross-validation. K-fold validation using either approach (i.e., division into k parts or a kth part k times) seems most reasonable, with the testing set being 10–20% of the training set. For small sample sizes, the leave-one-out method may be the only practicable solution.

In the present case, I used a randomly selected 10% of the data to be the testing set and repeated this 1000 times. To implement this type of cross-validation we can proceed as follows (coding in Appendix C.6.1):

(1) Randomly select 10% of the data set to use as the test set.

(2) Fit the simplest and most complex models (in other cases different models could be selected) to the remaining 90% of the data.

(3) For each model, calculate the predicted values for the test set and compute the two residual sums of squares $\sum_{i=1}^{N} (\hat{y}_{i,k} - y_i)^2$. Retain these values as pairs.

(4) Repeat the above steps a large number of times (e.g., 1000).

(5) Because, the paired residual sums of squares come from the same training and testing sets, the set of paired residual sums of squares can be compared using a paired t-test.

Application of this procedure to the cricket data shows that the more complex model produces significantly worse predictions than the simpler model (see output in Appendix C.6.1). A plot of the paired residual sums of squares shows that in eight cases there were extreme outliers from the full model (Figure 6.1): deletion of these from the analysis does not change the conclusions (paired t-test, $t = -3.5178$, df $= 991$, $P = 0.0005$). Interestingly, the results for the training set for these eight cases gives no indication of the extreme deviation in prediction (an example is provided in Figure 6.1).

As shown by the above example, cross-validation provides a simple means of distinguishing between competing models in those cases in which step-wise regression results differ depending on the particular protocol adopted. As will be shown elsewhere in this chapter, cross-validation is a general tool for the comparison of different models. It may be also used to assess the fit of a given model, by comparing the R^2 of the fitted model with that for the R^2 between observed and predicted values of the testing set (see locally weighted regression smoothing and Appendix C.6.3).

Local smoothing functions

One predictor variable

The simplest case to consider is that in which there is a single response variable and a single predictor variable. This is the simple linear regression model

$$y_i = \theta_0 + \theta_1 x_i + \varepsilon \tag{6.6}$$

where ε is the error variance, assumed to be normally distributed with a mean of zero and a standard deviation that is constant across the range of measurements (i.e., ε is distributed as $N(0,\sigma)$). We have already discussed the use of

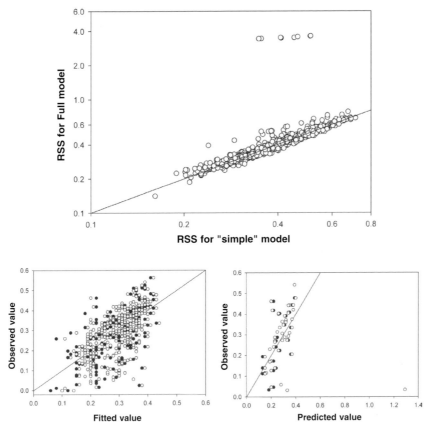

Figure 6.1 Results of cross-validation analysis of cricket data described in text.
Top panel shows the 1000 pairwise comparisons of the residual sums of squares (RSS)
for the two models. The lower plots show the results for a single comparison in which
an extreme outlier occurred (Left plot: results for the training set. Right plot: results
for test set. ● = "simple" model, ○ = full model).

randomization to deal with the problem of non-normal errors. Here, I am
concerned with the second assumption of linear regression, namely that the
equation is actually linear. Suppose that we have evidence that the line is not
linear (a common method to detect non-linearity is to include a quadratic term):
what options are available to describe the relationship? One option that we have
already discussed in all previous chapters is the use of a nonlinear equation, e.g.,
the logistic equation. Another is the use of a polynomial regression. The use of a
nonlinear or polynomial equation is probably the most appropriate route to take
if one has a reason to propose a particular equation or can find an equation that
fits the data. On the other hand, it could arise that there is no a priori model
or no clear empirical nonlinear model that is suitable, a situation that could
arise if the shape of the curve is complex.

Consider, for example the curve shown in Figure 6.2. This curve could represent some diurnal activity pattern, a relationship between some fitness measure and a trait, the time course of development or an evolutionary sequence in species numbers within a taxon. The curve is clearly not linear, is not suitable for a polynomial fit and there is no obvious nonlinear model (the curve is actually formed from two Gaussian functions, as described in the figure caption). To fit a smooth curve through the data, we can adopt one of the number of smoother algorithms, the ones to be discussed here being: (1) locally weighted regression smoothing (loess), (2) locally weighted regression smoothing with cross-validation (super-smoothing), and (3) the cubic spline.

Locally weighted regression smoothing

This is also known as **loess**, a name chosen by analogy with the fine deposit of silt laid down along a river valley, which itself forms a surface. This method, as with all the methods, fits a curve to a local set of points, which I shall refer to as the **window**, and creates a continuous curve by moving the window, point by point, through the entire data range. To describe this method, I shall assume that errors are normally distributed: under this assumption the local curve can be fitted using least squares regression. It is possible to drop this assumption and use a robust fit (the option `family="symmetric"` in S-PLUS) but the general principle is the same as with the least squares fit.

Consider the expanded portion of the curve shown in the lower plot of Figure 6.2 and focus upon the point x_0. We define a **neighborhood** as the set of points falling within a specified distance from x_0, this distance being determined by the user-determined value of **span**. This window could be defined as a particular distance or, as in S-PLUS, as a percentage of the total data set; in the example the number of data points in the window is 10% of the total. Suppose there are $n+1$ points in the neighborhood: these are ranked in ascending order according to their absolute distance from the focal point x_0 (column 2, Table 6.2). Next, we compute the distance between x_0 and each other point as a fraction of the maximum distance (column 3 in Table 6.2) and assign a weight to each point that is a function of this relative distance, the weight decreases as the relative distance increases, with the furthest point (and all points outside the window) receiving a weight of zero. A weighted least squares regression function is then calculated using the estimated weights and data points: this regression can either be linear or quadratic, which allows for local curvature. Having calculated the function, say $f(x)$, the predicted response value at x_0 is calculated as $f(x_0)$. This procedure is applied to all values of x and the response function plotted by connecting the predicted values.

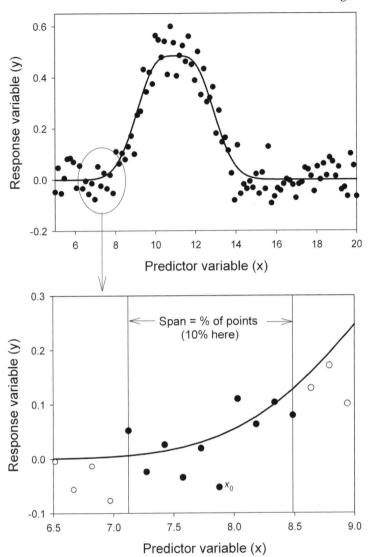

Figure 6.2 Hypothetical function to illustrate the use of smoothing routines when linear and nonlinear regression approaches are not workable. The function was generated using the sum of two Gaussian curves.

```
set.seed(1)                                       # Set random number seed
n           <- 100                                # Sample size
Curves      <- matrix(0,n,5)                      # Matrix for data
x           <- seq(5,20,length=n)                 # values of x
Curves[,1]  <- x                                  # Store x
error       <- rnorm(n,0,0.06)                    # Errors
Curves[,2]  <- dnorm(x, 10,1 ) + dnorm(x,12,1)    # Curve
Curves[,3]  <- dnorm(x, 10,1 ) + dnorm(x,12,1)+error  # Add error to curve
```

Table 6.2 *Illustration of the method of fitting a locally weighted regression function. The values of the predictor variable,* x *are ranked according to the distance from the focal point,* x_0

Predictor variable	Ranked distances, d_i	Relative distance, D_i	Weight, W_i
x_0	$d_0 = \lvert x_0 - x_0 \rvert = 0$	$D_0 = d_0/d_{\max} = 0$	$W_0 = (1 - D_0^3)^3 = 1$
x_1	$d_1 = \lvert x_0 - x_1 \rvert$	$D_1 = d_1/d_{\max}$	$W_1 = (1 - D_1^3)^3$
x_2	$d_2 = \lvert x_0 - x_2 \rvert$	$D_2 = d_2/d_{\max}$	$W_2 = (1 - D_2^3)^3$
.	.	.	.
.	.	.	.
x_i	$d_i = \lvert x_0 - x_i \rvert$	$D_i = d_i/d_{\max}$	$W_i = (1 - D_i^3)^3$
.	.	.	.
.	.	.	.
x_n	$d_n = \lvert x_0 - x_n \rvert = d_{\max}$	$D_{\max} = d_{\max}/d_{\max}$	$W_{\max} = (1 - D_{\max}^3)^3 = 0$

The fit between the observed points and the loess curve depends upon the span and the **degree** of the function (linear or quadratic). Decreasing the span and increasing the degree will get a "better" fit in the sense that the curve will pass through more points but this can lead to gross **overfitting**, whereby the curve is being fitted to the error about the true line. On the other hand, span values that are too large or a degree that is too small (i.e., linear rather than quadratic) can give very poor fits. These points are illustrated in Figure 6.3 in which is plotted three fits. The first fit shows the result of using a span of 1 (i.e., 100% of data points are used at each data point) and a degree of 1 (see Appendix C.6.2 for coding to produce fits). A measure of the amount of smoothing is given by the **equivalent number of parameters** (**ENP**), which equals 2.3 in this fit. The equivalent number of parameters is defined as

$$\text{ENP} = \frac{\sum_{i=1}^{N} \text{Variance}(\hat{y}_i)}{\sigma^2} \tag{6.7}$$

where N is the number of observations, $\hat{y}_i$ is the ith predicted value, and σ^2 is the error variance (it is calculated as the sum of the squared elements of the hat matrix). A measure of fit is the usual R^2, which is here 0.36, though the fit is quite abysmal! The lack of fit is also shown by the plot of residuals on predicted values, which shows a highly non-linear pattern (lower plot in Figure 6.3). Changing from a linear local fit to a quadratic local fit produces a somewhat better fit ($R^2 = 0.58$), but the distribution of residuals with respect to the predicted values is still unacceptable. Keeping the quadratic fit and decreasing the span to 0.3 produces an acceptable fit ($R^2 = 0.92$) and an acceptable distribution of residuals (rightmost plot of Figure 6.3). Note that the smoothing function tends to overfit the points lying beyond $x = 14$.

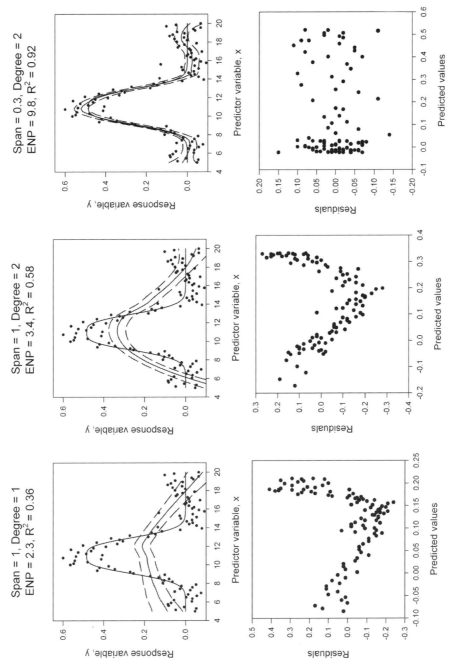

Figure 6.3 Results of fitting a locally weighted regression smoothing function (loess) to the data shown in Figure 6.2. The three fitted curves in the upper plot are the fitted prediction ±1SE. Lower plots show residuals as a function of the predicted values.

167

As with multiple regression, increasing the number of equivalent parameters by decreasing the span will increase the amount of variance accounted for, though the amount may not be more than expected by chance. There is no simple means of deciding when a fit is as good as can be expected but it is at least possible to compare two models using an approximate analysis of variance test. An approximate F-statistic can be constructed from (see Chapter 2 "Comparing models")

$$F = \frac{(RSS^{(n)} - RSS)/(\delta_1^{(n)} - \delta_1)}{RSS/\delta_1} \tag{6.8}$$

where $RSS^{(n)}$ is the residual sums of squares of the null model, RSS is the residual sums of squares of the alternate model and the δs play a similar role to the degrees of freedom. However, the degrees of freedom for the F-statistic are more complex functions of the δs and the reader is referred to Cleveland *et al.* (1992, p. 369) for their definition and a description of their calculation. The test is available in S-PLUS using the routine `anova.loess(model1, model2, test="F")` or simply `anova(model1, model2)`, where `model1` and `model2` are the two fitted models (e.g., `L.smoother1` in Appendix C.6.2). Comparing fits with span = 0.3 and 0.2 gives the output:

```
anova.model
Model 1:
loess(formula = Curves[, 3] ~ Curves[, 1], span = 0.3, degree = 2)
Model 2:
loess(formula = Curves[, 3] ~ Curves[, 1], span = 0.2, degree = 2)
Analysis of Variance Table
      ENP    RSS       Test    F Value    Pr(F)
1     9.8    0.29954   1 vs 2  1.38       0.23467
2    14.6    0.27289
```

There is thus no premium in decreasing the span to 0.2. Increasing the span to 0.4 gives

```
anova.model
Model 1:
loess(formula = Curves[, 3] ~ Curves[, 1], span = 0.3, degree = 2)
Model 2:
loess(formula = Curves[, 3] ~ Curves[, 1], span = 0.4, degree = 2)
Analysis of Variance Table
      ENP    RSS       Test    F Value    Pr(F)
1     9.8    0.29954   1 vs 2  3.23       0.030976
2     7.6    0.32990
```

which suggests that the model with a span of 0.3 is a statistically better fit than that with a span of 0.4.

An alternative method to assess the fit is cross-validation, as described above. Coding for K-fold cross-validation for the loess model is given in Appendix C.6.3. For the loess model selected above as being satisfactory, there is, in general, a good correspondence between the multiple R for the training set and that obtained from the testing set (see output in Appendix C.6.3). In one set, that indexed by 5, the correspondence is poor, with the multiple correlation obtained using the testing set being only 0.185. This illustrates the need to do several cross-validation runs, as a single run could itself be an outlier.

Super-smoothing

In the foregoing method, the span is kept constant, which may not be optimal. For example, in a region of little curvature a large span will produce a good fit. Equally, when the error variance is high a large span is to be preferred as it prevents over-fitting. The super-smoother approach uses local leave-one-out cross-validation to adjust the size of the span as the window is passed over the data range. In S-PLUS, the fit using the super-smoother can be obtained with the routine `supsmu`. Applying the super-smoother to the hypothetical data produces a fit that is virtually indistinguishable from that produced by the loess fit (Figure 6.4) giving us confidence that the loess fit is quite acceptable.

The cubic spline

In the previous two methods, the fitted function was estimated pointwise: a somewhat different approach to fitting a smooth curve through a set of data is to fit a series of polynomials. It has been found that sufficient flexibility in the amount of curvature required can be achieved by using third order polynomials, hence the term "cubic." The word "spline" referred originally to thin flexible rods used by draughtsmen to draw smooth curves and thus the term "cubic spline" is a smooth curve constructed using a "virtual" flexible rod that is mathematically described by a third order polynomial. The curve is constructed of a set of third-order polynomials that change at a series of control points called **knots**. To prevent the curve from having sharp changes at the knots, the polynomials are constrained to have matching first and second order derivatives at the knots. The cubic splines are estimated using the penalized residual sum of squares, *PRSS*,

$$PRSS = \sum_{i=1}^{N} (y_i - f(x_i))^2 + \lambda \int (f''(t))^2 dt \tag{6.9}$$

where N is the number of data points, $f(x_i)$ is a cubic polynomial and λ is a smoothing parameter that corresponds, conceptually though not

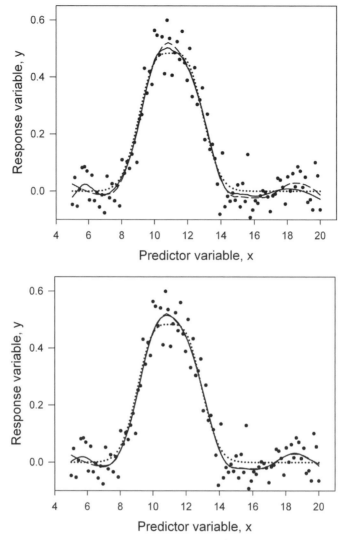

Figure 6.4 A comparison of three smoothing functions. See caption to Figure 6.2 for coding. Dotted line = true function; Dashed line = loess fit, solid line in top panel = super-smooth fit, solid line in bottom panel = cubic spline fit.

mathematically, to span in the loess function. The value of λ can be found by trial and error or by cross-validation. The cubic spline is available in S-PLUS as the routine `smooth.spline`. Using the default cross-validation approach to the estimation of λ gave the fit to the hypothetical data shown in Figure 6.4: as with the super-smoother, the cubic spline fit is indistinguishable from the loess fit.

While there is little to choose in terms of fit among the three methods (and also several others not discussed here), I recommend the loess method, because it permits an easy comparison among models and has a method for calculating confidence regions about the curves. Schluter (1988) suggested the use of the bootstrap method for the calculation of such regions but the validity of this approach has yet to be verified by simulation. Given the ease with which the three models can be implemented, it is preferable to try all three to see if the fitted curve is consistent across methods.

Two predictor variables

The easiest approach to deal with one response variable and two or more predictor variables is multiple regression. Provided that the terms are indeed additive, this is an appropriate strategy. However, there are a large number of possible functional relationships with two predictor variables: for example, $y = \theta_0 + \theta_1 x_1 + \theta_2 x_2$, $y = \theta_0 + \theta_1 x_1 + \theta_2 x_2 + \theta_3 x_1 x_2$, $y = \theta_0 + \theta_1 x_1 + \theta_2 x_2^2$, $y = \theta_0 + \theta_1 x_1^2 + \theta_2 x_2 + \theta_3 x_1 x_2$, etc. With a single predictor, the general shape of the relationship can be discerned by plotting the response variable on the predictor variable and using one of the local regression models described in the previous section. Such functions are available for three-dimensional (3D) plots. Here, I shall discuss the multivariate extension of the loess method described above. To illustrate the approach, I shall use an example later used in the description of regression tree analysis: there are two predictor variables, X1 (e.g., temperature) and X2 (e.g., habitat structure), and a response variable Y (e.g., density) that is determined according to the following rules:

if $X1 < 17$ then $Y = 5 + \varepsilon_1$
if $X1 > 17$ and $X2 < 10$ then $Y = 10 + \varepsilon_2$
if $X1 > 17$ and $X2 > 10$ then $Y = 20 + \varepsilon_3$

where ε_1 is $N(0,2)$, ε_2 is $N(0,4)$, ε_1 is $N(0,8)$. In the absence of the error terms (ε_1, ε_2, ε_3), the response variable, Y, is a step function (Figure 6.5): the error terms tend to smooth out the relationship. One hundred data points were generated using the above rules (see Figure 6.5 for the coding) and examined by stepwise regression with the most complex (full or saturated) model being $Y = \theta_0 + \theta_1 X1 + \theta_2 X2 + \theta_3(X1)(X2)$. Regardless of the stepwise procedure adopted, the best fitting model was the full model $Y = -7.536 + 0.786X1 - 4.311X2 + 0.275(X1)(X2)$. The predicted surface of this model captures the general shape of the true function but necessarily omits the steeply changing gradients (Figure 6.5).

Before subjecting the data to a local smoothing routine, it is useful to plot the data as a 3D surface "as is": because most 3D routines require an equally spaced grid of points on the x,y surface (X1,X2 in this case), this will generally require using some sort of interpolation or extrapolation. The surface so generated for the density data is quite rugged but does give some hints of curvature (Figure 6.6). A loess generated surface using a degree of one (see Appendix C.6.4 for the coding)

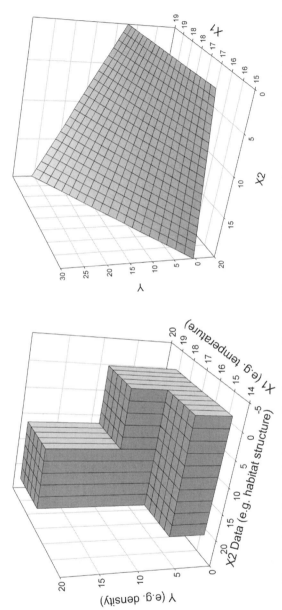

Figure 6.5 Hypothetical relationship between Y and two predictor variables, $X1$ and $X2$. Plot on left shows the deterministic relationship. The plot on the right shows the surface for the best fitting linear model, $Y = -7.536 + 0.786(X1) - 4.311(X2) + 0.275(X1)(X2)$. S-PLUS coding to generate the data is shown below.

```
set.seed(1)                          # Initialize random number

N <- 100                             # Number of data points

X1 <-runif(N,15,19)                  # Generate values of X1

X2 <-runif(N,0,20)                   # Generate values of X2

# Nest density at these sites

Y <- matrix(0,N)                     # Set up matrix for Y values

for (i in 1:N)

{

if (X1[i]<17)                   Y[i] <- 5 + rnorm(1,0,2)    # error N(0, 2)

if (X1[i]>=17 & X2[i] < 10)     Y[i] <- 10 + rnorm(1,0,4)   # error N(0, 4)

if (X1[i] >= 17 & X2[i] >= 10)  Y[i] <- 20 + rnorm(1,0,8)   # error N(0, 8)

}
```

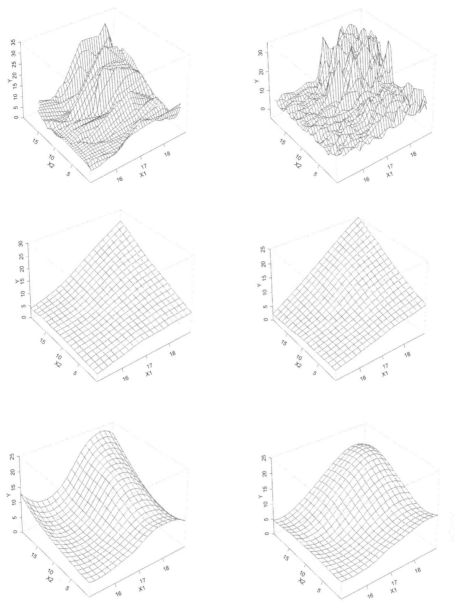

Figure 6.6   The three dimensional plots of density data. Top plots show plot produced from an equally spaced grid of X1, X2 using interpolation of the original data. Lower plots show surfaces generated using loess with either degree=1 (middle) or degree = 2 (bottom). Left-hand plots use 100 data points and right-hand plots use 1000 data points.

shows the same pattern as found with the multiple linear regression model, though it appears to be slightly superior in not showing a rise towards the minimum values of X1, X2. Using a degree of two (quadratic surface) gives a visually better fit though in this case there appears to be some over-fitting at the highest values of X1 and X2. The two fits can be compared using the approximate F-statistic described previously (Appendix C.6.5). The quadratic fit produces a significantly better fit than the linear model (Appendix C.6.5). This is shown very clearly in the rightmost plots in Figure 6.6 which were produced using 1000 generated data points.

Generalized additive models

One predictor variable

An investigation into possible non-linearities in a relationship can be time consuming if one has to examine a range of transformations or polynomials of the predictor variables. An alternate approach is presented by generalized additive models. The generalized additive model for a single predictor variable is given by

$$y = \theta_0 + s(x) \tag{6.10}$$

where the function $s(x)$ is a smoother such as a loess function. Generalized additive models extend the standard linear regression equation by replacing the coefficient, θ_1 (Eq. (6.6)) by a smooth function. As with the local smoothing methods, the prime function of generalized additive models is to explore the nature of relationships. Consider the set of data points plotted in Figure 6.7. Although there is a clear linear relationship between the two variables there is also a visually discernible nonlinear component. The equation actually used to generate the data was $y = 10 + \sqrt{x^{-1}} - x + e^{0.2x} + \varphi(x) + \varepsilon$, where ε is the error term distributed as $N(0,1)$ and $\varphi(x) = 0.4e^{-\frac{1}{2}(0.995x)^2}$. A generalized additive model can be fit using any of the smoothing routines discussed previously. Here, I use the loess method, invoked in S-PLUS by `gam(y~x, data=Data)`, where "Data" is the data frame containing the data. The residual sums of squares is 214.11 with 4.22 degrees of freedom used in the fit. In comparison, a linear fit to the data gives a residual sums of squares of 293.85 with 2 degrees of freedom used in the fit. We can construct an approximate F-statistic in the usual manner as

$$F_{4.21-2,\,200-2.21} = \frac{(293.85 - 214.11)/(4.21 - 2)}{214.11/(200 - 4.21)} = 32.99 \tag{6.11}$$

which is highly significant ($P = 5.17 \times 10^{-14}$). This test can be automatically carried out in S-PLUS by invoking the `anova` routine (e.g., `Fit <- gam(y~lo(x),`

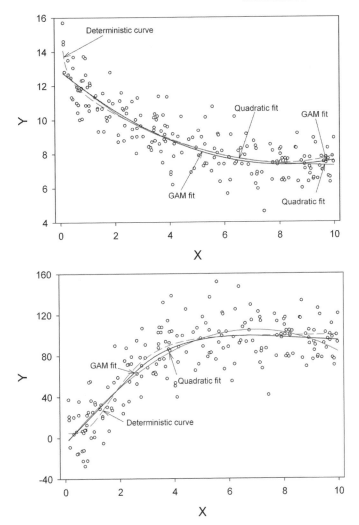

Figure 6.7 A comparison of a generalized additive and a quadratic regression model fitted to two hypothetical data sets. Data generated using coding shown below, with full analysis and output in Appendix C.6.6.

```
set.seed(1)                              # initiate random number generator

n        <- 200                          # Sample size

X        <- runif(n, 0.1,10)             # values of X

# Y values for first example

error    <- rnorm(n,0,1)                 # error terms

Y        <- 10+(1/sqrt(X))+exp(0.2*X)-X+pnorm(0.005*X,1)+error

# Y values for second example

error    <- rnorm(n,0,20)                # error terms

Y        <- 5+95*(1-exp(-1*X))^5 + error
```

`data=Data); anova(Fit))`. Also shown in Figure 6.7 is the fit achieved using a quadratic polynomial; there is very little to distinguish the two curves and in this instance one would certainly prefer the quadratic equation, because of the more rigorous statistical basis. Such is not the case with the second example, created from a four parameter Chapman equation $y = 5 + 95(1-e^{-x})^5 + \varepsilon$, where ε is distributed as $N(0,20)$. The quadratic fit indicates a declining response at high values of x (Figure 6.7); this is not visually justified from the distribution of points, and one would be likely to seek an asymptotic function, which could be tedious. The generalized additive model produces an excellent fit to the data, clearly indicates a non-linear component, and is significantly better than the quadratic fit (coding and output in Appendix C.6.6).

With a single predictor variable, the generalized additive model offers relatively little advantage over more "traditional" methods (transformations or polynomials) but when there are a number of predictor variables the number of possible alternative models greatly increases and generalized additive models can be very useful.

Several predictor variables

The extension to more than a single predictor variable is relatively straightforward, the general equation for k predictor variables being

$$y = \theta_0 + s(x_1) + s(x_2) + \ldots + s(x_k) \tag{6.12}$$

To illustrate the analysis, I shall use the two models considered in the previous section, $y = 15 + \sqrt{x_1^{-1}} - x_1 + e^{0.2x_1} + \varphi(x_1) + 95(1 - e^{-x_2})^5 + \varepsilon$, where ε is distributed as $N(0,5)$. The 3D pattern so created could represent population density as a function of two environmental or geographic variables (Figure 6.8). The interpolated perspective plot of the 400 data points generated (20 by 20 evenly spaced grid, assuming the highly unusual case of a perfectly executed experimental design) is very rugged and a clear pattern is not readily discernible, but a plot of the response variable on each predictor variable suggests a negative relationship with $X1$ and an asymptotic relationship with $X2$ (middle row, Figure 6.8). These data suggest three possible models: (1) $X1$ constant, $X2$ nonlinear, (2) $X1$ linear, $X2$ nonlinear, (3) $X1$ and $X2$ nonlinear. We commence by fitting the simplest model using the loess fitting routine (the data are in a file called "Curves")

```
Model.1 <- gam(Y~lo(X2), data=Curves)
anova(Model.1)
```

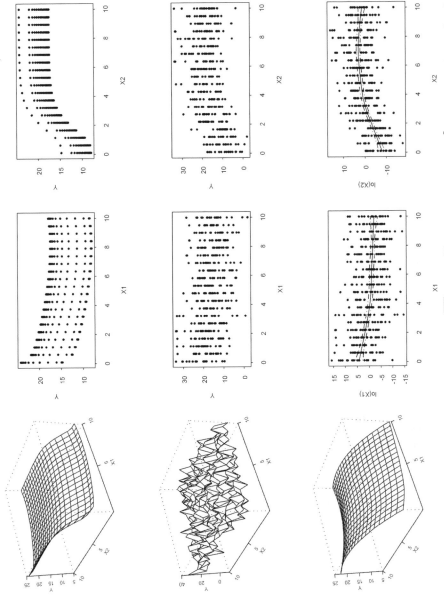

Figure 6.8 Top row: Deterministic plot of equation $y = 15 + \sqrt{x_1^{-1}} - x_1 + e^{0.2x_1} + \phi(x_1) + 95(1 - e^{-x_2})^5$; middle row: interpolated plot of data generated using above equation with an error term $N(0,5)$; bottom row: multivariate loess plot of the generated data. Side plots show partial fits (±SE) for the generalized additive model. The fitted values are scaled to a mean of zero and partial residuals shown.

The relevant output is

	Df	Npar Df	Npar F	Pr(F)
(Intercept)	1			
lo(X2)	1	2.4	20.40267	2.095264e-010

There is a highly significant effect attributable to the nonlinear component. We next fit the second model, entering X1 as a linear predictor and compare the two models

```
Model.2 <- gam(Y~X1+lo(X2), data=Curves)
anova(Model.1, Model.2, test="F")
```

which produces

```
Analysis of Deviance Table
Response: Y
```

	Terms	Resid.	Df Resid.	Dev Test	Df	Deviance	F Value	Pr(F)
1	lo(X2)	395.6118	12363.14					
2	X1 + lo(X2)	394.6118	11056.36	+X1	1	1306.783	46.64028	3.236633e-011

The addition of X1 produces a highly significant reduction in the residual sums of squares. Finally we compare models 3 and 2

```
Model.3 <- gam(Y~lo(X1)+lo(X2), data=Curves)
anova(Model.2, Model.3, test="F")
```

giving

```
Analysis of Deviance Table
Response: Y
```

	Terms	Resid.	Df Resid.	Dev Test	Df	Deviance	F Value	Pr(F)
1	X1 + lo(X2)	394.61	11056.36					
2	lo(X1) + lo(X2)	392.22	10859.78	1 vs. 2	2.39	196.58	2.97	0.04299

There is a marginally significant nonlinear effect attributable to X1. Because these tests are approximate, this result should be viewed cautiously. A plot of the partial fits for the generalized additive model for the two predictors suggests that the fits are reasonable and that the two effects are relatively independent (which they are). A multivariate loess plot broadly matches the deterministic surface (cf. top and bottom 3D plots in Figure 6.8).

Tree models

Multiple regression is limited by the assumption of additivity in the model (which includes polynomial and interaction terms). Provided that this assumption is upheld, multiple regression is a very useful statistical tool. However, when there are a large number of variables, or the likelihood that variables interact in complex ways, multiple regression may either not be able to discern the patterns in the data or may produce misleading answers. An alternative approach that is not as restrictive as multiple regression is that of classification and regression tree analysis. These two methods differ only in that in classification trees the endpoint is a category, such as a species, whereas in regression trees the endpoint is a predicted value such as nest density, body size, etc. The general aim of these methods is to produce a binary tree in which each **node** of the tree represents a binary division of the data present at that node, determined by some statistical criterion such as least squares. The **terminal nodes** are called **leaves**, the initial split is called the **root node** and the number of leaves is called the **size** of the tree. Each node is considered separately and all the available predictor variables are analyzed; thus, for example, at the first split the data may be best divided according to some predictor variable $X1$, while at a subsequent node the best split of the data passing through that node may be accomplished using some other predictor variable, say $X2$. Regression trees were first developed in the context of hypothesis generating routines rather than hypothesis testing routines. However, as will be shown, the use of randomization can turn them into hypothesis testing machines in the same manner as stepwise regression. Tree based models have a number of important attributes: they are easy to interpret when the predictors consist of both categorical and continuous variables, they are invariant to monotone transformations of the predictor variables, they can capture non-additive behavior, and they allow very general interactions between predictor variables. As illustrated in Table 6.3, tree models can be applied to a wide variety of biological questions.

How to split

At each node, tree models seek to maximize the probability of correctly assigning the response variable into two divisions. The majority of methods use a **one-step look-ahead**, which means that the statistic (such as least squares) is minimized (or maximized) only with respect to the node under scrutiny. The basic perspective of the classification tree is the multinomial distribution whereas that of regression trees is some continuous distribution such as the normal. Both of these models can be illustrated by a single example, the extinction probability of New Zealand avifauna as a function of body mass.

Table 6.3 *Examples illustrating the use of regression tree analysis to address biological phenomena in which there may be a large number of interacting factors*

Title	Reference
Ixodes ricinus (Acari: Ixodidae) infestation on roe deer (*Capreolus capreolus*) in Trentino, Italian Alps	Chemini *et al.* (1997)
Modelling the effects of environmental conditions on apparent photosynthesis of *Stipa bromoides* by machine learning tools	Dalaka *et al.* (2000)
Using regression trees to identify the habitat preference of the sea cucumber (*Holothuria leucospilota*) on Rarontonga, Cook Islands.	Dzeroski and Drumm (2003)
Genetic markers applied in regression tree prediction models.	Hizer *et al.* (2004)
Isolation vs. extinction in the assembly of fishes in small northern lakes	Magnuson *et al.* (1998)
Probability of infestation and extent of mortality associated with the Douglas-fir beetle in the Colorado Front Range	Negron (1998)
Tree regression analysis on the nesting habitat of smallmouth bass	Rejwan *et al.* (1999)
Predicting invasions of woody plants introduced into North America.	Reichard and Hamilton (1997)
A comparison of estimated proportional hazards models and regression trees.	Segal and Bloch (1989)
Stand and neighbourhood parameters as determinants of plant species richness in a managed forest.	Skov (1997)
Spatial distribution of developmental egg ages within a herring *Clupea harengus* spawning ground	Stratoudakis *et al.* (1998)
Catches per unit of effort of bigeye tuna: A new analysis with regression trees and simulated annealing.	Watters and Deriso (2000)

This model is discussed in detail later in this chapter; here, I present a simplified extract consisting of a single split (and reduced data set). In Table 6.4, 16 species of New Zealand birds are listed, some of which became extinct prior to the colonization of New Zealand by Europeans. Alongside each species is given the status as a categorical variable ("Extinct" or "Extant"), its status as a numerical variable (probability of extinction = 1 or 0), and its adult body mass. The birds have been ranked in the order of body size, beginning with the smallest. The object is to split the birds into two groups, based on their body mass, that maximizes the likelihood of assigning an individual to the correct group. Under the classification model framework, the likelihood of observing the two

Table 6.4 *A hypothetical example demonstrating how the best split at a node in regression tree analysis is calculated. The data consist of 16 species of New Zealand birds ranked according to body mass. "Observed extinction" is a categorical or binary numerical variable giving the probability of being extant (0 or 1). In the table, the species are divided into two groups at a body mass of 975 g, with species below this threshold given a predicted probability of being extinct of 1 and species above the threshold a value of 0. The probabilities and sample sizes to compute* $|D_L + D_R|$ *and* $|G_L + G_R|$ *for the classification model are shown in the lower table. The far right column shows the data necessary to calculate the deviances for the regression tree model*

Species	Observed extinction		Body mass (g)	$(y_j - \hat{\mu}_j)^2$
	Categorical	Numerical (y_j)		
Malacorhynchus scarletti	Extant	0	800	0.04[a]
Larus dominicanus	Extinct	1	850	0.64[a]
Mergus australis	Extant	0	900	0.04
Egretta alba	Extant	0	900	0.04
Corvus moriorum	Extant	0	950	0.04
Best split is here				
Botaurus poiciloptilus	Extinct	1	1000	0.033[b]
Anas superciliosa	Extinct	1	1000	0.033
Podiceps cristatus	Extinct	1	1100	0.033
Stictocarbo punctatus	Extinct	1	1200	0.033
Catharacta skua	Extinct	1	1950	0.033
Biziura delautouri	Extant	0	2000	0.669[b]
Phalacrocorax varius	Extinct	1	2000	0.033
Phalacrocorax carbo	Extinct	1	2200	0.033
Morus serrator	Extinct	1	2300	0.033
Leucocarbo carunculatus	Extant	0	2500	0.669
Leucocarbo chalconotus	Extinct	1	2500	0.033

	Predicted	
Observed	Extinct	Extant
Extinct	9/11=0.82	1/5=0.20
Extant	2/11=0.18	4/5=0.30

[a] $\hat{\mu}_1$ for this split=1/5=0.2. The two deviances are thus $(0-0.2)^2=0.04$ and $(1-0.2)^2=0.64$
[b] $\hat{\mu}_2$ for this split is 9/11. The two deviances are thus $(1-9/11)^2=0.033$ and $(0-9/11)^2=0.669$.

groups at some given split, L, is proportional to (see "Leaving normality" in Chapter 2)

$$L \propto p^r (1-p)^{n-r} \tag{6.13}$$

where r and $n-r$ are the numbers in the two groups and p is the probability that the response variable falls into this category. There will usually be more than a single split and hence more than two leaves. In this case, the above equation can be expanded to (see "From binomial to multinomial" in Chapter 2)

$$L \propto \prod_{i=1}^{N_{\text{Leaves}}} \prod_{j=1}^{N_{\text{Classes}}} p_{ij}^{n_{ij}} \tag{6.14}$$

where N_{Leaves} is the number of leaves, N_{Classes} is the number of classes (this will depend upon the number of categories into which the response variable is divided. For example, if there were three categories, say species A, species B, species C, then $N_{\text{Classes}}=3$. In the present example, $N_{\text{Classes}}=2$), n_{ij} is the observed number in class j within leaf i and p_{ij} is the probability. Recall that a comparison of models under the maximum likelihood framework can be accomplished by using the log-likelihood ratio of the likelihoods of the two models (see "Method 2: the log-likelihood ratio approach" in Chapter 2). This suggests the following approach to decide where to split. The **deviance for a tree**, D, is defined as

$$D = -2 \sum_{i=1}^{N_{\text{Leaves}}} \sum_{j=1}^{N_{\text{Classes}}} n_{ij} \ln(p_{ij}) \tag{6.15}$$

Now consider some node, which we label k. Prior to any division at this node, the deviance at this node, D_k is

$$D_k = -2 \sum_{j=1}^{N_{\text{Classes}}} n_{kj} \ln(p_{kj}) \tag{6.16}$$

The deviance of the two groups after the kth node is split is the sum of the deviances within each division

$$D_L + D_R = -2 \left(\sum_{j=1}^{N_{\text{Classes}}} n_{Lj} \ln(p_{Lj}) + \sum_{j=1}^{N_{\text{Classes}}} n_{Rj} \ln(p_{Rj}) \right) \tag{6.17}$$

where L and R refer to the split to the left and right, respectively. The reduction in deviance is thus

$$D_k - (D_L + D_R) = -2 \sum_{j=1}^{N_{\text{Classes}}} n_{kj} \ln(p_{kj}) + 2 \left(\sum_{j=1}^{N_{\text{Classes}}} n_{Lj} \ln(p_{Lj}) + \sum_{j=1}^{N_{\text{Classes}}} n_{Rj} \ln(p_{Rj}) \right)$$

$$= -2 \left(\sum_{j=1}^{N_{\text{Classes}}} n_{Lj} \ln(p_{kj}) + \sum_{j=1}^{N_{\text{Classes}}} n_{Rj} \ln(p_{kj}) \right)$$

$$+ 2 \left(\sum_{j=1}^{N_{\text{Classes}}} n_{Lj} \ln(p_{Lj}) + \sum_{j=1}^{N_{\text{Classes}}} n_{Rj} \ln(p_{Rj}) \right)$$

$$= 2 \sum_{j=1}^{N_{\text{Classes}}} \left(n_{Lj} \ln \left(\frac{p_{Lj}}{p_{kj}} \right) + n_{Rj} \ln \left(\frac{p_{Rj}}{p_{kj}} \right) \right)$$

$$(6.18)$$

The optimum split is where the reduction in deviance is the greatest (this is consistent with the principal of maximum likelihood). As we do not know the true value of the probabilities, we estimate these using the observed proportions

$$\hat{p}_{kj} = \frac{n_{kj}}{N}, \quad \hat{p}_{Lj} = \frac{n_{Lj}}{N_L}, \quad \hat{p}_{Rj} = \frac{n_{Rj}}{N_R} \qquad (6.19)$$

where N is the total number of observations (assuming no missing values) and $N = N_L + N_R$. Substituting in Eq. (6.18) gives

$$D_k - (D_L + D_R) = 2 \left[\sum_{j=1}^{N_{\text{Classes}}} n_{Lj} \ln(n_{Lj}) + n_{Rj} \ln(n_{Rj}) - n_{kj} \ln(n_{kj}) \right.$$

$$\left. + N \ln(N) - N_L \ln(N_L) - N_R \ln(N_R) \right] \qquad (6.20)$$

At any node, the value of D_k is constant and hence the split point can also be found by minimizing $|D_R + D_L|$.

Two alternate rules for deciding where to split a node are based on the idea of minimizing the average impurity. The first of these is the **entropy or information index**

$$E = \sum_{i=1}^{N_{\text{Leaves}}} \sum_{j=1}^{N_{\text{Classes}}} p_{ij} \ln(p_{ij}) \qquad (6.21)$$

This differs from the deviance by only a constant and hence will produce the same split.

The second is the **Gini index**, which before the kth node is split is

$$G_k = 1 - \left(\sum_{j=1}^{N_{\text{Classes}}} p_j^2 \right) \qquad (6.22)$$

The reduction in average impurity is then

$$G_k - (G_L + G_R) = 1 - \sum_{j=1}^{N_{\text{Classes}}} p_{kj}^2 - \left(1 - \sum_{j=1}^{N_{\text{Classes}}} p_{Lj}^2 + 1 - \sum_{j=1}^{N_{\text{Classes}}} p_{Rj}^2\right)$$

$$\sum_{j=1}^{N_{\text{Classes}}} p_{kj}^2 - \left(1 - \sum_{j=1}^{N_{\text{Classes}}} p_{Lj}^2 - \sum_{j=1}^{N_{\text{Classes}}} p_{Rj}^2\right)$$

(6.23)

As before, the split point can be found by minimizing $|G_L + G_R|$.

To find the optimal split, the data are ranked in ascending (or descending) order as shown in Table 6.4. We move down the rows and calculate the required statistic at each division; the table shows the division at a body mass intermediate between rows 5 and 6. The two relevant statistics are estimated as

$$|D_L + D_R| = -2\{(9)\ln(0.82) + (2)\ln(0.18) + (1)\ln(0.2) + (4)\ln(0.80)\} = 15.43$$

$$|G_L + G_R| = -(2 - 0.82^2 - 0.18^2 - 0.20^2 - 0.80^2) = 0.62$$

(6.24)

Plotting these statistics vs. the split point shows that this particular split gives the lowest value in both cases (Figure 6.9) and thus, for the classification model this split is optimal.

An alternate way of viewing these data is as a regression tree model in which species are given probabilities of being extinct (1=extinct and 0=extant).

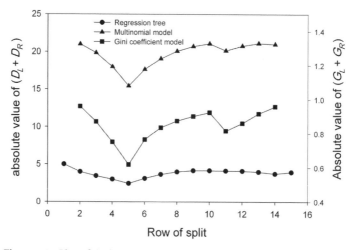

Figure 6.9 Plot of deviances vs. split point for the hypothetical example using the extinction data shown in Table 6.4.

As a species is either extinct or extant it can only be assigned a numerical value of 0 or 1. However, at any leaf, the mean value represents the probability of being extinct. In this case, we can view this model as a regression problem with binomial errors, though in the more general case we would assume a normally distributed error term. The obvious candidate to calculate the splits from the regression perspective is the sums of squares. We define the deviance at a node as

$$D_k = \sum_{j=1}^{n_k} (y_j - \mu_j)^2 \tag{6.25}$$

where n_k is the number of observations at the kth node, y_j is the jth observation of the response variable within this group and μ_j is the mean. As before, we can form a binary split at this node and define two deviances

$$D_L + D_R = \sum_{j=1}^{n_L} (y_{Lj} - \mu_L)^2 + \sum_{j=1}^{n_R} (y_{Rj} - \mu_R)^2 \tag{6.26}$$

We select the split such that $D_k - (D_L + D_R)$ is maximized, which is to say that we minimize $|D_L + D_R|$. As in least square estimation, we replace the unknown mean with its estimate (designated as $\hat{\mu}$ or $\bar{y}$) and hence minimize

$$D_L + D_R = \sum_{j=1}^{n_L} (y_{Lj} - \hat{\mu}_L)^2 + \sum_{j=1}^{n_R} (y_{Rj} - \hat{\mu}_R)^2 \tag{6.27}$$

For the extinction example, the sums of squares is minimized between rows 5 and 6 (Figure 6.9), as found using the classification approach.

To illustrate the regression tree model when the data are continuous and there is more than a single predictor variable, I shall use the example previously given in the discussion of the multivariate loess fit. Consider an organism whose nesting requirements are determined by two components of its environment, say X1 and X2. The interaction of these components is illustrated by the binary tree shown in Figure 6.10 (this example is inspired by an analysis of the nesting habitat of smallmouth bass given by Rejwan et al. 1999). If the value of X1 is less than 17 then the response variable, Y, has the value of 5, whereas if X1 is greater than 17 then the response variable depends upon the value of the second variable, X2. If X2 is less than 10 then Y takes the value 10, otherwise Y equals 20. The variable X1 could be temperature and X2 could be habitat structure. The interaction of these two habitat components produces the stepped 3D surface shown in the lower left of Figure 6.10. To make the situation more realistic, I generated values of Y based on the binary tree and added random

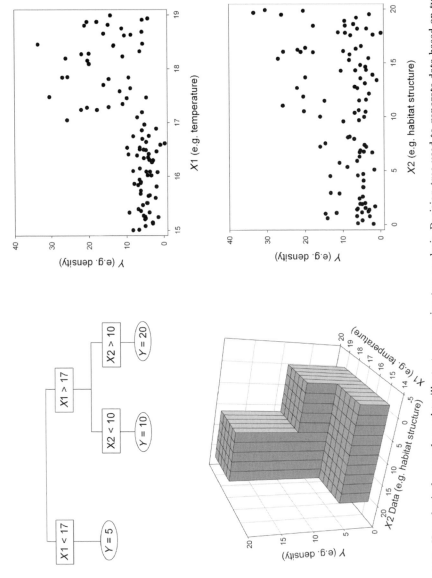

Figure 6.10 Hypothetical example used to illustrate regression tree analysis. Decision tree used to generate data based on two variables X1 and X2. The three dimensional plot shows a 3D plot of the response variable (Y) against the two predictor variables.

186

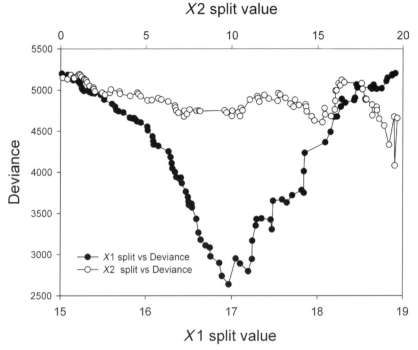

Figure 6.11 Plot of deviance vs. split point for the first node using the data illustrated in Figure 6.10.

normal variables with means of zero and standard deviations that increased with nest density (i.e., 2, 4, 8 for nest densities of 5, 10, 20, respectively; coding given in Figure 6.5). The rationale for using the sums of squares is that the variance at each leaf is the same (i.e., within each leaf the data are distributed as $N(\mu_i, \sigma)$): the foregoing model violates this assumption but appears a more reasonable assumption for real data. Plotting the simulated points on the two predictor variables produces patterns that are not readily discernible, though the plot of Y on X1 does suggest an abrupt change at about $X1=17$.

To find the appropriate split at the root node, we proceed as follows (Figure 6.11):

(1)   Rank all the values of X1.

(2)   Divide the data set into two parts, beginning at the first value of X1, and calculate the deviance, D (sums of squares) for the two parts

$$D = \sum_{i=1}^{n} (y_i - \bar{y}_{1,n})^2 + \sum_{i=n+1}^{N} (y_i - \bar{y}_{n+1,N})^2 \tag{6.28}$$

(3)   where n is the last row in the first group, y_i is the response variable in the ith row, $\bar{y}_{1,n}$ is the mean value of y in the first groups ($\hat{\mu}_j$ in

the extinction example), N is the total number of observations and $\bar{y}_{n+1,N}$ is the mean value in the second group. The optimal split point for X1 is the value of X1 that minimizes the deviance (i.e., least squares).

(4)   The process is repeated for X2.

(5)   The predictor variable chosen is the one that gives the smallest deviance. In the present example this is X1.

The process of splitting can continue until there is only one observation per node or the data are entirely homogeneous at a node. By default, in S-PLUS the splitting continues until the node is homogeneous or there are less than 5 observations to be split. A regression tree can be constructed using a dialogue box in S-PLUS or the coding shown in the caption to Figure 6.12. The lengths of the branches correspond to the reduction in deviance, it can be seen that although there are 13 terminal nodes (leaves) most of the reduction in deviance occurs in the first three branches (some branching is so close that the text merges together and cannot be separated close to the split points), which corresponds to the method by which the data were constructed (Figure 6.12).

Tree pruning: the cost-complexity measure

As the number of splits is increased, the deviance declines but the rate of decline becomes increasingly smaller (Figure 6.13). In the present example, there is very little decrease after three terminal nodes. It is desirable to have a more objective means of determining when to stop adding nodes or when to stop pruning. Breiman *et al.* (1984) developed a method for pruning trees to a given size that gives the smallest deviance of all possible pruned trees (to anticipate: it is possible that an optimal tree for a given size is not possible as can be seen by the lack of size 11 in Figure 6.12, in which case S-PLUS selects the next larger possible tree). Pruning is based on minimizing the **cost-complexity measure**,

$$D_k(T') = D(T') + k\text{Size}(T') \tag{6.29}$$

where $D(T')$ is the deviance for the subtree T', k is the cost-complexity parameter, and $\text{Size}(T')$ is the number of terminal nodes on the given subtree. A plausible measure for k is an approximation to Akaike's information criterion, $k = 2\hat{\sigma}^2$, where $\hat{\sigma}^2$ is the estimated variance within the leaves, which can be equated to the residual mean deviance in the full model. In our example, $\hat{\sigma}^2 = 10.71$ (see caption to Figure 6.12), giving $k = 21.42$. In S-PLUS, the tree that satisfies the cost-complexity measure can be found as shown in Figure 6.14. Although the number

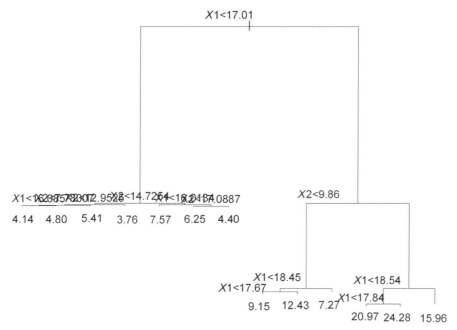

Figure 6.12 Regression trees fitted to hypothetical data from Figure 6.10. Top plot shows the fully fitted tree (nodes too crowded to display text properly). Coding in S-PLUS to generate tree and text (text output not shown).

```
Tree <- tree(Y~X1+X2, data=Data.df)    # Data in file called Data.df

Tree                                   # Print out results

summary (Tree)                         # Output summary of tree

plot(Tree); text(Tree)                 # Plot tree with value at splits
```

Summary output

```
Regression#tree:

tree(formula = Y~X1 + X2, data = Data.df)

Number of terminal nodes: 13

Residual mean deviance: 10.71 = 931.5/87

Distribution of residuals:

  Min.     1st Qu.   Median  Mean   3rd Qu.  Max.

 −11.6500  −1.7130  0.1132  0.0000  1.5780  9.4700
```

of terminal nodes is reduced from 13 to 8, there are still far more than actually used to generate the data. This illustrates the general finding that the cost complexity measure tends to be too generous and to over-fit.

Tree pruning: cross-validation

The technique employed here (as implemented by S-PLUS) is 10-fold cross-validation. The data set is to randomly split into 10 equal components, 1 retained for testing and the other 9 combined to generate the model.

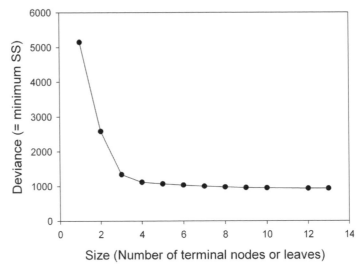

Figure 6.13 Deviance vs. the size of the tree (= number of terminal nodes = leaves) for the hypothetical example. In S-PLUS a similar graph can be produced by the coding

```
Tree        <- tree(Y~X1+X2, data=Data.df) # Generate tree
Tree.pruned <- prune.tree(Tree)            # prune tree
plot(Tree.pruned)                          # Plot deviance as a function of tree size
Tree.pruned$size                           # Output tree size
```

Note that there is no deviance for size = 11. This arises because of the method of pruning the tree (see text for further details).

Testing is done by calculating the deviance using the estimated tree as the predictor. Tree size is varied (in S-PLUS by varying k) and deviances calculated for each value. The set used for testing is then incorporated into the estimation set and one of these set aside for testing: thus this procedure yields 10 cross-validation results, which can then be averaged. The result can then be plotted as deviance vs. tree size. As a further test, the whole procedure can be repeated to yield more estimated curves. In principle, the deviance should be minimal at the best tree size. Results of applying the cross-validation routine 10 times to the present data set are shown in Figure 6.15 (coding in Appendix C.6.7): in 7 runs the best model had 3 terminal nodes and in 3 runs it had 4 terminal nodes. Taking the integral average to be 3 gives the tree shown in Figure 6.16. This tree agrees very well with the actual tree used to construct the data.

Tree pruning and testing: a randomization approach

The foregoing methods allow one to construct and prune a tree according to a particular criterion but the question remains whether the tree

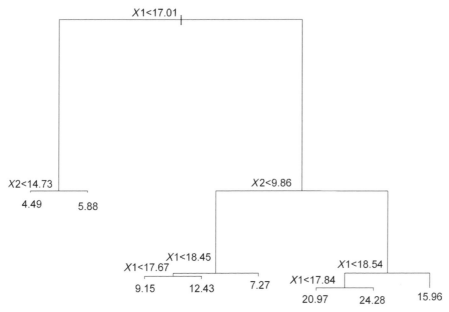

Figure 6.14 Regression tree fitted to hypothetical data from Figure 6.10 using the cost-complexity measure. Coding in S-PLUS to generate tree and text (text output not shown):

```
Tree <- tree(Y~X1+X2, data=Data.df)        # Data in file called Data.df
Prune.Tree <- prune.tree(Tree, k=21.42)    # Produce pruned tree
summary(Prune.Tree)                         # Output text results
plot(Prune.Tree); text(Prune.Tree)         # Plot tree
```

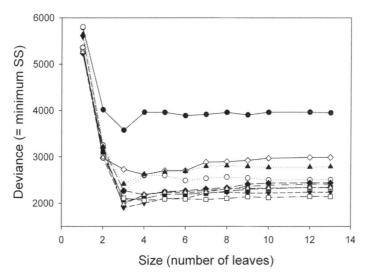

Figure 6.15 Cross-validation applied to the hypothetical example discussed in the text (coding to generate data given in Appendix C.6.7).

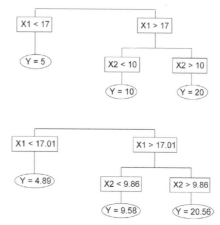

Figure 6.16 A comparison of the tree used to generate the data for the example analyzed in the text (upper) and the tree obtained from cross-validation (lower).

is statistically significant. We can use randomization to answer this question. Specifically, we can use randomization to test the tree generated by the above procedure. The null hypothesis is that there is no relationship between the response and predictor variables. We can test this hypothesis by randomly permuting the response variable, fitting a tree, and comparing the residual deviance from this tree with that from the observed data set. Tree fitting can be done using the given size of the tree as the default tree size (3 in this example): using the cost-complexity measure to prune the tree to the required size does not ensure that this size can be achieved. A conservative test is to use the tree that is the next larger if the desired size cannot be formed (this is the default for S-PLUS). Applying this test to the present data (Appendix C.6.8) shows that none of the randomized data sets fit the data as well or better than the observed data (note that the test is the probability of the residual deviance from a randomized data set being *as small or smaller* than the observed deviance). Most fitted trees had 3 terminal nodes, although one was as large as 10.

Putting it all together: the extinction of the avifauna of New Zealand

To illustrate the principles outlined in the previous sections, I shall present an analysis of data on the causes of extinction in the avifauna of New Zealand described in Roff and Roff (2003).

The large-scale extinction of the avifauna of New Zealand in the centuries following the colonization of the islands by the Maori is probably the best documented case of extinction caused by the direct or indirect actions of a stone-age people. The most frequently cited example is the extinction of the moas but numerous other taxa, including geese, ducks, rails, petrels, and passerines also became extinct in the period between Maori colonization and European contact.

On the North Island, there were 109 species prior to the Maori colonization of which 34 (31%) were extinct by 1770 (the time of European colonization). Of the 118 species on the South Island 37 (31%) were extinct by 1770. A wide variety of birds were exterminated: all eleven moas, most petrel species, some penguins, waterfowl, birds of prey, rails, and several passerines. This very disparate set of species lost suggests a variety of causes was responsible.

Frequently cited candidates as primary causes of extinction of the New Zealand avifauna are direct hunting by the Maori, destruction of the habitat by the Maori, and the impact of the pacific rat, *Rattus exulans*, through predation and/or habitat alteration. While it is accepted that the pacific rat was introduced into New Zealand by the Maoris, the exact date of introduction remains controversial. Climate has been discounted as unimportant, except in so much as it caused local reorganization of communities.

A fundamental problem in the analysis of patterns is that of erecting and testing hypotheses after inspection of the data. For example, visual inspection of the list of extinct and extant taxa suggests that flightless forms have a higher probability of extinction than volant (flight capable) forms. However, after such an inspection it is not statistically valid to then test this hypothesis using the same data, though of course it is typically the only data we have. What is required is a method of objectively finding patterns given a suite of potential candidate characteristics. The problem is that the factors underlying extinction probability may differ between taxa in highly non-linear manners. For example, birds laying small eggs may be more vulnerable to pacific rat predation (one, but not the only, possible impact of the pacific rat) than those laying large eggs, whereas large birds may be invulnerable to rat predation but be a focus of human hunting.

Regression tree analysis is ideally suited to addressing this question and Roff and Roff (2003) used the technique to discern the factors that correlate best with the probability of extinction in the New Zealand avifauna prior to European contact. Here, I concentrate on the extinction patterns in the North Island, though the same pattern was found also in the South Island. The response variable is the probability of extinction coded as 0 or 1 for each species. We considered 7 possible predictor variables; body mass, egg length, flight capability (volant or flightless), habitat type (3 classes), nesting site (4 classes), nest density (2 classes), and food (3 classes).

The fully fitted tree had 11 terminal nodes. A plot of the deviance vs. the number of terminal nodes shows a marked decrease in deviance up to 8 terminal nodes but further splitting produced little change in the model fit (Figure 6.17). The mean residual deviance was 0.06 giving a value of k of 0.12: pruning the tree using this value did not remove any terminal nodes. Figure 6.18 shows the result of applying 10 runs of the cross validation routine: there is considerable scatter

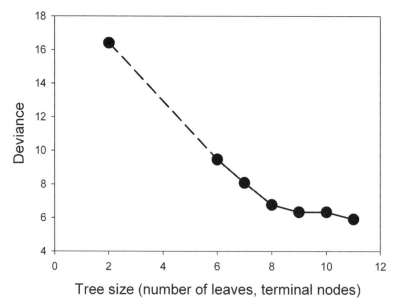

Figure 6.17 Deviance vs. tree size from the regression tree analysis of the extinction of the avifauna of New Zealand. The dashed line indicates the region (3–5) over which no optimal tree could be fitted.

but the optimal size appears to be 8 terminal nodes. To get a better fix on this value, I ran 50 runs of the cross-validation routine, which gave an optimal tree size of 8, whether one uses the mean, median or mode of the 50 values (see lower graph in Figure 6.18). Ninety randomizations of the response variable produced no tree of size 8 (or next largest) that had a smaller deviance than found in the observed data set. Hence, we can conclude that the fitted tree with 8 terminal nodes is significantly different from the null model.

This "best-fit" tree includes 3 of the 7 predictor variables: body mass, flight ability, and nesting site (a = in a cavity within the ground or in a fallen log etc. – e.g., petrels, kiwis; b = on the ground but not in a cavity – e.g., terns, most ducks; c = arboreal – e.g., most passerines, egrets, and herons; d = in a cavity not on the ground – e.g., some parrots. The analysis separated nesting site "a" from the rest; see Figure 6.19). The terminal and near-terminal nodes split the data according to body size and these can be grouped into four sets: leaves 1–3, leaves 4–5, leaves 6–7 and leaf 8 (Figure 6.19). Within each of these sets, we can investigate the relationship between extinction probability and body size using a continuous function such as logistic regression.

In the case of the grouping of nodes 1, 2, and 3, we added a quadratic term to test for the presence of a decreasing probability at the largest body masses (Roff and Roff 2003). Model fit was tested using log-likelihood (Chapter 2). The quadratic term was not significant ($\chi_1^2 = 1.06, P = 0.30$) and was dropped from

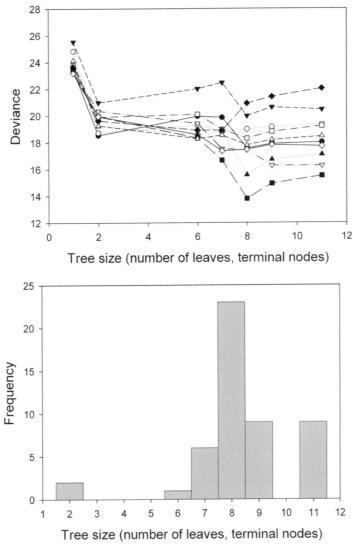

Figure 6.18 Top plot shows the results of 10 runs of the cross-validation routine for the New Zealand avifauna data. Bottom plot shows the distribution of the optimal tree size from 50 runs of the cross-validation routine.

the model. The model involving body mass was highly significant for all three groupings (nodes 1, 2, 3: $\chi^2_1=7.32$, $P=0.007$; nodes 4 and 5: $\chi^2_1=10.67$, $P=0.001$; nodes 6 and 7: $\chi^2_1=9.68$, $P=0.002$). In agreement with the regression tree analysis, the probability of extinction declines with body mass for two groupings and increases with body mass for the third (Figure 6.20). The final regression tree with the terminal node logistic regressions is shown pictorially in Figure 6.20.

North Island

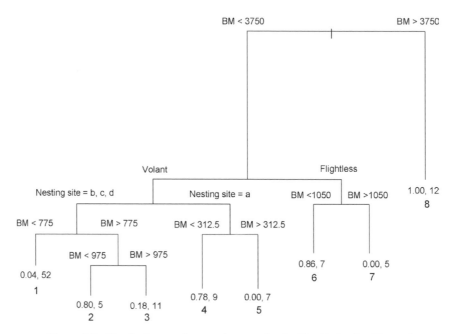

Figure 6.19 The final pruned regression tree for the New Zealand avifauna data. At each terminal node (leaf) is shown the probability of extinction and the sample size. For discussion, the terminal nodes are labeled 1–8. BM = body mass in grams.

Summary

(1)   A recurring problem with multiple regression is that of finding the best-fitting model. Stepwise regression methods frequently arrive at different solutions. Cross-validation can be used to distinguish among competing models.

(2)   In cross-validation a portion of the data is set aside and the remainder used to estimate the regression. This regression equation is then used to predict the values of the data points set aside; the fit can be judged by the correlation between the predicted and observed values in the "new" data set.

(3)   There are three "types" of cross-validation: the holdout method, K-fold cross-validation and leave-one-out cross-validation. Given a reasonable sample size K-fold cross-validation is to be preferred.

(4)   In the absence of an a priori function or an obvious phenomenological form, the form of a function can be estimated using a local smoothing function such as loess, super-smoothing or the cubic spline.

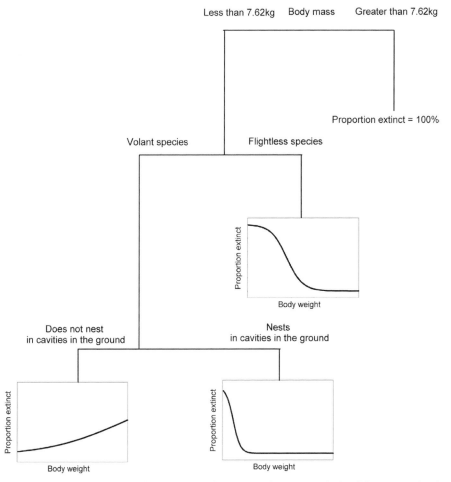

Figure 6.20   A pictorial summary of the regression tree analysis of the New Zealand avifauna data (redrawn from Roff and Roff 2003).

(5)   Differences among fitted models can be assessed approximately by use of an *F*-test or by cross-validation.

(6)   Generalized additive models extend the standard linear regression equation by replacing the coefficients of the regression equation by a smooth function.

(7)   As with smoothing functions, generalized additive models can be compared using either an *F*-test or cross-validation.

(8)   Tree models are particularly useful when there are many predictor variables and their interactions may be complex. The general aim of these methods is to produce a binary tree in which each node of the tree represents a binary division of the data present at that node, determined by some statistical criterion such as least squares.

(9)   Tree pruning can initially be done using the cost-complexity measure combined with cross-validation.

(10)   A randomization test can be used to determine if the regression tree accounts for more variation than expected by chance.

Further reading

Breiman, L., Friedman, J. H., Olshen, R. A. and Stone, C. G. (1984). *Classification and Regression Trees*. Belmont, California: Wadsworth International Group.

Chambers, J. M. and Hastie, T. J. (1992). *Statistical Models in S*. New York: Chapman and Hall/CRC.

De'ath, G. and Fabricius, K. E. (2000). Classification and regression trees: A powerful yet simple technique for ecological data analysis. *Ecology*, **81**, 3178–92.

Hastie, T. J. and Tibshirani, R. J. (1990). *Generalized Additive Models*. London: Chapman and Hall.

LeBlanc, M. and Crowley, J. (1992). Relative risk trees for censored survival data. *Biometrics*, **48**, 411–25.

Marshall, R. J. (2001). The use of classification and regression trees in clinical epidemiology. *Journal of Clinical Epidemiology*, **54**, 603–9.

Schluter, D. (1988). Estimating the form of natural selection on a quantitative trait. *Evolution*, **42**, 849–61.

Schluter, D. and Nychka, D. (1994). Exploring fitness surfaces. *American Naturalist*, **143**, 597–616.

Venables, W. N. and Ripley, B. D. (2002). *Modern Applied Statistics with S*. New York: Springer.

Exercises

(6.1) Use the data below to compare by an *F*-test the two regression models, $y = \theta_0 + \theta_1 x$ vs. $y = \theta_0 + \theta_1 x + \theta_2 x^2$

#	1	2	3	4	5	6	7	8	9	10
x	0.33	0.85	0.63	1.29	0.17	0.17	0.41	1.96	0.88	0.54
y	1.27	0.19	1.32	1.09	−0.06	−0.93	−0.10	2.48	1.53	−0.26
#	11	12	13	14	15	16	17	18	19	20
x	1.94	1.58	0.04	1.82	1.81	1.12	0.75	1.60	0.77	1.64
y	3.94	1.56	−0.18	1.72	4.15	1.03	0.62	1.75	0.11	2.30

(6.2) Use the data in question 6.1 to compare the two models $y = \theta_0 + \theta_1 x$ and $y = \theta_0 + \theta_1 x^2$ using leave-one-out cross-validation.

(6.3) Use 10-fold cross-validation to examine the fit of the regression model $y = \theta_0 + \theta_1 x + \theta_2 x^2$ to the data generated by the coding

```
set.seed(1)
n       <-100
x       <-runif(n,0,2)
error   <-rnorm(n,0,1)
y       <-x^2 +error
```

(Hint: consult Appendix C.6.1.)

(6.4) In a study of introgression between two salmonid species the following data were collected on the environmental characteristics of the streams from which the samples were taken and the surrounding forest. Forward and backwards stepwise regression give different models. Which model is the better predictor? Because of the relatively low number of observations, use 100 runs in which the testing set consists of 20% of the data. (Hint: consult Appendix C.6.1.) Is there any evidence of overfitting by the "better" model (Hint: modify coding in Appendix C.6.1 to calculate r^2 for single models)?

INTR. INDEX	S. LENGTH	FOREST	S. COND	S. ABLE	S. TEMP
0.54	25.3	70.7	0.8	0.30	17.0
0.41	25.3	70.7	0.8	0.30	18.1
0.37	8.0	90.4	0.2	1.00	13.6
0.26	55.6	79.5	1.2	0.27	11.2
0.48	27.4	84.8	0.9	0.57	17.5
0.03	81.8	26.5	1.0	1.00	14.6
0.03	228.5	62.5	1.0	1.00	10.8
0.05	45.0	20.2	0.3	0.41	10.6
0.06	97.2	46.7	1.5	0.51	17.8
0.34	311.2	51.3	1.3	0.14	12.6
0.23	28.8	81.0	2.7	1.00	12.6
0.06	36.0	45.9	1.4	0.59	12.2
0.31	80.5	29.0	1.7	1.00	17.3
0.03	343.2	50.0	2.2	1.00	10.8
0.30	113.2	87.8	2.8	0.27	11.9
0.30	33.1	83.5	0.8	1.00	13.7
0.35	102.0	65.2	1.0	0.18	19.1
0.00	618.6	70.2	1.4	1.00	12.6
0.25	16.9	38.5	2.0	1.00	14.4
0.16	101.2	77.3	1.7	1.00	15.3

0.08	151.4	24.1	1.9	1.00	10.2
0.04	58.0	64.5	1.1	1.00	17.2
0.13	24.8	19.7	4.3	1.00	16.8
0.02	86.8	54.8	1.2	1.00	13.9
0.04	32.2	46.0	1.2	1.00	11.0
0.09	581.1	35.4	0.8	0.24	12.6
0.05	262.6	18.4	1.7	1.00	13.3
0.02	618.6	70.2	1.4	0.39	13.8
0.03	80.2	55.1	2.1	1.00	16.7
0.11	122.7	48.7	1.0	0.67	11.4
0.04	41.6	71.7	0.7	1.00	14.3

INTR. INDEX=introgression index, S. LENGTH=stream length, FOREST= index of forest quality, S. COND= stream conductivity, S. ABLE=index of stream quality, S. TEMP=stream temperature.

(6.5) Fit two loess functions to the data in Q6.1, using span=0.2, degree=1 and span=1, degree=1. Does the former model give a significantly better fit to the data? (Also try plotting the data and the residuals making use of the coding given in Appendix C.6.2).

(6.6) The number of data points in Q6.1 is too few to do 10-fold cross-validation. There is sufficient to do 3-fold cross-validation. Using the coding in Appendix C.6.3 as a guide, do 3-fold cross-validation.

(6.7) Modify the coding used in Q6.5 to randomly select one-third of the data as the test set and perform 100 cross-validations. Compare the r^2 between predicted and observed with the multiple R^2 from the fit on the training sets.

(6.8) Use the following generalized additive models to analyze the data in the table below:

(1)   All predictor variables entered as functions.
(2)   X1 entered as a linear function.
(3)   X1 not entered.

#	X1	X2	X3	Y	#	X1	X2	X3	Y
1	1	9	2	90	16	5	0	3	44
2	4	9	9	813	17	3	0	2	11
3	3	8	0	86	18	7	9	0	111
4	6	2	0	11	19	3	2	3	35
5	0	5	1	29	20	8	6	5	167
6	0	4	0	7	21	5	2	7	368

7	2	7	2	81	22	8	4	1	20
8	9	0	9	761	23	4	6	4	117
9	4	0	0	9	24	3	5	3	51
10	2	3	8	536	25	1	0	9	732
11	9	6	1	44	26	2	5	6	260
12	7	4	3	77	27	5	3	6	226
13	0	2	3	40	28	8	3	3	64
14	9	1	7	350	29	8	3	1	24
15	9	5	9	773	30	9	4	3	67

(6.9) The tree diagram shown below gives the factors that govern the probability of extinction in a group of birds.

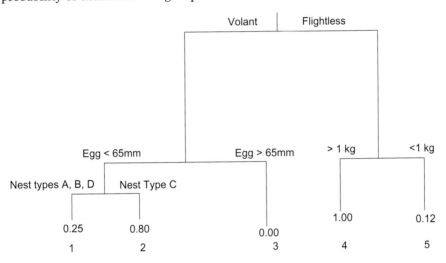

A hypothetical data set of 1000 species was constructed using the above tree assuming 50% in each binary category (flight capability = WING, nest type = NEST). A binary habitat category was also created that was unrelated to the probability of extinction. Body and egg sizes were drawn from uniform random distributions. The appropriate coding is

```
set.seed(1)
N <- 1000   # Number of species
# Create a vector with 50% 0s and 50% 1s
M       <- N/2
Dummy   <-c(rep(0,M),rep(1,M))
# Create vectors for the three binary variables with a randomized
# Dummy
Wing    <- sample(Dummy)   # 0=flightless, 1=volant
```

```
Nest      <- sample(Dummy)   # 0 nest type A,B,D, 1 = C
Habitat   <- sample(Dummy)   # Not connected to survival
# Create vectors of egg size and body size from random uniform
# distribution
Egg     <- runif(N,0,130)   # 0 < 65 , 1 > 65 egg size
Body    <- runif(N,0,2)     # 0 < 1kg, 1 > 1kg
# Create expected response vector
P  <- matrix(0,N,1)
# Cycle through conditions
for ( i in 1:N)
{
  if(Wing[i]==0 & Body[i]<1)   P[i]   <- 0.12
  if(Wing[i]==0 & Body[i]>=1)  P[i]   <- 1
  if(Wing[i]==1 & Egg[i] > 65) P[i]   <- 0
  if(Wing[i]==1 & Egg[i] <= 65 & Nest[i] == 0)   P[i] <- 0.25
  if(Wing[i]==1 & Egg[i] <= 65 & Nest[i] == 1)   P[i] <- 0.80
}
# Now test to see if species is extinct or extant
# Generate uniform random numbers 0-1 to see if species survives
Prand <- runif(N,0,1)
for ( i in 1:N){if (Prand[i] < P[i] )P[i] <- 1 else P[i] <- 0}
# Combine variables into a single dataframe
Q7.Data <- data.frame(Wing,Egg,Body,Nest,Habitat,P)
```

Use regression tree analysis to determine the "best" tree. Use the response variable as a numeric variable (setting it as a factor gives the same answer. The advantage in using P as a numeric is that the leaves give the predicted probabilities of extinction). Compare the "best" tree with the one used to construct the data. Are they different? Be careful in your assessment: consider how the tree given above would look if the first division was as in the output.

Hint: Use the following steps

Step 1: Create tree and plot deviance against possible tree sizes

Step 2: Cross-validation of tree to find optimal size (Appendix C.6.6)

Step 3: Randomization test for tree with 8 leaves (Appendix C.6.7).

List of symbols used in Chapter 6

Symbols may be subscripted

δ Parameter that plays the same role as degrees of freedom in approximate F-statistic.

ε	Error term
θ	Parameter
μ	Mean
σ	Standard deviation
σ^2	Variance
$\varphi(\ )$	Function

AIC	Akaike's information criterion
D	Deviance
E	Entropy or information index
ENP	Equivalent number of parameters
G	Gini coefficient
L	Likelihood
LL_{max}	Maximum log-likelihood
N	Sample size
$PRSS$	Penalized residual sums of squares
RSS	Residual sums of squares
$X1, X2$	Variables
$f(\ )$	Function
k	(1) Number of parameters, (2) Number of divisions in cross-validation
P	Probability
r, n	Number in two groups at tree node
$s(x)$	Smoother function
x, y	Observations

7

Bayesian methods

Introduction

The approaches we have examined thus far belong to a school called the **frequentist school**. Frequentists use the likelihood function to calculate the probability of observing a particular set of data for a given value of the statistic, that is, given some set of observations, say x, the frequentist approach is to calculate the probability of occurrence of the observed data for some given statistic θ. Symbolically, this is written as $P(x|\theta)$, that is "the probability of x given θ."

The **Bayesian approach** is different in that it reverses the probability statement and asks "what is the probability of θ given x," which can be written as $P(\theta|x)$. To apply the Bayesian approach we require a **prior probability distribution** for the statistic θ. The current data are then used to modify this probability, thereby forming a **posterior probability distribution**. Where frequentists and Bayesians disagree is the case in which there is no prior probability, though as we shall see this disagreement in many cases is actually rather insignificant.

Both approaches can be useful in gaining a perspective on a set of data, in particular, the Bayesian approach can be highly informative in cases of decision making. For example, a physician may ask "what is the probability that this patient actually has a disease given that she has a set of symptoms that occur in 99% of people with the disease?" This seemingly simple question can have a surprising answer in that it is possible for the probability that a person with the symptoms has the disease may be very low. This example is discussed later in a more general context concerning the correct assignment of species given some observable characteristic. The difficulty with the Bayesian approach is in assigning the prior probability and operationally in calculating the posterior probability. In this chapter, I shall present some simple examples to illustrate the

Bayesian approach and then some more complex "real-life" examples to illustrate further the utility of the method and the problems of its implementation.

Derivation of Bayes' theorem

To understand the Bayesian perspective, we begin by considering **conditional probability**. Thus far, we have only considered the case in which $P(A)=p$, that is, the probability of event A is p. We now consider the conditional probability $P(A|B)=p$, which in words is "given that event B is observed, the probability of observing event A is p." Quite clearly, this is a different statement from the former ($P(A)=p$) and the value of p will be different. We can represent this situation as a Venn diagram in which A and B are intersecting circles within the universal set (Figure 7.1). Now, given that B is observed, B represents the entire set of possible outcomes, and the probability of A is the proportion of the area of B contained within the circle (set) labeled A, which by simple geometry is

$$P(A|B) = \frac{P(A \cap B)}{P(B)} \tag{7.1}$$

where $P(A \cap B)$ is the conventional means of denoting the intersection of A and B. Rearranging Eq. (7.1) to give the probability of A and B

$$P(A \cap B) = P(A|B)\,P(B) \tag{7.2}$$

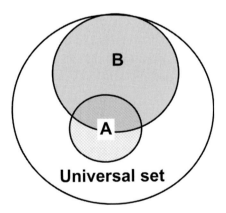

Figure 7.1 A diagrammatic representation of the occurrence of event A relative to event B. The outer circle represents all possible events (by convention called the universal set). The proportion of occurrences of B is shown by the inner grey circle. Event A is the hatched inner circle and the intersection of A and B, denoted by $A \cap B$, is the joint event A and B.

which must also be equal to

$$P(A \cap B) = P(B|A) P(A) \tag{7.3}$$

So now we can write Eq. (7.1) as

$$P(A|B) = \frac{P(B|A) P(A)}{P(B)} \tag{7.4}$$

The denominator can itself be written as a set of conditional probabilities

$$P(B) = P(B|A) P(A) + P(B|A^C) P(A^C) \tag{7.5}$$

where "A^C" is the conventional method of denoting "not A." Substituting in Eq. (7.4) gives **Bayes' Theorem**

$$P(A|B) = \frac{P(B|A) P(A)}{P(B|A) P(A) + P(B|A^C) P(A^C)} \tag{7.6}$$

The term $P(A)$ is the prior probability and $P(A|B)$ is the posterior probability, that is, after event B is observed the probability of A can be updated based on this observation. The above can be extended to more than two possible events and in the terminology generally adopted in this book rewritten as

$$L(\theta|x) = \frac{L(x|\theta) P(\theta)}{\int L(x|\theta) P(\theta) \, d\theta} \tag{7.7}$$

where θ is the statistic of interest, x is the set of observations and $L(\theta)$ is the likelihood function (it is perhaps better to call it the **loss function** as, in principle, one does not have to use the likelihood, though this is the general approach), and $P(\theta)$ is the prior probability for θ.

Two simple Bayesian models

A simple classification problem

This is the more "ecological" version of the example presented in the introduction. Suppose a fraction p_A of species A possess a characteristic B. In the population of non-A species, the fraction of species that possess characteristic B is p_{A^C}. Given that we have in our hand an organism that has characteristic B, what is the probability that it belongs to species A, which is to say, $P(A|B)$?

The probability of A, $P(A)$, is the proportion of species A in the fauna. Hence, the probability of non-A species, $P(A^C)$ is $1 - P(A)$. The probability of characteristic B

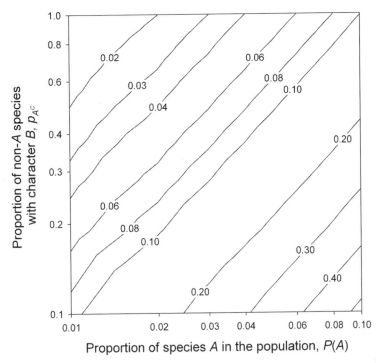

Figure 7.2   Contour plot showing the probability of a randomly selected species bearing characteristic B being species A when the probability of species A showing the character is 0.99. Letting the horizontal and vertical axes be x and y, respectively, the equation used is

$$P(A|B) = \frac{0.99x}{0.99x + y(1-x)}$$

given that the species is A, $P(B|A)$, is p_A, and the probability of characteristic B given that the species is not A, $P(B|A^C)$, is p_{A^C}. From these data, we can write

$$P(A|B) = \frac{p_A P(A)}{p_A P(A) + p_{A^C}[1 - P(A)]} \qquad (7.8)$$

The above equation is not particularly impressive until one realizes that it tells us that even if 99% of species A has the characteristic B, the probability that a randomly chosen species from the population with characteristic B is actually species A can be very small (Figure 7.2). This arises because the probability of A given B has to take into account the proportion of A in the population and the proportion of non-A that bear character B: when only 1% of the fauna is species A and 50% of other species also bear the characteristic B then the probability that a randomly chosen species with character B is species A is only 0.02 (Figure 7.2).

Estimating the mean of a normal distribution

First, we shall consider the problem of estimating the mean, θ, given that the variance is known (say σ^2). Recall from Chapter 2 that the likelihood of observing a value x from a normal distribution is given by

$$L(x|\theta) = \frac{1}{\sqrt{2\pi\sigma}} e^{-\frac{1}{2}\left(\frac{x-\theta}{\sigma}\right)^2} = \varphi(x - \theta, \sigma) \tag{7.9}$$

We shall assume that the prior probability distribution is also normal with mean (μ_0) and variance, (σ_0^2). The unfixed parameters of the prior probability distribution are known as **hyperparameters**. Using Bayes' theorem, we can compute the likelihood value for θ given the observation x as

$$L(\theta|x) = \frac{\varphi(x - \theta, \sigma)\,\varphi(\theta - \mu_0, \sigma_0)}{\int \varphi(x - \theta, \sigma)\,\varphi(\theta - \mu_0, \sigma_0)\, d\theta} \tag{7.10}$$

The numerator is the product of two exponential equations and after some algebra can be written as

$$L(\theta|x) \propto \varphi(\theta - \mu_1, \sigma_1) \tag{7.11}$$

where $\mu_1 = ((1/\sigma_0^2)/\mu_0 + (1/\sigma^2)x)/((1/\sigma_0^2) + (1/\sigma^2))$, $(1/\sigma_1^2) = (1/\sigma_0^2) + (1/\sigma^2)$. The inverse of the variance is termed the **precision** and hence, the Bayes' estimate of the mean is equal to a weighted average of the prior mean and the observation, where the weights are the prior and data precisions. An interesting way of looking at the result is to rewrite the posterior mean as

$$\mu_1 = x - (x - \mu_0)\frac{\sigma^2}{\sigma^2 + \sigma_0^2}$$

$$\mu_1 = \mu_0 + (x - \mu_0)\frac{\sigma_0^2}{\sigma^2 + \sigma_0^2} \tag{7.12}$$

Any single data point is itself an estimate of the mean θ and hence the above equations emphasize that the data estimate of θ is "shrunk" towards the prior mean (bottom equation) or adjusted towards the observed value by the observed datum (top equation).

The above result can be applied to multiple observations by treating $\bar{x}$ as a single observation:

$$L(\theta|x_1, x_2, \ldots, x_n) = L(\theta|\bar{x}) \propto \varphi(\theta - \mu_n, \sigma_n) \tag{7.13}$$

where $\mu_n = ((1/\sigma_0^2)\mu_0 + (n/\sigma^2)\bar{x})/((1/\sigma_0^2) + (n/\sigma^2))$, $(1/\sigma_n^2) = (1/\sigma_0^2) + (n/\sigma^2)$.

The application of the above formulae requires that we supply values for the parameters of the prior. This is discussed in detail in the next section but it is worth noting here that it is the issue of "subjectivity" in the decision of the prior that troubles many researchers.

Deciding on the prior distribution

Noninformative priors

The simplest assumption that can be made about the prior is that all values are equally likely: such a prior is termed a **noninformative, vague** or **indifferent prior**. In the case of the normal distribution, this means letting the variance go to infinity. While this is an **improper prior** (in the sense that it does not have a unit integral), it can be justified by the use of limits and gives a **proper posterior**, which is consistent with the likelihood approach. The assumption of a uniform noninformative prior has disturbed many non-Bayesians, particularly as it can arise that a one-to-one transformation, which should not affect the answer, can change the effect of the prior (for a relatively accessible mathematical discussion on this issue refer to pp. 48–52 in Press (1989). This has lead to searches for noninformative priors that are invariant under transformation. For example, suppose θ is a binomial variable. By definition, the parameter space is 0–1 and the noninformative, uniform prior is $P(\theta)=1$. (Throughout the remainder of this chapter, I shall use the function $P(\bullet)$ to denote the prior.) Three other priors that have been proposed are: $P(\theta) = \theta^{-1}(1-\theta)^{-1}$, $P(\theta) \propto [\theta(1-\theta)]^{-\frac{1}{2}}$, and $P(\theta) \propto \theta^{\theta}(1-\theta)^{(1-\theta)}$ (Berger 1985, p. 89). All are reasonable, and if it makes a difference as to which is used then there is something wrong with the model, because the prior is clearly not noninformative!

Using a noninformative prior for the estimation of the mean of a normal distribution based on a single observation leads simply to the simple likelihood

$$L(\theta|x) = \varphi(x - \theta, \sigma) \tag{7.14}$$

The posterior distribution for a given $\sigma = 0.5$ is shown in Figure 7.3.

Natural conjugate priors

For some probability distributions, there are "natural" priors in the sense that the probability distributions suggest at least the form of the prior distribution. The class of priors that have the same parametric form as the posterior distribution are known as **conjugate priors**. These priors are frequently

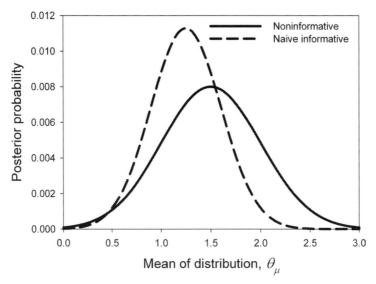

Figure 7.3 Posterior probabilities for the estimation of the mean of a normal distribution based on a known variance, a single observation x and naïve noninformative or informative priors. S-PLUS coding is

```
x                        <- 1.5                          # Value of x
sd                       <- 0.5                          # σ
Posterior1               <- dnorm(x,Theta,sd)           # Unscaled posterior
Posterior1               <- Posterior1/sum(Posterior1)  # Scaled posterior
plot(Theta, Posterior1)                                 # Plot
mu0                      <- 1                            # μ0
mu1                      <- (mu0+x)/2                    # μ0
sd1                      <- 0.5/sqrt(2)                  # σ1
Posterior2               <- dnorm(Theta,mu1,sd1)        # Unscaled posterior
Posterior2               <- Posterior2/sum(Posterior2)  # Scaled posterior
plot(Theta, Posterior2)
```

selected because of their convenient mathematical properties with respect to solving the Bayesian formula. For example, if the case in which we are interested follows the binomial distribution (e.g., the simple classification problem given earlier) the likelihood function is

$$L(\theta) = {}^nC_x\theta^x(1-\theta)^{n-x} \tag{7.15}$$

where n is the number of observations and x is the number of "successes." This likelihood can be written in the form of a **beta distribution**

$$L(\theta) = \frac{\theta^x(1-\theta)^{n-x}}{\int\theta^x(1-\theta)^{n-x}d\theta} = \frac{\theta^x(1-\theta)^{n-x}}{B(x+1, n-x+1)} \tag{7.16}$$

where $B(x, n)$ is a beta function. By Bayes' theorem, we have

$$L(\theta|x) = \frac{{}^nC_x\theta^x(1-\theta)^{n-x}P(\theta)}{\int {}^nC_x\theta^x(1-\theta)^{n-x}P(\theta)d\theta} \tag{7.17}$$

where $P(\theta)$ is the prior probability distribution. Given a prior set of experiments in which x_0 "successes" in n_0 trials were observed, we could take the prior to be the likelihood

$$P(\theta) = L(\theta) = \frac{\theta^{x_0}(1-\theta)^{n_0-x_0}}{B(x_0+1,\, n_0+1)} \tag{7.18}$$

Now, we do not have prior values for x_0 and n_0 and hence, we use two arbitrary parameters α, β, permitting them to take positive noninteger values and rewrite the formula as

$$P(\theta) = L(\theta) = \frac{\theta^{\alpha-1}(1-\theta)^{\beta-1}}{B(\alpha,\, \beta)},\quad \alpha > 0,\ \beta > 0 \tag{7.19}$$

The useful feature of this distribution is that it is very flexible with a shape that ranges from a bell-shaped distribution through a uniform distribution to a U-shaped distribution (Figure 7.4). Further, because it has the same parametric form as the likelihood of the observed data, the posterior distribution is readily obtained analytically

$$L(\theta|x) \propto \theta^x(1-\theta)^{n-x}\theta^{\alpha-1}(1-\theta)^{\beta-1}$$
$$L(\theta|x) \propto \theta^{x+\alpha-1}(1-\theta)^{n-x+\beta-1} \tag{7.20}$$

Table 7.1 gives examples of other natural conjugate priors.

Naïve Informative

It may be that we can use information or make an assumption to more precisely define the prior distribution and reduce the number of parameters in the posterior probability function. For example, in the estimation of the mean of a normal distribution, we might assume that $\sigma = \sigma_0$, that is, the variances of the prior and posterior distributions are the same. Given this, the mean and variance of the posterior distribution is

$$\mu_1 = \frac{\left(\frac{1}{\sigma_0^2}\right)\mu_0 + \left(\frac{1}{\sigma^2}\right)x}{\left(\frac{1}{\sigma_0^2}\right) + \left(\frac{1}{\sigma^2}\right)} = \frac{\mu_0 + x}{2}$$
$$\frac{1}{\sigma_1^2} = \frac{1}{\sigma_0^2} + \frac{1}{\sigma^2} = \frac{2}{\sigma^2} \tag{7.21}$$

Table 7.1 *Some natural conjugate priors (from Press 1989)*

Sampling distribution	Natural conjugate prior
Binomial	"Success" probability is beta
Negative binomial	"Success" probability is beta
Poisson	Mean is gamma
Exponential with mean λ^{-1}	λ is gamma
Normal with known variance but unknown mean	Mean is normal
Normal with known mean but unknown variance	Variance is inverted gamma

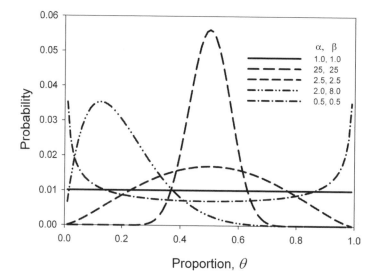

Figure 7.4 Probability densities for the beta distribution for various values of α and β. In S-PLUS, a curve can be generated using the coding:

```
x         <- seq(0.01,0.99,.01)    # Generate proportions
alpha     <- 24                     # Set alpha
beta      <- 25                     # Set beta
y         <- dbeta(x,alpha,beta)    # Generate densities
plot(x,y)                           # Plot data
```

The posterior distribution is still normal, but with a different mean and variance:

$$L(\theta_\mu|x) = \varphi(\theta - \mu_1, \sigma_1) \tag{7.22}$$

To apply this model, we require μ_0. For illustrative purposes, I have assumed $\mu_0 = 1$. The new observation shifts the mean and reduces the variance of the posterior probability distribution (Figure 7.3).

Hierarchical Bayesian

In the foregoing analysis, I assumed that both μ_0 and σ_0 were known: in general, the parameters of the prior distribution are unlikely to be known. There are two possible routes to solving this problem. First, and most simply, they can be estimated from the data, in which case the estimator is known as an **Empirical Bayes' estimator**. Second, we can address the prior probability in the same manner as the posterior probability, namely via Bayes' theorem, in which case the estimator is termed **Hierarchical Bayesian**. Prior probabilities for the parameters of the prior are termed hyperpriors. The general approach is relatively straightforward but the implementation can be horrendously complex and involve considerable numerical simulation.

In the case of estimating the mean of a normal distribution, we have two unknown hyperpriors, μ_0 and σ_0. For simplicity, I shall assume that these two parameters are independent. Under independence, the joint probability density function can be written as the product of two separate functions, $p(\mu_0, \sigma_0) = (\mu_0) p_2(\sigma_0)$. This allows us to construct a prior that is itself a conditional probability distribution, a hyperprior. The prior distribution on θ is

$$P(\theta) = \int L(\theta | \mu_0, \sigma_0) \, p_1(\mu_0) \, p_2(\sigma_0) \, \mathrm{d}\mu_0 \, \mathrm{d}\sigma_0 \qquad (7.23)$$

where $L(\theta | \mu_0, \sigma_0) = L(\theta) = \varphi(\theta_\mu - \mu_0, \sigma_0)$. The above equation states that the joint likelihood (prior) is obtained by integrating over the probability distributions of μ_0 and σ_0. The posterior distribution can now be written as

$$L(\theta | x) = \int L(\theta | x, \mu_0, \sigma_0) \, p_1(\mu_0 | x) \, p_2(\sigma_0 | x) \mathrm{d}\mu_0 \, \mathrm{d}\sigma_0 \qquad (7.24)$$

To obtain the probability functions p_1 and p_2 we again make use of Bayes' theorem

$$p_i(v | x) = \frac{L(x | v) \, p_i(v)}{\int L(x | v) \, p_i(v) \mathrm{d}v} \qquad (7.25)$$

where v is either μ_0 or σ_0. We are now left with the question of deciding upon the distribution functions for μ_0 and σ_0 and the likelihood function. The latter has actually already been defined as normal and so

$$L(x | \mu_0, \sigma_0) = \varphi(x - \mu_0, \sigma_0) \qquad (7.26)$$

Given appropriate choices for the probability functions p_1 and p_2, it may be possible to arrive at an analytical solution, but in many cases one may have to

Table 7.2 *Probability values for assumed values of μ_0 and σ_0 and the associated likelihoods for the 16 combinations given an observed value of x=1.5*

σ_0		$\mu_0=$	0.0	0.5	1.0	3.0
		$p_1(\mu_0)=$	0.1	0.2	0.5	0.2
σ_0	$p_2(\sigma_0)$		$L(\mu_0, \sigma_0\|x=1.5)$			
0.25	0.01		10^{-10}	4.1×10^{-6}	0.0042	2×10^{-10}
0.30	0.05		9.5×10^{-8}	2.0×10^{-4}	0.0319	1.9×10^{-7}
0.50	0.90		0.0031	0.0747	0.8371	0.0061
0.75	0.04		0.0011	0.0067	0.0327	0.0022

resort to numerical methods. To illustrate the procedure, suppose, based on prior information, we assign to each unknown parameter (μ_0 and σ_0) four possible values and probabilities of these values as shown in Table 7.2. We are quite sure about the value of σ_0 but only moderately confident in our estimate of μ_0.

Because the two parameters are independent of each other, there are a total of 16 combinations, with the probability of a particular combination, given the value of x, being determined by Eq. (7.25)

$$L(\mu_0,\sigma_0|x) = \frac{\varphi(x-\mu_0,\sigma_0)\,p_1(\mu_0)\,p(\sigma_0)}{\sum \varphi(x-\mu_0,\sigma_0)\,p_1(\mu_0)\,p(\sigma_0)} \tag{7.27}$$

where the summation is taken over the 16 combinations. The result of the calculations shows that the combination $\mu_0=1.0$, $\sigma_0=0.50$ is overwhelmingly the most likely. To obtain the posterior distribution for θ we select a value of θ, calculate the likelihood for the given values of μ_0, σ_0, and x and sum over all 16 combinations

$$L(\theta|x) = \sum L(\theta|\mu_0,\sigma_0,x)L(\mu_0,\sigma_0|x) \tag{7.28}$$

The prior probabilities are then obtained by dividing through by the overall sum (Appendix C.7.1).

Further examples of Bayesian analyses

Effects of different priors: the difference between two means

The data consists of two samples of sizes n_i with means $\bar{x}_i(i=1,2)$. The issue under investigation is the difference between the two means, θ_1 and θ_2 and hence, we are interested in the distribution of $\theta=\theta_1-\theta_2$. The likelihood function for each population, subscripted as i ($=1,2$) can be written as (Chapter 2)

$$L(x_i|\theta_i) = \varphi(x_i - \theta_i, \sigma_i) \tag{7.29}$$

Assuming the prior to be normal and, in the absence of information to the contrary, that there is a common mean, μ_0, and variance, σ_0^2, the likelihood for the mean of each population is

$$L(\theta_i) = \varphi(\theta_i - \mu_0, \sigma_0) \tag{7.30}$$

The prior distribution for the difference between the two means is therefore

$$P(\theta) = \varphi(0 - \theta, \sigma_0\sqrt{2}) = \varphi(\theta, \sigma_0\sqrt{2}) \tag{7.31}$$

The likelihood function of $X = \bar{x}_1 - \bar{x}_2$ conditional on θ is

$$L(X|\theta) = \varphi(X - \theta, \sigma) \tag{7.32}$$

where $\sigma^2 = \frac{\sigma_1^2}{n_1} + \frac{\sigma_2^2}{n_2}$. What we wish to calculate is the likelihood of θ conditional on $X = \bar{x}_1 - \bar{x}_2$

$$L(\theta|X) = \frac{\varphi(X - \theta, \sigma)\,\varphi(\theta, \sigma_0\sqrt{2})}{\int\varphi(X - \theta, \sigma)\,\varphi(\theta, \sigma_0\sqrt{2})\mathrm{d}\theta} \tag{7.33}$$

$$= \varphi(Y - \theta, \ V)$$

where $Y = \dfrac{2\sigma_0^2 X}{2\sigma_0^2 + \sigma^2}$ and $V^2 = \dfrac{2\sigma_0^2\sigma^2}{2\sigma_0^2 + \sigma^2}$.

The problem is now to decide upon the prior distribution. Using a noninformative prior gives the posterior probability to be $\varphi(X - \theta, \sigma)$, i.e.,

$$L(\theta|X) = \varphi(X - \theta, \sigma) \tag{7.34}$$

A researcher is likely to be interested not so much in the probability of a difference of θ but that the difference is equal to or greater than θ, which is simply the cumulative normal. For example, suppose the observed difference between the two samples is 1.5 units and the samples are large enough that the two variances can be substituted in the above formula (say $\sigma = 0.5$), then the posterior probability of a difference at least as large as θ is $\int\varphi\left(\frac{1.5-\theta}{0.5}\right)\mathrm{d}x$ (Figure 7.5). Given some knowledge of what constitutes a biologically significant difference the researcher can decide if such a difference has a "high" probability. Suppose a difference of 2 units were considered biologically important, the probability of a difference at least as large as 2 is approximately 0.16, which many researchers would certainly consider a value that cannot be considered "insignificant."

Rather than using a noninformative prior, we can use a naïve informative prior by assuming that the two variances in Eqs (7.31) and (7.32) are equal

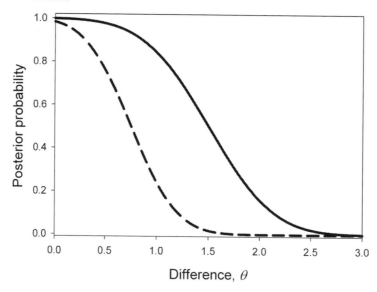

Figure 7.5 Posterior probability distributions for a difference between two normally distributed variables using a noninformative prior (——), or a naïve informative prior (— — —). The data for the naïve noninformative were generated using S-PLUS coding

```
Theta            <- seq(0,3,0.01)        # Vector of theta values
cum              <- pnorm(1.5-Theta, 0, 0.5)   # Cumulative probability
plot(Theta, cum)                         # Plot data
```

(both are variances of the difference between two means), i.e., $\sigma^2 = 2\sigma_0^2$. Substituting for the mean and variance in Eq. (7.33) gives

$$Y = \frac{2\sigma_0^2 X}{2\sigma_0^2 + \sigma^2} = \frac{\sigma^2 X}{\sigma^2 + \sigma^2} = 0.5X$$

$$V^2 = \frac{2\sigma_0^2 \sigma^2}{2\sigma_0^2 + \sigma^2} = \frac{\sigma^2 \sigma^2}{\sigma^2 + \sigma^2} = 0.5\sigma^2$$

(7.35)

which leaves us with only the "known" σ^2 (as before, say $\sigma = 0.5$). The posterior distribution is now

$$L(\theta|X) = \varphi\left(\frac{0.5X - \theta}{\sigma\sqrt{0.5}}\right)$$

(7.36)

The probability of obtaining a difference at least as large as 2 is now 0.0002, which is highly unlikely (Figure 7.5). Thus, a simple change in the assumption about two parameters produces a dramatic difference in the assessment of the probability of a difference between the two groups. This emphasizes the importance of very carefully evaluating the premises underlying the probability distributions.

Thus far, in this example, we have assumed a prior in which the means and variances of the two normal distributions are the same (θ_1 and θ_2 had a common normal distribution). Whereas this might be an appropriate null hypothesis, a more general assumption to make in the face of no knowledge is that they could have come from two different normal distributions. The prior now consists of the mixture of two normal distributions, which is itself normal with mean μ and variance v^2, but for which there is uncertainty about the parameter values. To incorporate this complexity into the estimation procedure, we must take a hierarchical Bayesian approach. Assuming that these two parameters are independent, the joint probability density function can be written as the product of two separate functions, $p(\mu, v) = p_1(\mu)\, p_2\,(v)$. The prior distribution on θ_1 and θ_2 is

$$P(\theta_1, \theta_2) = \int L(\theta_1, \theta_2 | \mu, v)\, p_1(\mu)\, p_2(v)\, d\mu\, dv \tag{7.37}$$

where $L(\theta_1, \theta_2 | \mu, v) = L(\theta_1 | \mu, v) L(\theta_2 | \mu, v)$, which is the product of the likelihoods given in Eq. (7.30) for given μ and v. The above equation states that the joint likelihood (prior) is obtained by integrating over the probability distributions of μ and v. The posterior distribution is

$$L(\theta_1, \theta_2 | x_1, x_2) = \int L(\theta_1, \theta_2 | x_1, x_2, \mu, v)\, p_1(\mu | x_1, x_2)\, p_2(v | x_1, x_2)\, d\mu\, dv \tag{7.38}$$

and the cumulative probability that we are interested in can now be written as

$$\int g(\theta | X) p_1(\mu | x_1, x_2) p_2(v | x_1, x_2) d\mu\, dv \tag{7.39}$$

where $g(\theta | X) = \int L(\theta | X)$. The cumulative probability does not depend upon $p_1(\mu | x_1, x_2)$ but is a function of $p_2(v | x_1, x_2)$. To obtain the latter, we again make use of Bayes' theorem

$$p_2(v | x_1, x_2) = \frac{L(x_1, x_2 | v)\, p_2(v)}{\int L(x_1, x_2 | v)\, p_2(v)\, dv} \tag{7.40}$$

and make use of the two conditional probabilities

$$L(x_1, x_2 | v) = \int L(x_1, x_2 | \mu, v) p_1(\mu)\, d\mu$$

$$L(x_1, x_2 | \mu, v) = \int L(\bar{x}_1, \bar{x}_2 | \theta_1, \theta_2) L(\theta_1, \theta_2 | \mu, v)\, d\theta_1 d\theta_2 \tag{7.41}$$

$$= \prod_{i=1}^{2} \varphi\left(\frac{\bar{x}_i - \mu}{\sqrt{(v^2 + \sigma_i^2)/n_i}} \right)$$

Substituting in Eq. (7.40), we arrive at

$$p_2(v|x_1, x_2) = \frac{\prod_{i=1}^{2} \varphi\left(\frac{\bar{x}_i - \mu}{\sqrt{(v^2 + \sigma_i^2)/n_i}}\right) p_2(v)}{\int \prod_{i=1}^{2} \varphi\left(\frac{\bar{x}_i - \mu}{\sqrt{(v^2 + \sigma_i^2)/n_i}}\right) p_2(v) \mathrm{d}v} \qquad (7.42)$$

It is still necessary to assign a functional form to $p_2(v)$ and values to σ_1 and σ_2. Once this is done the posterior probability can be calculated. The important message to be gained from this example is that the construction of the prior, even in a very simple case, can become quite complex and still require assumptions on distributions and parameter values that can be challenged. This is not meant to denigrate the Bayesian approach merely to make clear that its assumptions must always be clearly stated, particularly with respect to the construction of the prior distribution.

Sequential Bayes' estimation: survival estimates

Suppose we are interested in estimating the probability associated with some binomial event such as survival. We subject n *Daphnia* to an encounter with a predator, such as a stickleback and find that there are x survivors. As discussed in Chapter 2, the likelihood for this situation is

$$L(\theta) = {}^nC_x\theta^x(1 - \theta)^{n-x} \qquad (7.43)$$

where θ is the probability of "survival" at each trial. The observed value of θ is $8/10 = 0.8$. In the absence of any prior information, we make the usual Bayesian assumption that θ is equally likely to lie between zero and 1, i.e., it has a uniform prior. The prior probability is then a constant equal to 1 (for a uniform probability distribution between 0 and 1, we have $\int c \, \mathrm{d}\theta = 1$, $[c\theta]_0^1 = 1$). The posterior distribution is then given by

$$L(\theta|8) = \frac{{}^{10}C_8\theta^8(1 - \theta)^2}{\int {}^{10}C_8\theta^8(1 - \theta)^2 \mathrm{d}\theta} \qquad (7.44)$$

The posterior probability distribution can be readily obtained by numerical methods (Appendix C.7.2. I introduce these methods here, because in most cases the distributions are too complex to be solved by analytical means, though the above could be). The posterior distribution is simply the likelihood function scaled so that the cumulative probability sums to 1. The information we have in this equation is the same as we arrive at using the likelihood (Figure 7.6).

Suppose we repeat the experiment and obtain $x=2$, applying the Fisher exact test we find a significant difference between the two trials ($P=0.023$).

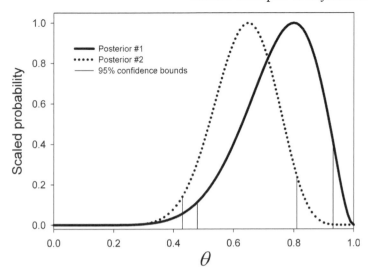

Figure 7.6 Posterior probabilities for the survival probability estimate θ, obtained using a binomial likelihood function. The first posterior was obtained from $x=8$ (8/10) and the second was obtained from the second set of trials in which $x=5$ (5/10) using the first posterior as the new prior.

At this point, we cannot combine the data and we must consider each experiment separately. But suppose in the second trial we obtain $x=5$, which is not significantly different from the first experiment ($P=0.3498$). We can apply Bayes' theorem using the previous posterior distribution as our prior, we have

$$L(\theta|5) = \frac{^{10}C_5\theta^5(1-\theta)^5\,{}^{10}C_8\theta^8(1-\theta)^2}{\int {}^{10}C_5\theta^5(1-\theta)^5\,{}^{10}C_8\theta^8(1-\theta)^2 d\theta} \qquad (7.45)$$
$$\propto \theta^{13}(1-\theta)^7$$

which is simply the scaled likelihood for the combined data (i.e., $L(\theta|13)$ using a uniform prior. Figure 7.6). The difference between the frequentist and Bayesian approaches is that the Bayesian approach focuses upon the posterior probability as a source for decision making whereas the frequentist perspective is typically on confidence intervals and hypothesis testing as the basis for decision making. As I stated above, and wish to strongly emphasize, both approaches have their merits and there is no reason not to analyze the data from both perspectives.

An application of sequential Bayes' estimation: population estimation using mark-recapture

Suppose we wish to estimate the number of animals in a closed population (no immigration or emigration). One method is the method of "mark-recapture," in which an initial sample of size M is taken, the animals marked and released and the population then resampled at random, the second sample of n animals comprising m marked animals. Assuming equal catchability, no births and no deaths between the two samples

$$\frac{M}{N} = \frac{m}{n} \tag{7.46}$$

where N is the population size. Rearranging the above gives the Lincoln or Petersen estimate of population size

$$\hat{N} = \frac{Mn}{m} \tag{7.47}$$

Assuming that sampling can be modeled by the binomial (the hypergeometric distribution is a plausible alternate because sampling is done without replacement), we can write the likelihood of obtaining m marked animals in a sample of n

$$L(\theta|m) \propto {}^{n}C_m \theta^m (1 - \theta)^{n-m} \tag{7.48}$$

where $\theta = M/N$. We are interested in estimating the population size N. Given that M is known, then for any given θ the population size is also given and we can write

$$L(N|m) \propto {}^{n}C_m \theta^m (1 - \theta)^{n-m} \tag{7.49}$$

where $N = M/\theta$. As before, in the absence of any information we assume a uniform distribution of population sizes, with the smallest being no less than the total number of unique animals sampled, $M + n - m$. If further samples are taken, the process can be repeated with the preceding posterior distribution taken as the prior. Unlike the previous example, because the number of marked animals changes between sampling (assuming that all animals captured are marked before being released), the true value of θ changes at each resampling.

Figure 7.7 shows the results of applying this model to simulated data in which the actual population size was 10 000 "animals" and the number sampled and recaptured as given in the figure caption (coding given in Appendix C.7.3). Unlike several "traditional methods," the sequential Bayes' method does not underestimate the population size (Gazey and Staley 1986), and gives comparable 95% confidence limits (Table 7.3). Bayes' estimation has been applied to more complex mark-recapture scenarios (e.g., Casteldine 1981; George and Robert 1992; Madigan and York 1997; Bartolucci *et al.* 2004).

Table 7.3 *Population estimates from mark-recapture methods applied to a simulation model in which population size remained stable at 10000 animals. Modified from Gazey and Staley (1986)*

Estimator	Estimated N	95% interval
Schnabel	8688	5256−25035
Modified Schnabel	8019	4964−16814
Schumacher and Eschmeyer	8498	5596−17652
Bayesian (mean)	10355	5650−18600

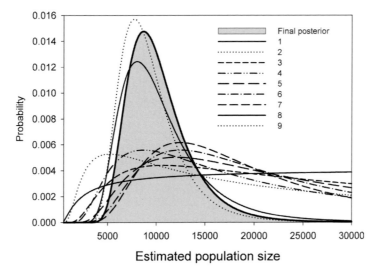

Figure 7.7 Sequential posteriors for a simulated mark-recapture analysis of a population of 10000 "animals." Mark-recapture history shown in table below, coding given in Appendix C.7.3.

Sample	n	M	M
1	34	50	0
2	42	84	1
3	43	125	0
4	40	168	1
5	32	207	0
6	56	239	1
7	42	294	1
8	44	335	4
9	56	375	3
10	44	428	1

Empirical Bayes' estimation: the James–Stein estimator of the mean

There are many circumstances in which we wish to estimate some character of an organism but have a relatively small sample size, which results in a large uncertainty in the estimate. For example, suppose we wish to estimate the mean clutch size of a particular species of bird, say species A, belonging to taxon B. We have a sample for species A giving an empirical mean of $\hat{\mu}_A$ but considerably more data on the taxon as a whole; sufficient in fact that, excluding A, we can state that mean clutch size in taxon B is normally distributed as $N(\mu, \sigma^2)$, which for convenience we shall rescale such that the variance is unity, $N(\mu, 1)$. Assuming that we have no reason to suspect that species A is not a representative species of the taxon, we can use the distribution within the taxon as the prior distribution to modify our assessment of the mean clutch size of species A. The relevant formula has already been derived in the estimation of the mean of a normal distribution (Eqs. (7.10–7.12)): our Bayesian assessment of the mean clutch size of species A, treating the mean as a single datum is

$$\mu_A = \mu + (\hat{\mu}_A - \mu)\frac{\sigma_A^2}{1 + \sigma_A^2}$$

$$= \mu + (\hat{\mu}_A - \mu)\left(1 - \frac{1}{1 + \sigma_A^2}\right) \quad (7.50)$$

We do not know the value of σ_A^2 but an unbiased estimate of $1/(1 + \sigma_A^2)$ is

$$\frac{n - 2}{\sum_{i=1}^{n} (\hat{\mu}_i - \mu_i)^2} \quad (7.51)$$

where μ_i is the mean of the ith species and $\hat{\mu}_i$ is its observed estimate. Substitution of (7.51) in Eq. (7.50) gives

$$\mu_A = \mu + (\hat{\mu}_A - \mu)\left(1 - \frac{n - 2}{\sum_{i=1}^{n} (\hat{\mu}_i - \mu_i)^2}\right) \quad (7.52)$$

The above is an example of a **James–Stein estimator** (for a mathematical discussion of this estimator in a general setting see Efron and Morris 1973: for a more readable account see Efron and Morris 1977). The above estimator assumes that the true means, μ_i are known, which, in general, will not be the case. As an approximation, we can replace μ_i by the grand mean, $\mu = \sum \mu_i/n$, which can be estimated by the empirically derived value $\sum \hat{\mu}_i/n$ (Efron and Morris 1973). This gives the estimate

$$\mu_A = \hat{\mu} + (\hat{\mu}_A - \hat{\mu})\left(1 - \frac{n - 3}{\sum_{i=1}^{n} (\hat{\mu}_i - \hat{\mu})^2}\right) \quad (7.53)$$

where $n-2$ in the numerator has been replaced by $n-3$ to take into account the estimation of one parameter. I have assumed in the above that species A is not included in the overall mean estimate: in fact, we can include all species in the overall mean and perform the estimation on each species independently, though this can be criticized as confounding the prior and posterior probabilities.

Efron and Morris (1973) applied the above approach to the estimation of batting averages of 18 major league baseball players. Such data might represent survival rates among different species of a taxon, or parasitism rates (an example discussed below). To test the efficacy of the maximum likelihood estimator (MLE) and the James–Stein estimator, Efron and Morris determined the batting averages for these players based on the first 45 times at bat ($\hat{\mu}_i$, $i = 1, 2, 3, \ldots, 18$) and compared predictions based on these data with the batting averages achieved in the remainder of the season (roughly 300 more times at bat). The maximum likelihood predictions are the averages calculated from the original data. To determine the James–Stein estimates, the data, being proportions, were first arc-sine transformed, which gave approximately unit variance. Estimates were then made using Eq. (7.53) with $\hat{\mu}$ being estimated using the entire data set. The resulting equation was

$$\mu_i = 0.209\hat{\mu}_i - 2.59 \tag{7.54}$$

Efron and Morris compared the total squared prediction errors $\left(\sum (\hat{\mu}_{i,\,predicted} - \hat{\mu}_{i,\,observed})^2 \right)$ of the two estimators using the transformed values: the MLE the total squared prediction error was 17.56 but for the James–Stein estimator it was only 5.01. However, we might ask if the estimates were in fact any good at all, that is, "Is there a significant correlation between the predicted estimate and the observed value?" Because the James–Stein estimate is a simple coding transformation of the data, the correlation between the observed and predicted values is the same for both estimates, though the estimate of the slope and intercept changes. The correlation in the present case is not significant ($r=0.34$, $P=0.167$) and the two regression equations are

$$Y = -2.68 + 0.18X_{MLE}$$
$$Y = -0.38 + 0.89X_{James-Stein} \tag{7.55}$$

where Y is the actual observed batting averages and X is the predicted. Thus although the correlation is not improved by the James–Stein estimator, and the estimates are still poor predictors of future averages, the actual relationship between observed and predicted is improved in the sense of being closer to a 1:1 ratio: this is shown in Figure 7.8 using the back-transformed values.

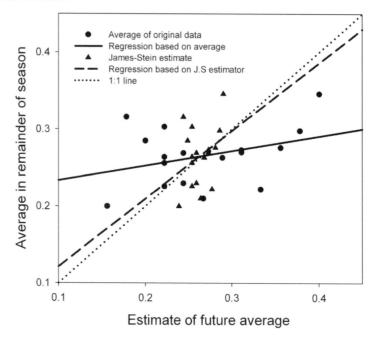

Figure 7.8 Plot of observed batting averages on values predicted using the average of 45 previous values (maximum likelihood estimate) or the James–Stein estimate based on these 45 values. Data from Efron and Morris (1973).

Predictive distributions: the estimation of parasitism rate in cowbirds

A similar approach as above was taken by Link and Hahn (1996) for the estimation of parasitism of nests by cowbirds. The basic data set was observations on the rates of parasitism in 26 host species. Instead of working with the normal distribution, Link and Hahn used the binomial distribution and derived an empirical Bayes' estimate directly from this. Before describing this study, we need to consider **predictive distributions**, which are concerned with predicting observable events rather than parameters, which cannot themselves be observed.

Let the likelihood of observing x be a function of a single parameter θ and be denoted as $L(x|\theta)$ with prior density $P(\theta)$. The posterior density for the n observations $x_1, x_2, \ldots, x_n$ is

$$L(\theta|x_1, x_2, \ldots, x_n) \propto \prod_{i=1}^{n} L(x_i|\theta)P(\theta) \tag{7.56}$$

We wish to predict a new observation y. As previously derived in the discussion on hyperpriors, the predictive density for y is

$$p(y|x_1, x_2, \ldots, x_n) = \int L(y|\theta)L(\theta|x_1, x_2, \ldots, x_n)\, d\theta \tag{7.57}$$

To illustrate the above in the context of the study of Link and Hahn (1996), recall that the posterior likelihood for an estimate of the proportion θ using the natural conjugate prior for the binomial is (Eq. (7.20))

$$L(\theta|x) \propto \frac{\theta^{x+\alpha-1}(1-\theta)^{n-x+\beta-1}}{B(\alpha, \beta)} \tag{7.58}$$

where α, β are the parameters of a beta distribution, x is the observed number of "successes" and n the number of observations. The likelihood of observing y successes in a sample of size N is

$$L(y|\theta) = {}^{N}C_{y}\theta^{y}(1-\theta)^{N-y} \tag{7.59}$$

Using Eq. (7.57), the predictive density of y is

$$p(y|x) = \int_{0}^{1} {}^{N}C_{y}\theta^{y}(1-\theta)^{N-y} \frac{\theta^{x+\alpha-1}(1-\theta)^{n-x-\beta-1}}{B(x+\alpha, n-x+\beta)} d\theta$$

$$= \frac{{}^{N}C_{y}}{B(x+\alpha, n-x+\beta)} \int_{0}^{1} \theta^{x+\alpha-1+y}(1-\theta)^{n-x-\beta-1+N-y} d\theta \tag{7.60}$$

$$= \frac{{}^{N}C_{y}B(x+y+\alpha, N-y+n-x+\beta)}{B(x+\alpha, n-x+\beta)}$$

which is known as the **probability mass function of a beta binomial distribution**. Now we can, without loss of generality, redefine α and β as $\alpha=\alpha+x$ and $\beta=n-x+\beta$ and write the formula in terms of the probability of y successes in N trials

$$p(y; N) = \frac{{}^{N}C_{y}B(y+\alpha, N-y+\beta)}{B(\alpha, \beta)} \tag{7.61}$$

The importance of this distribution in the present case is that it permits the empirical estimation of the two parameters α and β. With respect to parasitism by cowbirds, the above is the probability of selecting a host species at random from the list of N species and finding y parasitized nests. As with the James–Stein estimator discussed in the previous section, we estimate α and β from the data. Let the observed proportion of parasitized nests for species i be $\hat{\theta}_i = x_i/n_i$, where x_i is the number of parasitized nests and n_i is the number of nests examined. Then approximate estimates are

$$\hat{\alpha} = \hat{\theta}\left(\frac{\hat{\theta}(1-\hat{\theta})}{\hat{\sigma}_{\theta}^2}\right)$$

$$\hat{\beta} = (1-\hat{\theta})\left(\frac{\hat{\theta}(1-\hat{\theta})}{\hat{\sigma}_{\theta}^2} - 1\right) \tag{7.62}$$

where $\hat{\theta} = \sum_{i=1}^{N} \hat{\theta}_i/N$ and $\hat{\sigma}_{\hat{\theta}}^2 = \frac{1}{N}\sum_{i=1}^{N} \sqrt{\hat{\theta}_i(1-\hat{\theta}_i)/n_i}$. The posterior probability of observing a parasitism rate in the ith species given an observed rate of $\hat{\theta}_i$ is thus

$$L(\theta_i|\hat{\theta}_i) = \frac{\theta^{x_i+\alpha-1}(1-\theta)^{n_i-x_i+\beta-1}}{B(\alpha+x_i, \beta+n_i-x_i)} \tag{7.63}$$

The expected value of θ_i given $\hat{\theta}_i$, $E(\theta_i|\hat{\theta}_i)$, has the extraordinarily simple formula

$$E(\theta_i|\hat{\theta}_i) = \frac{\hat{\alpha}+x_i}{\hat{\alpha}+\hat{\beta}+n_i}$$

$$= \left(\frac{\hat{\alpha}}{\hat{\alpha}+\hat{\beta}}\right)\left(\frac{\hat{\alpha}+\hat{\beta}}{\hat{\alpha}+\hat{\beta}+n_i}\right) + \hat{\theta}_i\left(\frac{n_i}{\hat{\alpha}+\hat{\beta}+n_i}\right) \tag{7.64}$$

There is relatively little change in estimated parasitism rates, except where sample sizes are very small (Figure 7.9). Unlike the formula given in the last section for the mean, because of n_i, the above formula is a nonlinear function and ranking is not preserved. Perhaps the most dramatic change is that parasitism on the wood thrush, which ranked 8th in the original rankings moves up to the first rank (Figure 7.9), a move that can be attributed to the large sample size for the wood thrush and the low sample sizes of those previously ranked higher.

Hypothesis testing in a Bayesian framework: the probability of extinction from sighting data

Suppose, we have two competing models for which we can attach posterior probabilities $L_0(\theta_0|x)$ and $L_1(\theta_1|x)$. An obvious measure of how well model 1 fares in relation to model 2 is the simple ratio

$$BF = \frac{L_0(\theta_0|x)}{L_1(\theta_1|x)} \tag{7.65}$$

which is known as the **Bayes' factor**. Unlike hypothesis testing in a frequentist framework, there is no generally accepted Bayes' factor for which a hypothesis is accepted or rejected. The principle use of the Bayes' factor is to weigh the evidence in favor of one model or the other. We can write Bayes' factor in terms of competing hypotheses H_0 and H_1 by noting that the probability that H_0 is true given some observation x is by Bayes' theorem

$$L(H_0|x) = \frac{L(x|H_0)P(x)}{L(x|H_0)P(x) + L(x|H_1)(1-P(x))} \tag{7.66}$$

where $P(x)$ is the prior probability for x. The particular value of the Bayes' factor depends on which hypothesis is used as the numerator, it is generally easier to

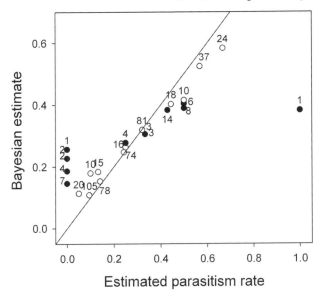

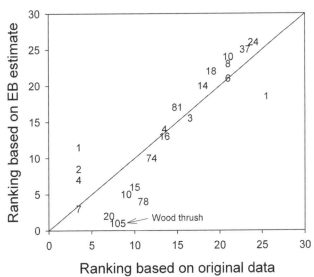

Figure 7.9 Correspondence between the average parasitism rates by cowbirds on other species and the Empirical Bayes' (EB) estimators. Numbers show sample sizes for each species (● < 10, ○ ≥ 10). Data from Link and Hahn (1996).

consider ratios greater than 1, and thus the likelihoods should be arranged accordingly, which involves no loss of generality. What ratio provides strong evidence for the hypothesis in the numerator? Blau and Neely (1975, p. 141) note that a ratio of 10:1 "is ordinarily taken as showing a real difference in plausibility, while 100 denotes strong preferences."

To illustrate this approach, I shall use the Bayesian analysis presented by Solow (1993) of extinction probability estimated from sighting data. The data set to be modeled consists of sightings of the Caribbean monk seal since 1915, the last sighting occurring in 1952. Taking the sighting in 1915 as the beginning of the series gives four sightings (1922, 1932, 1948, and 1952) denoted as x_1, x_2, x_3, $x_4 = \mathbf{x}$. The likelihood of observing $n = 4$ sightings in X years (1915–1992) is a Poisson process with some undetermined rate θ. Labeling the hypothesis that the monk seal is not extinct as H_0, we can write the likelihood for these sightings as

$$
\begin{aligned}
L(\mathbf{x}|H_0) &= \int_0^\infty L(\mathbf{x}|\theta)\mathrm{d}P(\theta) \\
&= \int_0^\infty \theta^n e^{-\theta T} \mathrm{d}P(\theta)
\end{aligned}
$$
(7.67)

Solow used the noninformative prior $\mathrm{d}P(\theta) = \mathrm{d}\theta/\theta$ $(0 \leq \theta \leq \infty)$. The alternate hypothesis, H_1 is that the monk seal became extinct in the period x_n to X (1952–1992). Letting the time at extinction be x_E, we have

$$
L(\mathbf{x}|H_1) = \int_{x_n}^X L(\mathbf{x}|x_E)\mathrm{d}P(x_E)
$$
(7.68)

The likelihood of $\mathbf{x}$ given x_E is equal to

$$
L(\mathbf{x}|x_E) = \int_0^\infty \theta^n e^{-\theta x_E} \mathrm{d}P(\theta)
$$
(7.69)

As with θ, Solow assumed the noninformative prior, $\mathrm{d}P(x_E) = \mathrm{d}x_E/X (0 \leq x_E \leq X)$. Integration of the above gives the Bayes' factor to be

$$
\mathrm{BF} = \frac{n-1}{(X/x_n)^{n-1} - 1} = \frac{3}{(77/37)^3 - 1} = 0.37
$$
(7.70)

Inverting this ratio gives 2.7, a result that can be worded as "the likelihood that the Caribbean monk seal is extinct, given the particular set of observations, is 2.7 times the likelihood that it is not extinct." Does this ratio provide strong evidence for the continued existence of the Caribbean monk? On the basis of the 10:1 criterion, we cannot distinguish between these alternatives. How long would we have to go without sightings until a ratio of 10:1 is achieved? From Eq. (7.69) the answer is 116 years (2031).

There is a "classical" hypothesis test for these two hypotheses, namely that the probability of not making a sighting given that the subject is not extinct is $(x_n/X)^n$, which gives a probability of 0.053 (Solow 1993). This test gives a marginal rejection of the null hypothesis that the Caribbean monk seal is not extinct. Failure to see the seal in 78 years gives $P=0.05$ and a probability of 0.01 is given by the failure to observe a seal in 117 years.

Summary

(1) In the previous chapters of this book, analyses were based upon the probability statement $P(x|\theta)$, which is to say, "the probability of x given parameter θ." Using the maximum likelihood principle, an estimate of θ can be obtained. Hypotheses can be similarly tested under this framework. The Bayesian perspective reverses the statement giving $P(\theta|x)$, which is to say "the probability of θ given x," shifting the focus of enquiry to the probability distribution of the parameter. A Bayesian analysis relies upon Bayes' theorem, $L(\theta|x) \propto L(x|\theta)P(\theta)$, where $L(\theta|x)$ is the posterior (likelihood) probability and $P(\theta)$ is the prior probability.

(2) The principle difficulty in Bayesian analysis is the selection of the prior. In the absence of any information, a noninformative prior is typically selected. For some probability distributions there are prior distributions, called natural conjugate priors, that have the useful property that they have the same parametric form as the proposed likelihood making it relatively easy to analytically derive the posterior distribution. A reduction in complexity of the problem can sometimes be achieved by making assumptions on the relationships between parameters, giving what Deely (2004) refers to as naïve informative priors. The most general approach to the construction of priors involves a hierarchical approach with the prior distribution itself being derived from a Bayesian analysis: these priors are called hyperpriors.

(3) A Bayesian analysis can be applied sequentially, with the preceding posterior used as the new prior. Two examples, presented in this chapter (survival estimates and population estimation by mark-recapture) are based on the analysis of the binomial distribution.

(4) The parameters of the prior distribution may be estimated from the data, the resulting estimate being called an empirical Bayes' estimator. The application of the James–Stein estimator can be done using the available data.

(5) The ratio of two posterior probabilities is known as the Bayes' factor and can be used to distinguish between competing hypotheses. A factor of 10 is considered by many to be good evidence for a difference between the models and a factor of 100 is very strong evidence.

Further reading

Berger, J.O. (1985). *Statistical Decision Theory and Bayesian Analysis.* New York: Springer-Verlag.

Gelman, A., Carlin, J.B., Stern, H.S. and Rubin, D.B. (1995). *Bayesian Data Analysis.* London: Chapman and Hall.

Gotelli, N. J. and Ellison, A.M. (2004). *A Primer of Ecological Genetics.* Sunderland: Sinauer Associates, Inc.

Leonard, T. and Hsu, J.S.J. (2001). *Bayesian Methods.* Cambridge: Cambridge University Press.

Press, S.J. (1989). *Bayesian Statistics: Principles, Models, and Applications.* New York: John Wiley & Sons.

Exercises

(7.1) Snails parasitized by a particular nematode show a characteristic "malaise" which is detectable by a behavioral test in 97% of infected snails. However, other factors can also cause this behavior and it is found that 67% of noninfected snails exhibit the behavior. Assuming that the proportion of noninfected snails in the population is 83%, what is the probability that a snail that exhibits the behavior is infected?

(7.2) A general distribution used for rare occurrences is the Poisson distribution $L(x) = \frac{\theta^x}{e^\theta x!}$, where $L(x)$ is probability of x occurrences given a mean of θ. Assuming a uniform prior for θ between 0 and c what is the posterior probability for θ?

(7.3) From prior information, the value of c in the above question is estimated to be 0.1. In the present sample one occurrence is noted. Using an interval of 0.001, plot the posterior distribution. Does it appear sensible? Repeat the analysis assuming no upper bound for c (Hint: Take the maximum value of θ to be 10).

(7.4) The data below show the mean values for five bird species normalized to a mean of 0 and variance of 1. Each sample consists of the mean of five individuals. Use an empirical Bayesian approach to "better" estimate these values. Are the new estimates an improvement?

Species #	1	2	3	4	5	6	7	8	9	10
True value	−1.88	−1.02	−0.36	−0.13	−0.04	−0.03	0.00	0.01	0.34	1.21
Observed	−3.85	−1.74	1.74	0.32	4.10	−1.47	1.80	2.03	1.81	0.46

(7.5) The coding below was used to generate the data in the preceding question. Use it to explore the consequences of the distribution of species not being normal. Is this an important assumption?

```
# Seed for random number generator
set.seed(0)
nspecies        <- 10                          # Number of species
nsample         <- 5                           # Sample size per species
N               <- nspecies*nsample            # Total sample size
# Create vector of species values
Species.means   <- rnorm(nspecies, mean=0, sd=1)
Species         <- sort(Species.means)   # Sort and store values
# Replicate species means to total sample size
Species.means   <- sort(rep(Species, nsample))
# Set up index for By routine
Index           <- sort(rep(seq(1, nspecies),nsample))
# Creates individual values from normal with mean = Species. means
# and sd=5
Species.values  <- rnorm(N,Species. means,sd=5)
# combine
Data0           <- data.frame(Index, Species.means, Species.values)
# Calculate means per species
Obs             <- unlist(by(Data0, Data0[,1], function (Data0)
                     mean(Data0[,3])))
# sums of squares
sum.var         <- (nspecies-1) *var(Obs)
# Grand mean
GM              <- mean(Obs)
# EB estimates
EB.estimate     <- GM +(Obs-GM) * (1-(nspecies-3)/sum.var)
# Bind for data set
Data            <- cbind(Species, Obs,EB.estimate)
# Print SS using obs means and SS using EB estimates
print(c(sum((Data[,1]-Data[,2])^2), sum((Data[,1]-Data[,3])^2)))
```

List of symbols used in Chapter 7

α, β	Parameters in the beta function
θ	Parameter
σ^2	Variance
$\varphi(x-\theta, \sigma)$	Probability density for normal with mean θ and standard deviation σ
μ	Mean
$\hat{\mu}$	Estimate of μ

ν	Symbol standing for either μ or σ
BF	Bayes' Factor
$B(\alpha, \beta)$	Beta function
$L(\theta)$	Likelihood given parameter θ
$L(x\,|\,\theta)$	Likelihood of x given θ
M	Number of marked animals in first sample
N	Population size
$\hat{N}$	Estimate of N
$P(A \cap B)$	Probability of A and B
$P(A^C)$	Probability of "not A." Upper case P is also used primarily in this chapter to denote the prior probability
$P(x\,|\,\theta)$	Probability of x given θ
^nC_x	$x!/[n!(n-x)!]$
X	$\bar{x}_1 - \bar{x}_2$
$X_{\mathrm{MLE}}, X_{\mathrm{James\text{-}Stein}}$	Estimated values
m	Number of marked animals in samples subsequent to the first
n	Sample size
p	Probability or probability function
p_A	Probability of A
x	Observed value
$\bar{x}$	Observed mean

References

Alatalo, R. V. (1982). Bird species distributions in the Galapagos and other archipelagoes: competition or chance? *Ecology*, **63**, 881–7.

Anderson, M. J. and ter Braak, C. J. F. (2003). Permutation tests for multi-factorial analysis of variance. *Journal of Statistical Computation and Simulation*, **73**, 85–113.

Anderson, T. W. (1958). *An Introduction to Multivariate Statistical Analysis*. New York: Wiley.

Arditi, R. (1989). Avoiding fallacious significance tests in stepwise regression; a Monte Carlo method applied to a meteorological theory for the Canadian lynx cycle. *International Journal of Biometeorology*, **33**, 24–6.

Armbruster, W. S. (1986). Reproductive interactions between sympatric *Dalechampia* species: are natural assemblages "random" or organized? *Ecology*, **67**, 522–33.

Armbruster, W. S., Edwards, M. E. and Debevec, E. M. (1994). Floral character displacement generates assemblage structure of Western Australian triggerplants (Stylidium). *Ecology*, **75**, 315–29.

Arvesen, J. N. and Schmitz, T. H. (1970). Robust procedures for variance component problems using the jackknife. *Biometrics*, **26**, 677–86.

Avise, J. C., Reeb, C. A. and Sanders, N. C. (1987). Geographic population and species differences in mitochondrial DNA of mouthbrooding catfishes (Ariidae) and dmersal spawning toadfishes (Batrachoididae). *Evolution*, **41**, 991–1002.

Bartolucci, F., Mira, A. and Scaccia, L. (2004). Answering two biological questions with latent class model via MCMC applied to capture-recapture data. In M. Di Bacco, G. D'Amore, and F. Scalfari, eds., *Applied Bayesian Statistical Studies in Biology and Medicine*, pp. 7–24. Boston: Kluwer Academic Publishers.

Begin, M. and Roff, D. A. (2004). The effect of temperature and wing morphology on quantitative genetic variation in the cricket, *Gryllus firmus*, with an appendix examining the statistical properties of the Jackknife-MANOVA method of matrix comparison. *Journal of Evolutionary Biology*, **17**, 1255–67.

Bentzen, P., Leggett, W. C. and Brown, C. G. (1988). Length and restriction site heteroplasmy in the mitochondrial DNA of American shad (*Alosa sapidissima*). *Genetics*, **118**, 509–18.

Berger, J. O. (1985). *Statistical Decision Theory and Bayesian Analysis*. New York: Springer-Verlag.

Besag, J. and Clifford, P. (1989). Generalized Monte Carlo significance Tests. *Biometrika*, **76**, 633–42.

Besag, J. and Clifford, P. (1991). Sequential Monte Carlo p-Values. *Biometrika*, **78**, 301–4.

Blau, G. E. and Neely, W. B. (1975). Mathematical model building with an application to determine the distribution of Dursban insecticide added to a simulated ecosystem. In A. Macfadyen, ed., *Advances in Ecological Research*, vol. 9, pp. 133–63. London: Academic Press.

Bliss, C. I. (1935). The calculation of the dosage-mortality curve. *Annals of Applied Biology*, **22**, 220–33.

Bowers, M. A. and Brown, J. H. (1982). Body size and coexistence in desert rodents: chance or community structure? *Ecology*, **63**, 391–400.

Brandl, R. and Topp, W. (1985). Size structure of *Pterostichus* spp. (Carabidae): aspects of competition. *Oikos*, **44**, 234–8.

Breiman, L., Friedman, J. H., Olshen, R. A. and Stone, C. G. (1984). *Classification and Regression Trees*. Belmont, California: Wadsworth International Group.

Buonaccorsi, J. P. and Liebhold, A. M. (1988). Statistical methods for estimating ratios and products in ecological studies. *Environmental Entomology*, **17**, 572–80.

Capone, T. A. and Kushlan, J. A. (1991). Fish community structure in dry-season stream pools. *Ecology*, **72**, 983–92.

Carpenter, J. R. (1999). Test inversion bootstrap confidence intervals. *Journal of the Royal Statistical Society, B*, **61**, 159–172.

Case, T. J., Faaborg, J. and Sidell, R. (1983). The role of body size in the assembly of West Indian bird communities. *Evolution*, **37**, 1062–74.

Castledine, B. J. (1981). A Bayesian analysis of multiple-recapture sampling for a closed population. *Biometrika*, **67**, 197–210.

Chambers, J. M. and Hastie, T. J. (1992). *Statistical Models*. New York: S. Chapman & Hall/CRC.

Chemini, C., Rizzoli, A., Merler, S., Furlanello, C. and Genchi, C. (1997). *Ixodes ricinus* (Acari: Ixodidae) infestation on roe deer (*Capreolus capreolus*). Trentino, Italian Alps. *Parassitologia*, **39**, 59–63.

Cleveland, W. S., Grosse, E. and Shyu, W. M. (1992). Local regression models. In J. M. Chambers, and T. J. Hastie, eds., *Statistical Models in S*. pp. 309–76 London: Chapman and Hall.

Cochran, W. G. (1954). Some methods for strengthening the common χ^2 tests. *Biometrics*, **10**, 417–51.

Cole, B. J. (1981). Overlap, regularity, and flowering phenologies. *American Naturalist*, **117**, 993–7.

Connor, E. F. and Simberloff, D. (1979). The assembly of species communities: chance or competition? *Ecology*, **60**, 1132–40.

Connor, E. F. and Simberloff, D. (1986). Competition, scientific method, and null models in ecology. *American Scientist*, **74**, 155–62.

Conover, W. J., Johnson, M. E. and Johnson, M. M. (1981). A comparative study of tests for homogeneity of variances, with applications to the outer continental shelf bidding data. *Technometrics*, **23**, 351–61.

Cordell, H. J. and Carpenter, J. R. (2000). Bootstrap confidence intervals for relative risk parameters in affected-sib-pair data. *Genetic Epidemiology*, **18**, 157–72.

Cordell, H. J. and Olson, J. M. (1997). Confidence intervals for relative risk estimates obtained using affected-sib-pair data. *Genetic Epidemiology*, **14**, 593–98.

Couteron, P., Seghieri, J. and Chadoeuf, J. (2003). A test for spatial relationships between neighbouring plants in plots of heterogeneous plant density. *Journal of Vegetation Science*, **14**, 163–72.

Cox, D. R. and Hinkley, D. V. (1974). *Theoretical Statistics*. London: Chapman and Hall.

Cox, D. R. and Snell, E. J. (1989). *Analysis of Binary Data*. London: Chapman and Hall.

Crowley, P. H. (1992). Resampling methods for computation-intensive data analysis in ecology and evolution. *Annual Review of Ecology and Systematics*, **23**, 405–48.

Dalaka, A., Kompare, B., Robnik-Sikonja, M. and Sgardelis, S. P. (2000). Modelling the effects of environmental conditions on apparent photosynthesis of *Stipa bromoides* by machine learning tools. *Ecological Modelling*, **129**, 245–57.

Damgaard, C. and Weiner, J. (2000). Describing inequality in plant size or fecundity. *Ecology*, **81**, 1139–42.

Davison, A. C. and Hinkley, D. V. (1999). *Bootstrap Methods and their Applications*. Cambridge: Cambridge University Press.

De'ath, G. and Fabricius, K. E. (2000). Classification and regression trees: a powerful yet simple technique for ecological data analysis. *Ecology*, **81**, 3178–92.

Deely, J. (2004). Comparing two groups or treatments - a Bayesian approach. In M. Di Bacco, G. D'Amore, and F. Scalfari, eds., *Applied Bayesian Statistical Studies in Biology and Medicine*. pp. 89–107. Boston: Kluwer Academic Publishers.

Diamond, J. M. and Gilpin, M. E. (1982). Examination of the "null" model of Connor and Simberloff for species co-occurrences on islands. *Oecologia*, **52**, 64–74.

Dietz, E. J. (1983). Permutation tests for association between two distance measures. *Systematic Zoologist*, **32**, 21–26.

Dillon, R. T. J. (1981). Patterns in the morphology and distribution of gastropods in Oneida lake, New York, detected using computer-generated null hypotheses. *American Naturalist*, **118**, 83–101.

Dixon, P. M., Weiner, J., Mitchell-Olds, T. and Woodley, R. (1987). Bootstrapping the Gini coefficient of inequality. *Ecology*, **68**, 1548–51.

Dobson, A. J. (1983). *An Introduction to Statistical Modelling*. London: Chapman and Hall.

Draper, N. R. and Smith, H. (1981). *Applied Regression Analysis*. New York: John Wiley & Sons.

Dzeroski, S., and Drumm, D. (2003). Using regression trees to identify the habitat preference of the sea cucumber (*Holothuria leucospilota*) on Rarontonga, Cook Islands. *Ecological Modelling*, **170**, 219–26.

Edgington, E. S. (1987). *Randomization Tests*. New York: Marcel Dekker, Inc.

Efron, B. (1979). Computers and the theory of statistics: thinking the unthinkable. *Siam Review*, **2**, 460–80.

Efron, B. (1981). Nonparametric standard errors and confidence intervals. *The Canadian Journal of Statistics*, **9**, 139–72.

Efron, B. (1982). *The Jackknife, the Bootstrap and Other Resampling Plans*. Philadelphia: Society for Industrial and Applied Mathematics.

Efron, B. (1987). Better bootstrap confidence intervals. *Journal of the American Statistical Association*, **82**, 171–200.

Efron, B., Halloran, E. and Holmes, S. (1996). Bootstrap confidence levels for phylogenetic trees. *Proceedings of the National Academy of Science USA* **93**, 13429–34.

Efron, B. and Morris, C. (1973). Stein's estimation rule and its competitors – an empirical Bayes approach. *Journal of the American Statistical Association*, **68**, 117–30.

Efron, B. and Morris, C. (1977). Stein's paradox in statistics. *Scientific American*, 119–28.

Efron, B. and Tibshirani, R. J. (1993). *An Introduction to the Bootstrap*. New York: Chapman and Hall.

Eliason, S. R. (1993). *Maximum likelihood estimation*. Newbury Park: Sage Publications.

Felsenstein, J. (1985). Confidence limits on phylogenies: an approach using the bootstrap. *Evolution*, **39**, 783–91.

Felsenstein, J. and Kishino, H. (1993). Is there something wrong with the bootstrap on phylogenies? A reply to Hillis and Bull. *Systematic Biology*, **42**, 193–200.

Flury, B. (1988). *Common Principal Components and Related Multivariate Models*. New York: Wiley.

Fong, D. W. (1989). Morphological evolution of the amphipod *Gammarus minus* in caves: quantitative genetic analysis. *American Midland Naturalist*, **121**, 361–78.

Garthwaite, P. H. (1996). Confidence intervals from randomization tests. *Biometrics*, **52**, 1387–93.

Gazey, W. J. and Staley, M. J. (1986). Population estimation from mark-recapture experiments using a sequential Bayes algorithm. *Ecology*, **67**, 941–51.

Gelman, A., Carlin, J. B., Stern, H. S. and Rubin, D. B. (1995). *Bayesian Data Analysis*. London: Chapman and Hall.

George, E. I. and Robert, C. P. (1992). Capture-recapture estimation via Gibbs sampling. *Biometrika*, **79**, 677–83.

Gilpen, M. E. and Diamond, J. M. (1982). Factors contributing to non-randomness in species co-occurrences on islands. *Oecologia*, **52**, 75–84.

Gonzalez, L. and Manly, B. F. J. (1998). Analysis of variance by randomization with small data sets. *Environmetrics*, **9**, 53–65.

Gotelli, N. J. (2000). Null model analysis of species co-occurrence patterns. *Ecology*, **81**, 2606–21.

Gotelli, N. J. and Ellison, A. M. (2004). *A Primer of Ecological Genetics*. Sunderland: Sinauer Associates, Inc.

Hanski, I. (1982). Structure in bumblebee communities. *Annales Zoologici Fennici*, **19**, 319–26.

Harvey, P. H., Colwell, R. K., Silvertown, J. W. and May, R. M. (1983). Null models in ecology. *Annual Reviews of Ecology and Systematics*, **14**, 189–211.

Hastie, T. J. and Tibshirani, R. J. (1990). *Generalized Additive Models*. London: Chapman and Hall.

Hellmann, J. J. and Fowler, G. W. (1999). Bias, precision, and accuracy of four measures of species richness. *Ecological Applications*, **9**, 824–34.

Hendrickson, J. A. J. (1981). Community-wide character displacement reexamined. *Evolution*, **35**, 794–810.

Hillis, D. M. and Bull, J. J. (1993). An empirical test of bootstrapping as a method for assessing confidence in phylogenetic analysis. *Systematic Biology*, **42**, 182–92.

Hizer, S. E., Wright, T. M. and Garcia, D. K. (2004). Genetic markers applied in regression tree prediction models. *Animal Genetics*, **35**, 50–2.

Holyoak, M. (1993). The frequency of detection of density dependence in insect orders. *Ecological Entomology*, **18**, 339–47.

Jackson, D. A., Somers, K. M. and Harvey, H. H. (1992). Null models and fish communities evidence of nonrandom patterns. *American Naturalist*, **139**, 930–51.

Jacobsen, N. O. (1984). Estimates of pup production, age at first parturition and natural mortality for hooded seals in the west ice. *Fiskeridirektoratet. Skrifter. Serie Havundersoekelser*, **17**, 483–98.

Jernigan, R. W., Culver, D. C. and Fong, D. W. (1994). The dual role of selection and evolutionary history as reflected in genetic correlations. *Evolution*, **48**, 587–96.

Joern, A. and Lawlor, L. R. (1980). Food and microhabitat utilization by grasshoppers from arid grasslands: comparisons with neutral models. *Ecology*, **61**, 591–9.

Kendall, M. G. and Buckland, W. R. (1982). *A Dictionary of Statistical Terms*. London: Longman Group Ltd.

Kennedy, P. E. and Cade, B. S. (1996). Randomization tests for multiple regression. *Communications in Statistics, Simulation and Computing*, **25**, 923–36.

Kimura, D. K. (1980). Likelihood methods for the von Bertalanffy growth curve. *Fishery Bulletin*, **77**, 765–76.

Knapp, S. J., Bridges, J. W. C. and Yang, M. (1989). Nonparametric confidence estimators for heritability and expected selection response. *Genetics*, **121**, 891–8.

Kochmer, J. P. and Handel, S. N. (1986). Constraints and competition in the evolution of flowering phenology. *Ecological Monographs*, **56**, 303–25.

Krause, A. and Olson, M. (1997). *The Basics of S and S-PLUS*. New York: Springer.

Lawlor, L. R. (1980). Overlap, similarity, and competition coefficients. *Ecology*, **6**, 245–51.

LeBlanc, M. and Crowley, J. (1992). Relative risk trees for censored survival data. *Biometrics*, **48**, 411–25.

Leonard, T. and Hsu, J. S. J. (2001). *Bayesian Methods*. Cambridge: Cambridge University Press.

Link, W. A. and Hahn, D. C. (1996). Empirical Bayes estimation of proportions with application to cowbird parasitism rates. *Ecology*, **77**, 2528–37.

Losos, J. B., Naeem, S. and Colwell, R. K. (1989). Hutchinsonian Ratios and Statistical Power. *Evolution*, **43**, 1820–6.

Lynch, M. and Walsh, B. (1998). *Genetics and Analysis of Quantitative Traits*. Sunderland, MA: Sinauer Associates.

Madigan, D. and York, J. C. (1997). Bayesian methods for estimation of the size of a closed population. *Biometrika*, **84**, 19–31.

Magnuson, J. J., Tonn, W. M., Banerjee, A., Toivonen, J., Sanchez, O. and Rask, M. (1998). Isolation vs. extinction in the assembly of fishes in small northern lakes. *Ecology*, **79**, 2941–56.

Manly, B. F. J. (1991). *Randomization and Monte Carlo Methods in Biology*. London: Chapman and Hall.

Manly, B. F. J. (1993). A review of computer-intensive multivariate methods in ecology. In G. P. Patil, and C. R. Rao, eds., *Multivariate Environmental Statistics*. pp. 307–46. Amsterdam: Elsevier Science Publishers.

Manly, B. F. J. (1995). A note on the analysis of species co-occurrences. *Ecology*, **76**, 1109–15.

Manly, B. F. J. (1997). *Randomization, Bootstrap and Monte Carlo Methods in Biology*. New York: Chapman and Hall.

Marshall, R. J. (2001). The use of classification and regression trees in clinical epidemiology. *Journal of Clinical Epidemiology*, **54**, 603–9.

Meyer, J. S., Ingersoll, C. G., McDonald, L. L. and Boyce, M. S. (1986). Estimating uncertainty in population growth rates: jackknife vs. bootstrap techniques. *Ecology*, **67**, 1156–66.

Miller, R. G. (1974). The jackknife – a review. *Biometrika*, **61**, 1–15.

Mingoti, S. A. and Meeden, G. (1992). Estimating the total number of distinct species using presence and absence data. *Biometrics*, **48**, 863–75.

Mooney, C. Z. and Duval, R. D. (1993). *Bootstrapping: A nonparametric approach to statistical inference*. Newbury Park: Sage Publications.

Mueller, L. D. (1979). A comparison of two methods for making statistical inferences on Nei's measure of genetic distance. *Biometrics*, **35**, 757–63.

Mueller, L. D. and Altenberg, L. (1985). Statistical inference on measures of niche overlap. *Ecology*, **66**, 1204–10.

Negron, J. F. (1998). Probability of infestation and extent of mortality associated with the Douglas-fir beetle in the Colorado Front Range. *Forest Ecology and Management*, **107**, 71–85.

Phillips, P. C. and Arnold, S. J. (1999). Hierarchical comparison of genetic variance-covariance matrices. I. Using the Flury hierarchy. *Evolution*, **53**, 1506–15.

Pleasants, J. M. (1990). Null-model tests for competitive displacement: the fallacy of not focusing on the whole community. *Ecology*, **71**, 1078–84.

Pollard, E. and Lakhani, K. H. (1987). The detection of density-dependence from a series of annual censuses. *Ecology*, **58**, 2046–55.

Potvin, C. and Roff, D. (1996). Permutation tests in ecology: A statistical panacea? *Bulletin of the Ecological Society of America*, **77**, 359

Press, S. J. (1989). *Bayesian Statistics: Principles, Models, and Applications*. New York: John Wiley & Sons.

Quenouille, M. (1949). Approximate tests of correlation in time series. *Journal of the Royal Statistical Society, Series B*, **11**, 18–84.

Ranta, E. (1982a). Animal communities in rock pools. *Annales Zoologi Fennici*, **19**, 337–47.

Ranta, E. (1982b). Species structure of North European bumblebee communities. *Oikos*, **38**, 202–9.

Reichard, S. H. and Hamilton, C. W. (1997). Predicting invasions of woody plants introduced into North America. *Conservation Biology*, **11**, 193–203.

Rejwan, C., Collins, N. C., Brunner, L. J., Shuter, B. J. and Ridgway, M. S. (1999). Tree regression analysis on the nesting habitat of smallmouth bass. *Ecology*, **80**, 341–8.

Ricklefs, R. E., Cochran, D. and Pianka, E. R. (1981). A morphological analysis of the structure of communities of lizards in desert habitats. *Ecology*, **62**, 1474–83.

Roff, D. A. (1997). *Evolutionary Quantitative Genetics*. New York: Chapman and Hall.

Roff, D. A. (2000). The evolution of the G matrix: selection or drift? *Heredity*, **84**, 135–42.

Roff, D. A. (2002). Comparing G matrices: a MANOVA method. *Evolution*, **56**, 1286–91.

Roff, D. A. and Bentzen, P. (1989). The statistical analysis of mitochondrial DNA polymorphisms: χ^2 and the problem of small samples. *Molecular Biological Evolution*, **6**, 539–45.

Roff, D. A. and Bradford, M. J. (1996). Quantitative genetics of the trade-off between fecundity and wing dimorphism in the cricket *Allonemobius socius*. *Heredity*, **76**, 178–85.

Roff, D. A. and Preziosi, R. (1994). The estimation of the genetic correlation: the use of the jackknife. *Heredity*, **73**, 544–8.

Roff, D. A. and Roff, R. J. (2003). Of rats and Maoris: a novel method for the analysis of patterns of extinction in the New Zealand avifauna prior to European contact. *Evolutionary Ecology Research*, **5**, 1–21.

Roff, D. A., Mousseau, T. A. and Howard, D. J. (1999). Variation in genetic architecture of calling song among populations of *Allonemobius socius*, *A. fasciatus* and a hybrid population: drift or selection? *Evolution*, **53**, 216–24.

Roff, D. A., Mousseau, T., Møller, A. P., Lope, F. D. and Saino, N. (2004). Geographic variation in the G matrices of wild populations of the barn swallow. *Heredity*, **93**, 8–14.

Sahai, H. and Ageel, M. I. (2000). *The Analysis of Variance*. Boston: Birkhauser.

Saitoh, T., Bjornstad, O. N. and Stenseth, N. C. (1999). Density dependence in voles and mice: a comparative study. *Ecology*, **80**, 638–50.

Schluter, D. (1988). Estimating the form of natural selection on a quantitative trait. *Evolution*, **42**, 849–61.

Schluter, D. and Nychka, D. (1994). Exploring fitness surfaces. *American Naturalist*, **143**, 597–616.

Schoener, T. W. (1984). Size differences among sympatric, bird-eating hawks: a worldwide survey. In D. R. Strong, D. Simberloff, L. G. Abele, and A. B. Thistle, eds., *Ecological Communities: Conceptual Issues and the Evidence*, pp. 245–81. Princeton, NJ: Princeton University Press.

Segal, M. R. and Bloch, D. A. (1989). A comparison of estimated proportional hazards models and regression trees. *Statistics in Medicine*, **8**, 539–50.

Shackell, N. L., Lemon, R. E. and Roff, D. A. (1988). Song similarity between neighbouring American redstarts (*Setophaga ruticilla*): a statistical analysis. *The Auk*, **105**, 609–15.

Shaw, R. G. (1991). The comparison of quantitative genetic parameters between populations. *Evolution*, **45**, 143–51.

Shaw, R. G. and Mitchell-Olds, T. (1993). ANOVA for unbalanced data: an overview. *Ecology*, **74**, 1638–45.

Silvertown, J. and Wilson, J. B. (1994). Community structure in a desert perennial community. *Ecology*, **75**, 409–17.

Simons, A. M. and Roff, D. A. (1994). The effect of environmental variability on the heritabilities of traits of a field cricket. *Evolution*, **48**, 1637–49.

Skov, F. (1997). Stand and neighbourhood parameters as determinants of plant species richness in a managed forest. *Journal of Vegetation Science*, **8**, 573–8.

Solow, A. R. (1993). Inferring extinction from sighting data. *Ecology*, **74**, 962–4.

Soltis, P. S. and Soltis, D. E. (2003). Applying the bootstrap in phylogeny reconstruction. *Statistical Science*, **18**, 256–67.

Stratoudakis, Y., Gallego, A. and Morrison, J. A. (1998). Spatial distribution of developmental egg ages within a herring *Clupea harengus* spawning ground. *Marine Ecology-Progress Series*, **174**, 27–32.

Strong, D. R. J. (1979). Tests of community-wide character displacement against null hypotheses. *Evolution*, **33**, 897–913.

Strong, D. R. J. (1982). Null hypotheses in ecology. In E. Saarinen, ed., *Conceptual Issues in Ecology*, Dordrecht: D. Reidel, pp. 245–59.

Strong, D. R. J. and Simberloff, D. S. (1981). Straining at gnats and swallowing ratios character displacement. *Evolution*, **35**, 810–12.

Strong, D. R., Simberloff, D., Abele, L. G. and Thistle, A. B. (1984). *Ecological communities: conceptual issues and the evidence.* Princeton, N.J: Princeton University Press.

Stuart, A., Ord, K. and Arnold, S. (1999). Kendall's Advanced Theory of Statistics. In *Classical Inference and the Linear Model.* Vol. 2A. London: Arnold.

ter Braak, C. J. F. (1992). Permutation versus bootstrap significance tests in multiple regression and ANOVA. In G. R. K. –H. Jöckel, W. Sendler, eds., *Bootstrapping and related techniques: proceedings of an International Conference held in Trier, Germany,* June 4–8, 1990, pp. 79–86. New York: Springer-Verlag.

Tibshirani, R. J. (1988). Variance stabilization and the bootstrap. *Biometrika*, **75**, 433–44.

Tokeshi, M. (1986). Resource utilization, overlap and temporal community dynamics: a null model analysis of an epiphytic chironomid community. *The Journal of Animal Ecology*, **55**, 491–506.

Tukey, J. W. (1958). Bias and confidence in not quite large samples. *Annals of Mathematical Statistics*, **29**, 614.

Venables, W. N. and Ripley, B. D. (2002). *Modern Applied Statistics with S*. New York: Springer.

Vitt, L. J., Sartorius, S. S., Avila-Pires, T. C. S., Esposito, M. C. and Miles, D. B. (2000). Niche segregation among sympatric Amazonian teiid lizards. *Oecologia*, **122**, 410–20.

Watters, G. and Deriso, R. (2000). Catches per unit of effort of bigeye tuna: a new analysis with regression trees and simulated annealing. *Inter-American Tropical Tuna Commission Bulletin*, **21**, 531–71.

Willis, J. H., Coyne, J. A. and Kirkpatrick, M. (1991). Can one predict the evolution of quantitative characters without genetics? *Evolution*, **45**, 441–4.

Wilson, J. B. (1987). Methods for detecting non-randomness in species co-occurrences: a contribution. *Oecologia*, **73**, 579–82.

Zhou, X.-H., Gao, S. and Hui Siu, L. (1997). Methods for comparing the means of two independent log-normal samples. *Biometrics*, **53**, 1129–35.

Appendix A

An overview of S-PLUS methods used in this book

Data storage methods

There are three storage methods in S-PLUS pertinent to the programs in this book.

Data frames

These are "matrices" in which columns can be of varying types. For example, one column might be a numeric and another might be a character. They are equivalent to the general data sets in most statistical packages, spreadsheets, graphics packages, etc. Consider the following example data frame, which I shall call "Data," consisting of three columns, labeled X, Y, and GROUP:

X	Y	GROUP
1	7	A
2	9	B
4	10	C
3	2	D

The entries in a data frame can be accessed by several methods. Suppose we wish to access the datum denoted in bold in the above data frame, we can use `Data$Y[2]` or `Data[2,2]`. All of the values in column 2 can be accessed by `Data$Y` or `Data[,2]`. Functions that require data in several formats (e.g., ANOVA where the dependent variable is numeric and the independent variable is categorical) require data frames but those that do not may be able to use either data frames or matrices (e.g., linear regression).

Matrices

A matrix can contain data of only one type (e.g., all numeric). The usual way to access elements of a matrix is to specify their row and column number, thus X[3,5] means the element in matrix X occupying the cell on row 3 and column 5. A vector is simply a matrix consisting of a single row or column.

Lists

A list consists of the concatenation of objects of different classes. Their primary relevance in the programs presented in this book is that the output of statistical functions can be saved in the form of a list. To obtain a relevant variable it is necessary to access the appropriate component of the list. For illustration, consider a one-way analysis of variance using data contained in a data frame named `data.df`. The dependent variable is labeled X (numeric) and treatment is labeled GROUP (factor). The command

```
ANOVA.model <- aov(X~GROUP, data=data.df)
```

produces an ANOVA object from which the ANOVA table can be constructed using the summary command, stored in a new object ANOVA.S

```
ANOVA.S <- summary(ANOVA.model, ssType=3)
```

Typing ANOVA.S produces the output

```
> ANOVA.S
    Type III  Sum of Squares
              Df    Sum of Sq    Mean Sq     F Value     Pr(F)
    GROUP      1     0.18212     0.182116    0.130879    0.7191101
    Residuals 48    66.79133    1.391486
```

Suppose we need the mean squares and degrees of freedom to calculate variance components, we can extract this information from the object ANOVA.S. To determine the names of the list items in ANOVA.S we type names(ANOVA.S), which gives

```
[1] "Df" "Sum of Sq" "Mean Sq" "F Value" "Pr(F)"
```

The first list item is the degrees of freedom "DF", which can be accessed by typing ANOVA.S$Df, producing the response

```
> ANOVA.S$Df
    GROUP    Residuals
      1          4
```

To access the mean squares, we can type `ANOVA.S$"Mean Sq"` or `ANOVA.S$Mean` (the shortened version is permitted, because there is no semantic conflict with other component names). The variance component among groups is given by $(MS_{GROUP} - MS_{Residuals})/25$, where MS denotes means squares. This can be calculated directly in S-PLUS by

```
(ANOVA.S$Mean[1]-ANOVA.S$Mean[2])/25
```

Assigning and comparing values

S-PLUS differs from many other packages in how values are assigned to the variables. To set, for example, the variable X equal to 5, we use `X <- 5`, rather than $X=5$. The equal sign is used to assign data within function calls: thus to generate 10 random normal deviates with a mean of zero and a standard deviation of 1 we use the function `rnorm`, passing the information as `rnorm(n=10, mean=0,sd=1)`.

Some symbols for logical operators are also "non-standard:"

Symbol	Function
<	Less than
>	Greater than
<=	Less than or equal to
>=	Greater than or equal to
==	Equal to
!=	Not equal to

Some examples

(1) Assign a value, 5, to a variable named X1:

```
X1 <- 5
```

(2) Assign three values, 5, 3, and 9, to a vector named Xvector:

```
Xvector <- c(5,3,9)
```

(3) Assign three values, 5, 3, and 9, to a 3×1 (rows by columns) matrix named Xmatrix.1:

```
Xmatrix.1 <- matrix(c(5,3,9),nrow=3,ncol=1)
```

(4)   Assign six values, column 1=5, 3, 9, column 2 = 2, 4, 1 to a 3×2 matrix
        named Xmatrix.2:

```
# Note that rows are filled first
  Xmatrix.2 <- matrix(c(5,3,9,2,4,1),3,2)
```

(5)   Generate 25 random normal deviates with mean zero and standard
        deviation=1 and place in Xnormal:

```
Xnormal <- rnorm(n=25,mean=0,sd=1)
```

(6)   Assign the 8th entry of Xnormal to a variable named y:

```
# Note that positions in datasets are denoted by square brackets
  y <- Xnormal[8]
```

(7)   Assign the value from cell 2, 1 (row, column) of Xmatrix2 to a variable
        named y:

```
y <- Xmatrix.2[2,1]
```

(8)   Generate two sets of 25 random normal deviates, both with zero mean
        and unit SD and assign column-wise to matrix called DATA

```
X1      <- rnorm(n=25,mean=0,sd=1)
X2      <- rnorm(n=25,mean=0,sd=1)
DATA    <- matrix(c(X1,X2),ncol=2)
```

(9)   Generate two sets of 25 random normal deviates, both with zero mean
        and unit SD and assign column-wise to a data frame called DATA.DF.
        Note that the columns will be automatically labeled X1 and X2.

```
X1      <- rnorm(n=25,mean=0,sd=1)
X2      <- rnorm(n=25,mean=0,sd=1)
DATA.DF <- data.frame(X1,X2)
```

(10)   Generate two sets of 25 random normal deviates, both with unit SD but
        different means of 0 and 0.5 (mean= and SD= can be omitted). Assign
        both to data frame called data.df consisting of two columns, the first
        containing the 50 data points and the second column indicating group
        membership (1 or 2).

```
X       <- c(rnorm(25,0.5,1),rnorm(25,0.0,1))
GROUP   <- c(rep(1,times=25),rep(2,times=25))
data.df <- data.frame(X,GROUP)
```

(11)   Simulate the linear regression model $Y = a + bX + \varepsilon$, where $a = 2$, $b = 3$ and ε is a random normal deviate with mean zero and SD$=10$. Use 20 equally spaced predictor values from 1 to 20. Place data in two vectors called X.lin and Y.lin.

```
X.lin   <- seq(from=1, to=20, length=20)   # X values
error   <- rnorm(20,0,10)                   # Vector of errors
Y.lin   <- 2 + 3*X +error                   # Y values
```

Manipulating data

(1)   Calculate the mean of Xnormal:

```
mean(Xnormal)
```

(2)   Calculate the mean of column 2 of Xmatrix.2 (two methods shown)

```
mean(Xmatrix.2[1:3,2])
mean(Xmatrix.2[,2])
```

(3)   Select all values in Xnormal that are greater than zero and assign to y:

```
y <- Xnormal[Xnormal>0]
```

(4)   Find the number of values in Xnormal:

```
length(Xnormal)
```

(5)   Find the number of values in Xnormal that are greater than zero:

```
length(Xnormal[Xnormal>0])
```

(6)   Find the number of values in the 2nd column of Xmatrix.2 that are greater than 1:

```
y <- Xmatrix.2[,2]   # Assign column to y
length(y[y>1])        # Find number > 1
```

(7)   Apply a two-sample t-test to data in DATA.DF, comparing X1 with X2 (alternate methods shown):

```
t.test(DATA.DF$X1,DATA.DF$X2)
t.test(DATA.DF[,1],DATA.DF[,2])
```

Output

```
Standard Two-Sample t-Test
data: DATA.DF[, 1] and DATA.DF[, 2]
t = -1.1368, df = 48, p-value = 0.2613
alternative hypothesis: difference in means is not equal to 0
95 percent confidence interval:
-1.0501346 0.2915413
sample estimates:
mean of x mean of y
-0.100205 0.2790916
```

(8)   Compare groups in data.df using a oneway ANOVA and type 3 sums of
       squares. First convert group membership value in data.df to a factor.
       Note the use of "~" in the ANOVA model description, this is the general
       means of indicating relationships in function formulae.

```
data.df <- convert.col.type(data.df, "GROUP", "factor")
ANOVA.model <- aov(X~GROUP, data=data.df)
summary(ANOVA.model,ssType=3)
```

Output

```
Type III Sum of Squares
```

	Df	Sum of Sq	Mean Sq	F Value	Pr(F)
GROUP	1	0.18212	0.182116	0.130879	0.7191101
Residuals	48	66.79133	1.391486		

(9)   Perform a linear regression analysis using X.lin and Y.lin

```
Lin.Model <- lm(Y.lin~X.lin)
summary(Lin.Model)
```

Output

```
Call: lm(formula = Y ~ X)
Residuals:
```

Min	1Q	Median	3Q	Max
-16.23	-6.459	2.202	3.347	14.41

```
Coefficients:
```

	Value	Std. Error	t value	Pr(>|t|)
(Intercept)	-1.1979	3.8007	-0.3152	0.7562
X.lin	3.0854	0.3173	9.7245	0.0000

```
Residual standard error: 8.182 on 18 degrees of freedom
```

```
Multiple R-Squared: 0.8401
F-statistic: 94.57 on 1 and 18 degrees of freedom, the p-value is
1.37e-008
Correlation of Coefficients:
(Intercept)
X.lin -0.8765
```

(10)   Extract the regression coefficients from the above results (2 methods shown)

```
Lin.Model$coeff
Lin.Model[1]
```

Output

```
(Intercept)     X.lin
-1.197929    3.085367
```

(11)   Extract the slope

```
Lin.Model$coeff[2]
```

Output

```
X.lin
3.085367
```

Appendix B

Brief description of S-PLUS subroutines used in this book

General routines such as "log", "print" etc. are omitted. Arguments set at the default values are not displayed. For a complete description, see the language reference section in S-PLUS.

anova(Model) Compute an ANOVA table for **Model**

anova(Model1, Model2, test="F") Compare **Model1** and **Model2**

aov(formula, data=) Fit an analysis of variance model using **formula** on **data**

as.numeric(x) returns a vector like **x**, but with storage mode "double," if **x** is a simple object of mode "numeric." Otherwise, **as.numeric** returns a numeric object of the same length as **x** and with data resulting from coercing the elements of **x** to mode "numeric."

bootstrap(data=, statistic, B=, trace=F) Performs bootstrap for observations in **data=** using the statistic specified by **statistic**. The number of replicates is **B** and **trace** determines printing of replicate number during computation.

by(X, Indices, function) Split dataset **X** by **Indices** and apply **function** to each part.

c Concatenates objects into a vector or a list (e.g., c(0.5, 0.2)).

cbind() Returns a matrix that is pieced together, column-wise, from several vectors and/or matrices.

ceiling() Creates integers from floating point numbers by going to the next larger integer.

chisq.test(x, correct=F) Chi-square contingency test using **x**. Apply Yates correction if **correct=T**

choose(n,k) Is the binomial coefficients, $n!/(k!(n-k)!)$

contourplot(Z ~ X*Y, data=, at=0, xlab="X", ylab="Y") Contour plot of Z on X, Y. In above example, a single contour line of zero (at=0) is drawn.

convert.col.type(target=X, column.spec=, column.type=) Converts column(s), column.spec, in a 1- or 2-dimensional dataset, **X**, to a particular data type, **column.type**.

cor(X, Y, na.method = "omit") Returns the correlation of a vector or the correlation matrix of a data matrix. Missing values can be omitted or the calculation set to "fail."

dimnames Returns or changes the dimnames attribute of an array.

dnorm(x, mean=0, sd=1) Density for the normal distribution with mean and standard deviation parameters mean and sd.

factor(x) Makes **x** a factor.

fitted(Model) Extract fitted values from **Model**.

floor(x) Converts x to smallest integer $\leq$ x.

gam(formula, data=) Fit a generalized additive model according to **formula** using supplied **data**.

is.random States or changes whether or not a factor is considered random by the varcomp function.

jackknife(data=, statistic) Jackknife the observations given by **data** =, using the statistic specified by **statistic**.

length(X) Returns an integer that describes the length of the object, **X**, e.g., if **X** is a vector or matrix length(X) is the number of elements in **X**.

lines(x, y) Add lines to current plot.

lm(formula, data=) Fit linear regression using observations in **data** and equation given by **formula**.

loess(formula, span=, degree=) Fit a local regression model using **formula** with given **span** and **degree**.

manova(formula, data=) Produces a fit by MANOVA according to **formula** using **data**=.

matrix(data, nrow=, ncol=) Creates a matrix. **data** is a vector containing the data values for the matrix (if a single value, such as 0, is given the matrix is filled with this value). Missing values (NAs) are allowed. **nrow** is the number of rows. ncol is the number of columns.

mean(X) Mean of X (missing values are allowed).

menuTable(varnames=, data=, print.p= F, save.name=) Calls the table function to tabulate variables, **varnames**, in data set designated by **data** =, with the result not printed out (**print.p**=F) and saved as specified by **save.name**=.

ncol(X) Gives the number of columns of matrix **X**.

nrow(X) Gives the number of rows of matrix **X**.

nlmin(F,V) Finds a local minimum of a nonlinear Function, F, using a general quasi-Newton optimizer, given a vector, V, of starting values. Examples: C2.2, C2.4, C2.6, C2.7, C2.8, C2.9.

nlminb(F, V, L, U) Local minimizer for smooth nonlinear functions subject to bound-constrained parameters. F is the function, V a vector of starting values, L the lower bound and U the upper bound. Examples: C2.1, C2.3.

nls(F, data=D, start=list(....)) Fits a nonlinear regression model via least squares. The errors are assumed Gaussian and independent with constant variance. F is the function, D is the data matrix, start is a vector of starting values. Examples: C2.10, C2.11.

numerical.matrix(x) Convert **x** into a numerical matrix.

pchisq(q, df) Quantile for the chi-square distribution.

pf(q, df1, df2) Cumulative probability for the F distribution.

pnorm(x, mean=, sd=) Probability for the normal distribution with mean and standard deviation parameters mean and sd.

pointwise(Model, coverage=0.95) Computes pointwise confidence limits (coverage =) for predictions computed by the function predict.

predict(Model, newdata) Predict values using **newdata** and fitted **Model**.

predict.loess(Model, newdata) Returns the surface values of local regression surfaces and/or standard errors.

prune.tree(Model, best=) Cost-complexity pruning of fitted tree **Model**. The final size of the tree can be specified by **best**.

qnorm(x, mean=, sd=) Quantile for the normal distribution with mean and standard deviation parameters mean and sd.

qt(p, df) Quantile for t distribution for given **p** and **df** degrees of freedom.

rep(X, times =) Replicates the input **X** a certain number of times (or to a certain length, argument not shown).

residuals(Model) Extract residuals from **Model**.

rgamma(n, shape=, rate=) Generate **n** gamma-distributed variables with parameters **shape** and **rate**.

rnorm(n, mean=, sd=) Generates **n** random normal deviates with specified mean and standard deviation.

runif(n, min=, max=) Generates **n** random uniform deviates between **min** and **max**.

sample(X, size=, replace=T) Produces a vector of length **size** of objects randomly chosen from **X**. If **replace=T** sampling is done with replacement.

scatter.smooth(X, Y, span=, degree=) Produces a scatter plot and adds a smooth curve using the loess fitting method.

seq(from=, to=, by=, length=) Creates a vector of evenly spaced numbers. The start, end, spacing, and length of the sequence can be specified.

set.seed(i) Puts the random number generator in a reproducible state, thereby allowing the production of the sequence of random numbers. i is an integer between 0 and 1023.

sort(X)   Sort vector **X** into ascending order.

sum(X)   Returns the number that is the sum of all of the elements of all of the arguments. Missing values may optionally be removed before the computations.

summary   Provides a synopsis of an object. Classes which already have methods for this function include: aov, aovlist, data.frame, factor, gam, glm, lm, loess, mlm, ms, nls, ordered, terms, tree.

supsmu(x, y)   Supersmoothing routine for **x, y**.

tree(formula, data=)   Fit a regression or classification tree to data.

t.test(X, Y)   Performs a two sample *t*-test between numeric vectors **X** and **Y**. One sample, paired *t*-test, or a Welch modified two-sample *t*-test are also possible.

unlist   Returns a vector or a list which is the result of simplifying the recursive structure of the input list.

varcomp(formula, data=, method=)   Estimation of variance components using the model, **formula**, with **data=**, and **method=** (restrictive maximum likelihood in examples in this book).

write(X, file=, ncolumns=, append= T)   Write data **X** to ascii file **file=**, consisting of **ncolumns**. If **append= T** data appended to file, otherwise the file is overwritten.

Appendix C

S-PLUS codes cited in text

S-PLUS codes discussed in the text. Codes are delineated by Chapter and number; e.g., C.2.3. refers to Chapter 2, code set 3. In many cases, part or all of the calculations can be performed via the dialog boxes (which are not available in R). A WORD file containing this appendix is available at http://www.biology.ucr.edu/people/faculty/Roff.html. Comments, suggestions and questions can be sent to Derek.Roff@ucr.edu.

C.2.1 Calculating parameter values for a threshold model

```
# Set up function to calculate negative of the log likelihood (minus
# constants)
# THETA is a vector containing the two parameters to be estimated.
# THETA[1] is p, THETA[2] is the heritability
# r is a vector containing the two values of r
# n is a vector containing the two values of n
    LL <- function(THETA)
    {
# Calculate log likelihood for the initial sample
    L0 <- r[1]*log(THETA[1])+(n[1]-r[1])*log(1-THETA[1])
# Calculate the initial population mean of the liability μ0
    mu0 <- qnorm(THETA[1],0,1)
# Calculate the mean liability of the offspring μ1
    mu1 <- mu0*(1-THETA[2])+THETA[2]*(mu0+dnorm(mu0,0,1)/THETA[1])
# Calculate predicted proportion, p2, of the designated morph in the offspring
    p2 <- pnorm(mu1,0,1)
# Calculate log likelihood (minus constants) for the second sample
    L1 <- r[2]*log(p2)+(n[2]-r[2])*log(1-p2)
```

```
# Return negative of the sum of the two log "likelihoods"
    return (-(L0+L1))
    }
# Main Program
# Set values for r and n
    r <- c(50,68)
    n <- c(100,100)
# Set initial estimates for THETA
    THETA <- c(0.8,0.1)
# Call minimization routine setting lower and upper limits to 0.0001 and
# 0.999, respectively
    min.func <- nlminb(THETA, LL,lower=0.0001, upper=0.9999)
# Print out estimates
    min.func$parameters
```

Output

```
min.func$parameters
0.5000000 0.5861732
```

C.2.2 Estimation of parameters of a simple logistic curve

For an alternative approach using the routine glm see C.2.8.

```
# Data (see Figure 2.5) are in a matrix or data frame called D.
# Col 1 is dose (x), col 2 is n, col 3 is r
  Dose    <- c(1.69,1.72,1.76,1.78,1.81,1.84,1.86,1.88)
  n       <- c(59,60,62,56,63,59,62,60)
  r       <- c(6,13,18,28,52,53,61,60)
  D       <- data.frame(Dose,n,r)
# Define function LL that will calculate the loss function
# b is a vector with the estimates of θ₁ and θ₂
```

```
  LL <- function(b){-sum(D[,3]*(b[1]+b[2]*D[,1])-D[,2]*log(1+exp(b[1]+
  b[2]*D[,1]))))}
  b           <- c(-50, 20)    # Create a vector with initial estimates
  min.func <- nlmin(LL, b)    # Call nonlinear minimizing routine
  min.func$x                  # Print out estimates
```

Output

```
min.func$x
-60.10328  33.93416
```

C.2.3 Locating lower and upper confidence limits for heritability of a threshold trait given offspring data

```
# Set up function to calculate negative of the log likelihood (omitting
# constants)
    LL <- function(h2)
    {
# Calculate the mean liability of the offspring μ₁
    mu1 <- mu0*(1-h2)+h2*Parental.mean
# Calculate predicted proportion, p2, of the designated morph in the offspring
# using library routine pnorm
    p2 <- pnorm(mu1,0,1)
# Calculate log likelihood for the offspring sample using the library routine
# dbinom
    L1 <- log(dbinom(r, n, p2))
# Return negative of the log-likelihood
    return (-L1)
    }
# MAIN PROGRAM
# Set values for r and n, the offspring sample
    r <- 68
    n <- 100
# Set initial proportion and calculate the mean liability
    p <- 0.5
    mu0 <- qnorm(p,0,1)
# Calculate Parental mean
    Parental.mean <- mu0+dnorm(mu0,0,1)/p
# Set initial estimates for h2
    h2 <- 0.5
# Call minimization routine setting lower and upper limits to 0.0001 and
# 0.999, respectively
    min.func <- nlminb(h2, LL, lower=0.0001, upper=0.9999)
# Save estimate
    MLE.h2     <- min.func$parameters
# Calculate Log-Likelihood at MLE
    Global.LL   <- -LL(MLE.h2)
# Create a function to square Diff so that minima are at zero
    Limit <- function(h2){(Global.LL+LL(h2)-0.5*3.841)^2}
# Find lower limit by restricting upper value below MLE.h2
    h2 <- 0.01
```

```
    min.func <- nlminb(h2, Limit, lower=0.0001, upper=0.9999*MLE.h2)
# Save estimate
    Lower.h2 <- min.func$parameters
# Find upper limit by restricting lower value above MLE.h2
    h2 <- 0.99
    min.func <- nlminb(h2, Limit, lower=1.0001*MLE.h2, upper=0.9999)
# Save estimate
    Upper.h2 <- min.func$parameters
# Print out results
    print(c(Lower.h2,MLE.h2,Upper.h2))
```

Output

```
print(c(Lower.h2, MLE.h2, Upper.h2))
0.2685602 0.5861735 0.9097963
```

C.2.4 95% confidence interval for parameters L_{max} and k of the von Bertalanffy equation conditioned on t_0 and variance

```
# Data are in file called D (see Figure 2.3)
# Col 1 is Age, Col 2 is length of females, which is the only sex analyzed here
    Age      <-c(1.0,2.0,3.3,4.3,5.3,6.3,7.3,8.3,9.3,10.3,11.3,12.3,13.3)
    Length <-c(15.4,28.0,41.2,46.2,48.2,50.3,51.8,54.3,57.0,58.9,59,60.9,61.8)
    D        <-data.frame(Age, Length)
# Create function to calculate sums of squares for three variable parameters
    LL <- function(b) {sum(((D[,2]-b[1]*(1-exp(-b[2]*(D[,1]-b[3])))))^2)}
# Calculate parameters for all three parameters
    b          <- c(60, 0.3, -0.1)    # Set initial estimates in vector b
    min.func <- nlmin(LL,b)          # Find minimum sums of squares
    MLE.b    <- min.func$x           # Save estimates
    t0         <- MLE.b[3]            # Set t0 to its MLE value
    n          <- nrow(D)            # Get sample size n
    var        <- LL(min.func$x)/n    # Calculate MLE variance, called var
# Calculate log-likelihood at MLE and subtract 1/2 chi-square value for k=2
    Chi.Contour <- (-n*log(sqrt(2*pi*var))-(1/(2*var))*LL(min.func$x))-
    0.5*(5.991)
# Condition on var and t0
# Create a matrix with values of Lmax and k, the two parameters of interest
# Set number of increments
    Nos.of.inc <- 20
```

```
# Set values of Lmax
    Lmax <- rep(seq(from=58,to=64,length=Nos.of.inc), times=Nos.of.inc)
# Set values of k
    k <- rep(seq(from=0.25,to=0.35,length=Nos.of.inc), times=Nos.of.inc)
    k <- matrix(t(matrix(k,ncol=Nos.of.inc)), ncol=1)
# Place Data in cols 1 and 2 of matrix Results
    Results        <- matrix(0,Nos.of.inc*Nos.of.inc,3)
    Results[,1]    <- Lmax
    Results[,2]    <- k
# Set number of cycles for iteration
    Nreps <- Nos.of.inc*Nos.of.inc
    for (I in 1:Nreps)
    {
# Calculate LL for this combination
    LL.I <- (-n*log(sqrt(2*pi*var))-(1/(2*var))*LL(c(Results[I,1],
    Results[I,2],t0)))
# Subtract Chi.Contour this from value
    Results[I,3] <- Chi.Contour - LL.I
    }
# Now plot contour
    contourplot(Results[,3]~Results[,2]*Results[,1], at=0, xlab="k",
    ylab="Lmax")
```

C.2.5 Output from S-PLUS for von Bertalanffy model fit using dialog box

```
*** Nonlinear Regression Model ***
Formula: LENGTH ~ Lmax * (1 - exp(- k * (AGE - t0)))
Parameters:
                Value        Std. Error         t value
Lmax        61.2333000       1.2141000        50.435300
   k         0.2962530       0.0287412        10.307600
  t0        -0.0572662       0.1753430        -0.326595

Residual standard error: 1.69707 on 10 degrees of freedom

Correlation of Parameter Estimates:
           Lmax            k
  k       -0.843
 t0       -0.544        0.821
```

C.2.6 Estimation of parameters of a simple logistic curve and calculation of the deviance

```
# Data are in a matrix or data frame called D (see C.2.2 for data)
# Col 1 is dose (x), col 2 is n, col 3 is r
# Define function LL that will calculate the loss function
# b is a vector with the estimates of θ₁ and θ₂
```

b is a vector with the estimates of θ_1 and θ_2

```
    LL <- function(b){-sum(D[,3]*(b[1]+b[2]*D[,1])-D[,2]*log(1+exp(b[1]
    +b[2]*D[,1])))}

    b              <- c(-50, 20)       # Create a vector with initial estimates
    min.func       <- nlmin(LL, b)     # Call nonlinear minimizing routine
    b              <- min.func$x       # Save estimates

# Create function to calculate expected frequencies

    Expected <- function(x) {exp(b[1]+b[2]*x)/(1+ exp(b[1]+b[2]*x))}

# Calculate expected frequencies
    Exp.Freq <- Expected(D[,1])
# Create vectors with observed and expected cell numbers
# Add 0.0000001 to observed values to prevent undefined logs

    r.obs                  <- D[,3]+0.0000001
    n.minus.r.obs          <- (D[,2]-D[,3])+0.0000001
    r.exp                  <- Exp.Freq*D[,2]
    n.minus.r.exp          <- (1-Exp.Freq)*D[,2]

# Calculate Deviance
    Deviance <- (r.obs*log(r.obs/r.exp)+n.minus.r.obs*log(n.minus.r.obs/
    n.minus.r.exp))
# Print out estimate and Deviance
    print(c(b, 2*sum(Deviance)))
```

Output
```
print(c(b, 2 * sum(Deviance)))
-60.10328 33.93416 13.63338
```

C.2.7 Comparing a 3-parameter with 2-parameter von Bertalanffy model using nlmin routine

For an alternative approach using the nls routine see C.2.10 and C.2.11. Data are in file called D (see Figure 2.3 and C.2.4 for actual data).

```
# Set up function to calculate sums of squares for 3 parameter model
    LL.3 <- function(b) {sum((D[,2]-b[1]*(1-exp(-b[2]*(D[,1]-b[3]))))^2)}
# Calculate parameters for all three parameters

    b                <- c(60, 0.3, -0.1)       # Initial estimates
    min.func         <- nlmin(LL.3,b)          # Call nlmin
    MLE.b3           <- min.func$x             # Save Estimates
    SS.3             <- LL.3(min.func$x)       # Save Sums of squares

# Set up function to calculate sums of squares for 2 parameter model
    LL.2 <- function(b) {sum((D[,2]-b[1]*(1-exp(-b[2]*D[,1])))^2)}
# Calculate parameters for all two parameters

    b                <- c(60,0.3)              # Set initial values
    min.func         <- nlmin(LL.2,b)          # Call nlmin
    MLE.b2           <- min.func$x             # Save Estimates
    SS.2             <- LL.2(min.func$x)       # Save Sums of squares
    n                <- nrow(D)                # Get sample size n
    F.value          <- (SS.2-SS.3)/(SS.3/(n-3))   # Compute F value
    P                <- 1 - pf(F.value, 1, n-3)    # Compute probability

# Print out results
    MLE.b3
    MLE.b2
    print(c(F.value, P))
```

Output

```
MLE.b3
61.21610737 0.29666467 -0.05492771
MLE.b2
60.9913705 0.3046277
print(c(F.value, P))
0.09273169 0.76697559
```

C.2.8 Comparing one (= constant proportion) and two parameter logistic model

Two alternative methods are given. The first uses the data as shown in Figure 2.5, whereas the second converts the data set to the outcome for each individual (0,1 data) and then uses glm to fit the model and test for significance.

```
# Data are in a matrix or data frame called D. See Figure 2.5 for data.
# Function to calculate LL for 2 parameter model
    LL.2 <- function(b){-sum(D[,3]*(b[1]+b[2]*D[,1])-D[,2]*
    log(1+exp(b[1]+b[2]*D[,1])))}

# Function to calculate LL for 1 parameter model
    LL.1 <- function(b){-sum(D[,3]*(b[1]*D[,1])-D[,2]*log(1+exp(b[1]*
    D[,1])))}

# Function to calculate predicted proportion
    Expected <- function(x){exp(b[1]+b[2]*x)/(1+exp(b[1]+b[2]*x))}

# Function to calculate Deviance
    D.fit.function <- function()
    {
    Exp.Freq        <- Expected(D[,1])              # Expected frequencies
# Add small amount to avoid zeros
    r.obs               <- D[,3]+0.0000001
# Add small amount to avoid zeros
    n.minus.r.obs   <- (D[,2]-D[,3])+0.0000001
    r.exp               <- Exp.Freq*D[,2]
    n.minus.r.exp   <- (1-Exp.Freq)*D[,2]
    return(2*sum(r.obs*log(r.obs/r.exp)+n.minus.r.obs*log(n.minus.r.
    obs/n.minus.r.exp)))
    }

# Calculate stats for 2 parameter model
    b               <- c(-50,20)            # Initial estimates
    min.func        <- nlmin(LL.2,b)        # Call nlmin
    b               <- min.func$x           # Extract final estimates of b
    D.2             <- D.fit.function()     # Calculate deviance

# Calculate stats for 1 parameter model
    b               <- 30                   # Initial estimate
    min.func        <- nlmin(LL.1,b)        # Call nlmin
    b[1]            <- min.func$x           # Estimate
# Set b[2]=0 before using deviance function
    b[2]            <- 0
    D.1             <- D.fit.function()     # Calculate deviance
    n               <- nrow(D)              # Calculate n
    D.value         <- (D.1-D.2)V           # Calculate "D"
```

```
P              <- 1-pchisq(D.value,1)    # Compute probability
print(c(D.value, P))                     # Print out results
```

Output

```
print(c(D.value, P))
273.5865 0.0000
```

Alternative approach using glm and 0, 1 data

```
Successes   <- D[,2]-D[,3]  # Calculate a vector giving n-r for each row
Outcome     <- NULL         # Set up vector to take 0,1 data
# Iterate over rows making a vector with appropriate numbers of 0s and 1s
    for (i in 1:nrow(D))Outcome <- c(Outcome,rep(0,Successes[i]),
    rep(1,D[i,3]))
# Create a vector of doses for each individual
    Dose    <- rep(D[,1],D[,2])
    D       <- data.frame(Dose,Outcome)    # Convert to dataframe
# Use glm to fit logistic model
    Model <- glm(Outcome~Dose,data=D, family=binomial)
# Print out summary stats
    summary(Model)
```

Output

```
Call: glm(formula = Outcome ~ Dose, family = binomial, data = D)
Deviance Residuals:
         Min          1Q      Median           3Q          Max
    -2.474745   -0.5696173   0.2217815    0.4297788    2.373283

Coefficients:
                  Value    Std. Error    t value
  (Intercept)   -60.10328    5.163413   -11.64022
        Dose    33.93416     2.902441    11.69159

(Dispersion Parameter for Binomial family taken to be 1)

    Null Deviance: 645.441 on 480 degrees of freedom

Residual Deviance: 374.872 on 479 degrees of freedom

Number of Fisher Scoring Iterations: 5

Correlation of Coefficients:
        (Intercept)
Dose     -0.9996823
```

C.2.9 Comparing two von Bertalanffy growth curves (males and females) using nlmin function

For an alternative approach using the nls routine see C.2.10 and C.2.11. Data are in file called D. See Figure 2.9 for data: the actual form of the data matrix is three columns, col 1 = age, col 2 = female length, col 3 = male length.

```
Age       <- c(1.0,2.0,3.3,4.3,5.3,6.3,7.3,8.3,9.3,10.3,11.3,12.3,13.3)
Female.L <- c(15.4,28.0,41.2,46.2,48.2,50.3,51.8,54.3,57.0,58.9,59,60.9,61.8)
Male.L   <- c(15.4,26.9,42.2,44.6,47.6,49.7,50.9,52.3,54.8,56.4,55.9,
             57.0,56.0)
D         <- data.frame(Age, Female.L, Male.L)
# Set up function to calculate sums of squares for 3 parameter von Bertalanffy
# model
LL.3     <- function(b) {sum((length-b[1]*(1-exp(-b[2]*(Age-b[3]))))^2)}

# Calculate parameters for Females, which are in col 2
# Create age and length vectors for function
length    <- D[,2]
Age       <- D[,1]
b         <- c(60,0.3,-0.1)
min.func  <- nlmin(LL.3,b)
b.Female  <- min.func$x       # Save MLE estimates
SS.F      <- LL.3(min.func$x) # Store sums of squares at MLE

# Calculate parameters for Males, which are in col 3
length    <- D[,3]
b         <- c(60, 0.3, -0.1)
min.func  <- nlmin(LL.3,b)
b.Male    <- min.func$x       # Save MLE estimates
SS.M      <- LL.3(min.func$x) # Store sums of squares at MLE
SS        <- SS.F+SS.M        # Calculate total SS

# Now Calculate parameters assuming no difference
Age       <- c(D[,1],D[,1])
length    <- c(D[,2],D[,3])
b         <- c(60, 0.3, −0.1)
min.func  <- nlmin(LL.3,b)
b.both    <- min.func$x       # Save MLE estimates
SS.FM     <- LL.3(min.func$x)  # Store sums of squares at MLE

n         <- nrow(D)          # Get sample size n
```

```
F.value    <- ((SS.FM-SS)/(6-3))/(SS/(2*n-6))    # Compute F value
P          <- 1 - pf(F.value,3,2*n-6)            # Compute probability
```

```
# Print out results
    print(c(b.Female,b.Male))
    b.both
    print(c(F.value, P))
```

Output

```
print(c(b.Female, b.Male))
61.21610737 0.29666467 -0.05492771 56.47060501 0.37243270 0.14277098
b.both
58.70733949 0.33307408 0.04991537
print(c(F.value, P))
4.7376264 0.0117886
```

C.2.10 Comparing two von Bertalanffy growth curves (males and females) using nls function

Two different methods are presented below. The first extracts the necessary information to perform the test, whereas the second uses the S-PLUS function anova to perform the model comparison.

Data are in file called D (see Figure 2.9). The data are stored in three columns: col 1 = AGE, col 2 = LENGTH, col 3 = Sex (0 = female, 1 = male).

```
# Create data set
    Age      <- c(1.0,2.0,3.3,4.3,5.3,6.3,7.3,8.3,9.3,10.3,11.3,12.3,13.3)
    Female.L <- c(15.4,28.0,41.2,46.2,48.2,50.3,51.8,54.3,57.0,58.9,59,60.9,61.8)
    Male.L   <- c(15.4,26.9,42.2,44.6,47.6,49.7,50.9,52.3,54.8,56.4,55.9,57.0,56.0)
    LENGTH   <- c(Female.L,Male.L)
    AGE      <- rep(Age, times=2)
    n        <- length(Age)
    Sex      <- c(rep(0, times=n), rep(1, times=n))
    D        <- data.frame(AGE, LENGTH, Sex)
# Fit von Bertalanffy function using dummy variable Sex (=0 for female,
# 1 for male)

    Model <- nls(LENGTH~(b1+b4*Sex)*(1-exp(-(b2+b5*Sex)*(AGE-
    (b3+b6*Sex)))), data=D, start=list(b1=60,b2=0.1,b3=0.1,b4=0,b5=0,
    b6=0))
    b.separate   <- Model$parameters       # Save parameter values
    SS           <- sum(Model$residuals^2) # Save residual sums of squares
```

```
# Fit model assuming no difference between males and females
    Model <- nls(LENGTH~b1*(1-exp(-b2*(AGE- b3))), data=D,
    start=list(b1=60,b2=0.1,b3=0.1))
    b.both    <- Model$parameters              # Save parameter values
# Save residual sums of squares
    SS.FM     <- sum(Model$residuals^2)
    n         <- nrow(D)                       # Get sample size n
    F.value   <- ((SS.FM-SS)/(6-3))/(SS/(n-6))  # Compute F value
    P         <- 1 - pf(F.value, 3, n-6)      # Compute probability

# Print out results
    b.separate
    b.both
    print(c(F.value, P))
```

Output

```
b.separate
          b1          b2          b3          b4          b5          b6
    61.21511   0.2966925  -0.05478805  -4.745226  0.07577554  0.1976811
b.both
          b1          b2          b3
    58.70635   0.3331111   0.05006876
print(c(F.value, P))
4.7376262 0.0117886
```

Alternative coding using the anova function to compare models (results are identical)

```
# Fit von Bertalanffy function using dummy variable Sex (=0 for female,
    # 1 for male)
    Model.1 <- nls(LENGTH~ (b1+b4*Sex)*(1-exp((b2+b5*Sex)*(AGE-(b3
    +b6*Sex)))), data=D, start=list(b1=60,b2=0.1,b3=0.1,b4=0,b5=0,b6=0))
# Fit model assuming no difference between males and females
    Model.2 <- nls(LENGTH~b1*(1-exp(-b2*(AGE- b3))), data=D,
    start=list(b1=60,b2=0.1,b3=0.1))
# Compare models
    anova(Model.1,Model.2)
```

Output

```
Analysis of Variance Table
Response: LENGTH
```

	Terms	Resid. Df
1	(b1+b4*Sex)*(1-exp(-(b2+b5*Sex)*(AGE-(b3+b6*Sex))))	20
2	b1*(1-exp(-b2*(AGE-b3)))	23

	RSS	Test	Df	Sum of Sq	F Value	Pr(F)
1	49.50852	1				
2	84.69146	2	-3	-35.18293	4.737626	0.0117886

C.2.11 Comparing von Bertalanffy growth curves with respect to θ_3 $(=t_0)$, assuming differences between the sexes in other parameters

Two different methods are presented below. The first extracts the necessary information to perform the test, whereas the second uses the S-PLUS function anova to perform the model comparison.

Data are in file called D (see Figure 2.9 and C.2.10). The data are stored in three columns:

col 1 = AGE, col 2 = LENGTH, col 3 = Sex (0 = female, 1 = male).

```
# Fit von Bertalanffy function using dummy variable Sex (=0 for female,
    # 1 for male)
    Model <- nls(LENGTH~(b1+b4*Sex)*(1-exp(-(b2+b5*Sex)*(AGE-(b3+
    b6*Sex)))), data=D, start=list(b1=60,b2=0.1,b3=0.1,b4=0,b5=0,b6=0))

    b.separate    <- Model$parameters        # Save parameter values
    SS            <- sum(Model$residuals^2)   # Save residual sums of squares

# Fit model assuming no difference between males and females in t0
    Model <- nls(LENGTH~(b1+b4*Sex)*(1-exp(-(b2+b5*Sex)*(AGE))), data=D,
    start=list(b1=60,b2=0.1,b4=0,b5=0))

    b.both     <- Model$parameters        # Save parameter values
# Save residual sums of squares
    SS.FM      <- sum(Model$residuals^2)
    n          <- nrow(D)                  # Get sample size n
    F.value    <- ((SS.FM-SS)/(6-4))/(SS/(n-6))  # Compute F value
    P          <- 1 - pf(F.value, 2, n-6)       # Compute probability

# Print out results
    b.separate
    b.both
    print(c(F.value, P))
```

Output

b.separate

b1	b2	b3	b4	b5	b6
61.21511	0.2966925	-0.05478805	-4.745226	0.07577554	0.1976811

b.both

b1	b2	b4	b5
60.99053	0.3046447	-4.083912	0.04157999

print(c(F.value, P))

0.4794997 0.6260291

Alternative coding using the anova function to compare models (results are identical)

```
# Fit von Bertalanffy function using dummy variable Sex (=0 for female,
    # 1 for male)
    Model.1 <- nls(LENGTH~(b1+b4*Sex)*(1-exp(-(b2+b5*Sex)*(AGE-(b3+
    b6*Sex))))), data=D, start=list(b1=60,b2=0.1,b3=0.1,b4=0,b5=0,b6=0))
# Fit model assuming no difference between males and females in t0
    Model.2 <- nls(LENGTH~(b1+b4*Sex)*(1-exp(-(b2+b5*Sex)*(AGE))),
    data=D, start=list(b1=60,b2=0.1,b4=0,b5=0))
# Compare models
    anova(Model.1,Model.2)
```

Output

Analysis of Variance Table

Response: LENGTH

Terms	Resid Df
1 (b1+b4*Sex)*(1-exp(-(b2+b5*Sex)*(AGE-(b3+b6*Sex))))	20
2 (b1+b4*Sex)*(1-exp(-(b2+b5*Sex)*(AGE)))	22

	RSS	Test	Df	Sum of Sq	F Value	Pr(F)
1	49.50852		1			
2	22 51.88246	2	-2	-2.373932	0.4794997	0.6260291

C.3.1 An analysis of the Jackknife analysis of the variance using 1000 replicated data sets of 10 random normal, $N(0, 1)$, observations per data set

Note that the S-PLUS jackknife routine does not compute the jackknife estimate (only the mean of the delete-one estimates, called replicates). The pseudovalues can be calculated from the components of the routine.

```
set.seed(0)                        # Initialize random number seed
nreps    <- 1000                   # Set number of replicates
Output  <- matrix(0,nreps,4)       # Create matrix for output
n        <- 10                     # Sample size
Y        <- rnorm(nreps*n,0,1)     # Create nreps*n random normal values
X        <- matrix(Y,10,nreps)     # Put values into matrix with nreps columns
Tvalue  <- qt(0.975,9)             # Find appropriate t value for 95%

for (I in 1:nreps)                            # Iterate over nreps
{
x      <- X[,I]                               # Place data into vector
                                   # Jackknife data in Ith column of X
Out    <- jackknife(data=x, var(x))
Pseudovalues    <- n*Out$obs-(n-1)*Out$replicates  # Calculate pseudovalues
                                   # Store jackknife mean of variance
Output[I,1]      <- mean(Pseudovalues)
Output[I,2]      <- sqrt(var(Pseudovalues)/n)       # Store jackknife SE
}

Output[,3]   <- Output[,1] + Tvalue*Output[,2]    # Generate upper 95% limit
Output[,4]   <- Output[,1] - Tvalue*Output[,2]    # Generate lower 95% limit
N.upper      <- length(Output[Output[,3]<1,3])    # Find number that are < 1
N.lower      <- length(Output[Output[,4]>1,4])    # Find number that are > 1
                                                  # Print out results
print(c(mean(Output[,1]),N.upper/nreps, N.lower/nreps))
```

Output
```
print(c(mean(Output[, 1]), N.upper/nreps, N.lower/nreps))
1.009637 0.133000 0.001000
```

C.3.2 Testing for differences between the variances of two data sets using the jackknife

In this simulation, the null hypothesis is true, both samples of 10 observations per sample being drawn from a normal population with mean zero and unit standard deviation, $N(0,1)$.

```
# Initialize random number seed
set.seed(0)
nreps    <- 1000                        # Set number of replicates
n        <- 10                          # Sample size per population
```

```
Output   <- matrix(0,nreps,1)                       # Create matrix for output
# Put values in matrix with nreps cols
X        <- matrix(rnorm(nreps*n,0,1),n,nreps)
# Put values in matrix with nreps cols
Y        <- matrix(rnorm(nreps*n,0,1),n,nreps)

for (I in 1:nreps)                                  # Iterate over nreps
{
x                 <- X[,I]                          # Place data into x vector
Out.x             <- jackknife(data=x, var(x))      # Jackknife data in x
# Calculate pseudovalues
x.pseudovalues  <- n*Out.x$observed-(n-1)*Out.x$replicates
y                 <- Y[,I]                          # Place data into y vector
Out.y             <- jackknife(data=y, var(y))      # Jackknife data in y
# Calculate pseudovalues
y.pseudovalues  <- n*Out.y$observed-(n-1)*Out.y$replicates
# Perform t test
Ttest             <- t.test(x.pseudovalues, y.pseudovalues)
Output[I]         <- Ttest$p.value                  # Store probability
}
# Calculate proportion
p                 <- length(Output [Output[,]<0.05])/nreps
p
```

Output

```
p
0.035
```

C.3.3 Estimating the pseudovalues for the genetic variance – covariance matrix for full-sib data

The data consists of a set of full sib families (Data), generated using the program in C.3.4. The program uses MANOVA to compute the genetic variance – covariance matrix, here called Gmatrix. An identifier code is appended so that several data sets can be combined and the combined data set then analyzed using the MANOVA approach outlined in the text. Pseudovalues are stored in a matrix called Pseudovalues.

```
# The following two constants are set according to the particular data set
   Nos.of.Traits      <- 2                        # Number of traits
      k                <- 10                       # Number in each family
```

```
# Note also that the group designator is labeled FAMILY
# Number of (co)variances
    Nos.of.Covariances <- (Nos.of.Traits^2 +Nos.of.Traits)/2

# Create matrix for Genetic (co)variances
    Gmatrix             <- matrix(0,nrow=Nos.of.Traits,
                                ncol=Nos.of.Traits)

# Create matrix for G matrix estimated with one less group
    Gmatrix.minus.i     <- matrix(0,nrow=Nos.of.Traits,
                                ncol=Nos.of.Traits)

# Create matrix to take pseudovalues
    Gmatrix.Pseudo      <- matrix(0,nrow=Nos.of.Traits,
                                ncol=Nos.of.Traits)

# Set value of identifier variable
    Identifier          <- 1
# The Group designator is here labeled FAMILY. Ensure that FAMILY columns are
    # set as character
    Data                <- convert.col.type(target = Trait.df,
                                column.spec = list("FAMILY"),
                                column.type = "character")

# Extract Family codes (= Group designator)
    menuTable(varnames = "FAMILY", data = Data, print.p=F,
    save.name ="Family.Sizes")

# Set up matrices for storage of genetic pseudovalues
    Nos.of.Families <- length(Family.Sizes$FAMILY)
# Add extra col added for population identifier
    Pseudovalues    <- matrix(0,Nos.of.Families,
                                Nos.of.Covariances+1)

# Do MANOVA on entire data set
    Data.manova     <- manova(cbind(Trait.X,Trait.Y) ~ FAMILY, data=Data)
    Data.ms         <- summary(Data.manova) # Calculate sums of squares

# Calculate variance components as given on page 43 of Roff (1997)
    MS.AF           <- data.frame(Data.ms$SS[1])/(Nos.of.Families -1)
    MS.AP           <- data.frame(Data.ms$SS[2])/
                                (Total- Nos.of.Families)
# Genetic variance-covariance matrix
    Gmatrix         <- 2*(MS.AF-MS.AP)/k
```

```
# Now Jackknife the data
    for (i in 1:Nos.of.Families)
    {
# Name of ith family
    Ith.Family       <- Family.Sizes[i,1]
# Delete family from data
    Data.minus.one  <- Data[Data$FAMILY!=Ith.Family,]
# Do MANOVA on reduced data set
    Data.manova      <- manova(cbind(Trait.X,Trait.Y)
                          ~ FAMILY, data=Data.minus.one)
    Data.ms          <- summary(Data.manova)
    MS.AF            <- data.frame(Data.ms$SS[1])/(Nos.of.Families-2)
    MS.AP            <- data.frame(Data.ms$SS[2])/(Total-Nos.of.Families)
# Genetic variance-covariance matrix
    Gmatrix.minus.i <- 2*(MS.AF-MS.AP)/k
    Gmatrix.Pseudo  <- Gmatrix*Nos.of.Families - Gmatrix.minus.i*
                          (Nos.of.Families-1)

# Add pseudovalues to output matrix using diagonal elements and one set of
# covariances
    Jtrait                           <- 0
    for (Irow in 1:Nos.of.Traits) {
    for (Icol in Irow:Nos.of.Traits){
    Jtrait                           <- Jtrait+1
    Pseudovalues[i,Jtrait]           <- Gmatrix.Pseudo[Irow,Icol]}}
    Pseudovalues[i,Jtrait+1]         <- Identifier
    }
# Output information
    print(c(Identifier, Nos.of.Families,k))
    print(Gmatrix)          # Observed G matrix
```

Output

```
print(c(Identifier, Nos.of.Families, k))
          1       100       10
print(Gmatrix)
numeric matrix: 2 rows, 2 columns.
```

	FAMILY.Trait.X	FAMILY.Trait.Y
Trait.X	0.7981739	0.3943665
Trait.Y	0.3943665	0.4970988

C.3.4 Simulating data sets in which two characters are inherited according to a quantitative genetic algorithm

Assuming a full-sib pedigree structure, two trait values can be generated by the equations

$$X_{i,j} = a_{x,i}\sqrt{\frac{1}{2}h_x^2} + b_{x,i,j}\sqrt{1 - \frac{1}{2}h_x^2} \tag{1}$$

$$Y_{i,j} = r_A a_{x,i}\sqrt{\frac{1}{2}h_x^2} + a_{y,i}\sqrt{\frac{1}{2}(1 - r_A^2)h_y^2} + r_E b_{x,i,j}\sqrt{1 - \frac{1}{2}h_x^2} + b_{y,i,j}\sqrt{(1 - \frac{1}{2}h_y^2)(1 - r_E^2)} \tag{2}$$

where $X_{i,j}$, $Y_{i,j}$ are the trait values for the jth individual in family i, $a_{x,i}$, $a_{y,i}$ are random standard normal values, $N(0, 1)$, common to the ith family; $b_{x,i,j}, b_{y,i,j}$ are random standard normal values, $N(0, 1)$ of the jth individual from the ith family; h_X^2, h_Y^2 are the heritabilities of each trait; r_A is the genetic correlation between the two traits; r_E is the "environmental" correlation between the traits, calculated from the phenotypic correlation, r_P, as $r_E = (r_P - \frac{1}{2}h_X h_Y)/\sqrt{(1 - \frac{1}{2}h_X^2)(1 - \frac{1}{2}h_Y^2)}$. If only a single trait is required, use only the equation for trait X. For more details see Simons and Roff (1994).

```
# Create a population with two traits
# Nos.of Families = Number of families
# k = Number in family
# H2X, H2Y = Heritabilities
# rg, rp, re = Genetic, phenotypic and environmental correlations
# Total = Total number of individuals
# Initialize random number seed
    set.seed(0)
Nos.of.Families <- 100
    k        <- 10
    H2X      <- 0.5
    H2Y      <- 0.5
    rg       <- 0.5
    rp       <- 0.5
    re       <- (rp-0.5*rg*sqrt(H2X*H2Y))/sqrt((1-0.5*H2X)*(1-0.5*H2Y))
    if(re > 1) re <- 0.99     # Rounding error can generate this
    Total <- k* Nos.of.Families
```

```
# Set up random normal values
    a.xi    <- matrix(rep(rnorm(Nos.of.Families,0,1),k),Total,1)
    b.xij   <- matrix(rnorm(Total,0,1),Total,1)
    a.yi    <- matrix(rep(rnorm(Nos.of.Families,0,1),k),Total,1)
    b.yij   <- matrix(rnorm(Total,0,1),Total,1)
# Compute required constants
    Tx1     <- sqrt(0.5*H2X)
    Tx2     <- sqrt((1-.5*H2X))
    Ty1     <- rg*sqrt(0.5*H2Y)
    Ty2     <- sqrt(0.5*H2Y*(1-rg^2))
    Ty3     <- re*sqrt(1-0.5*H2Y)
    Ty4     <- sqrt((1-re^2)*(1-0.5*H2Y))
# Generate vector of family codes
    FAMILY  <-matrix(rep(seq(1,Nos.of.Families,1),k),Total,1)
# Generate values of traits X and Y and store in Trait
    Trait.X  <- a.xi*Tx1 + b.xij*Tx2
    Trait.Y  <- a.xi*Ty1 + a.yi*Ty2 + b.xij*Ty3 + b.yij*Ty4
# Convert Trait to a data.frame for analysis
    Trait.df    <- data.frame(FAMILY, Trait.X, Trait.Y)
# Convert Family codes into characters
    Trait.df    <- convert.col.type(target = Trait.df, column.spec = list
                    ("FAMILY"), column.type = "character")
```

C.3.5 Coding to estimate heritability for a full sib design using the routine "jackknife"

Data are in a file called Data. To illustrate the structure, suppose that there are 10 families with 4 individuals per family, with families labeled "1", "2", "3", ..., "10".

Cols with family codes					Cols with data for each family				
1	1	1	1	1	0.7	0.5	0.6	0.3	0.1
2	2	2	2	2	0.2	0.5	0.9	0.3	0.2
.	.	.	.	.	.	.	.	.	.
10	10	10	10	10	0.1	0.6	0.4	0.8	0.9

```
# Function to convert data into two column format with Family code in column 1
# and data in col 2 and calculate h2
H2.estimator <- function(d)
```

```
{
# d is in block format First convert it to two column format
# Find number of rows. This is necessary because of jackknife routine
    Nos.of.rows    <- nrow(d)
    Nos.of.cols    <- ncol(d)
# Set up constants setting range for variables
    n1    <- Nos.of.cols/2
    n2    <- n1+1
    n3    <- n1*2
# Set up 2 column matrix to take data
    D    <- matrix(0,Nos.of.rows*n1,2)
# Now pass data to matrix D
    D[,1]    <- d[,1:n1]
    D[,2]    <- d[,n2:n3]
# Convert Data to a data.frame for analysis
    D.df    <- data.frame(D)
# Convert Family codes into factors
    D.df    <- convert.col.type(target = D.df, column.spec = list("D.1"),
    column.type = "factor")
# Make Family a random effect for varcomp procedure
    is.random(D.df)    <- c(T,F)
# Call varcomp to estimate variance components
    Model    <- varcomp(D.2~D.1, data=D.df, method="reml")
# Calculate heritability
    h2.reml    <- (2*Model$variances[1])/sum(Model$variances)
    return(h2.reml)
}
# Call jackknife routine
    H2.jack    <- jackknife(data=Data, H2.estimator)
    n    <- nrow(Data)    # Find number of rows
# Calculate pseudovalues
    Pseudovalues    <- n*H2.jack$obs-(n-1) *H2.jack$replicates
    print(c(mean(Pseudovalues), sqrt(var(Pseudovalues)/n)))
```

C.3.6 Generation of data following the von Bertalanffy function

These lines generate a dataframe "Data" with three columns. Column 1 contains an identifier for each individual, column 2 contains the ages, and column 3 contains the simulated size at that age.

```
# Generate data
    set.seed(1)                          # Set seed for random number generator
# Create 5 age groups with 5 individuals in each
    Age    <- rep(seq(1:5),5)
    n      <- length(Age)                # Find total number of individuals
# Create a vector with individual names
    Ind    <- seq(1:n)
# Create vector of random normal errors N(0,10)
    Error <- rnorm(n,0,10)
# Generate length at age AGE
    Y      <- 100*(1-exp(-1*Age))+Error
    Data   <- data.frame(Ind,Age,Y)   # Bind 3 vectors, convert to dataframe
```

C.3.7 Jackknife estimation of parameter values for von Bertalanffy data generated by coding shown in C.3.6

Two methods are shown. The first uses explicit coding whereas the second uses the S-PLUS jackknife routine. Note that the S-PLUS jackknife routine does not compute the jackknife estimate (only the mean of the delete-one estimates, called replicates). The pseudovalues can be calculated from the components of the routine.

```
# Estimate parameter values by least squares using all data
    Out                 <- nls(Y~b1*(1-exp(-b2*Age)), data=Data,
                           start=list(b1=50, b2=.5))
    B.obs               <- Out$parameters    # Store parameter estimates
# Create matrix for storage of pseudovalues
    Pseudovalues  <- matrix(0,n,2)
# Jackknife the data
    for (i in 1:n)
    {
    Data.minus.one   <- Data[Data$Ind!=i,]     # Data set minus one individual
# Estimate parameter values by least squares using the reduceddata
    Out.pseudo          <- nls(Y~b1*(1-exp(-b2*Age)), data=Data.minus.one,
                           start=list(b1=50, b2=.5))
    B                   <- Out.pseudo$parameters  # Store parameter estimates
    Pseudovalues[i,] <- n*B.obs-(n-1)*B         # Calculate pseudovalue
    }
# Print out statistics for MLE estimators
    summary(Out)
```

```
# Print means and SEs for jackknife estimates
    print(c(mean(Pseudovalues[,1]), sqrt(var(Pseudovalues[,1])/n))) # b1
    print(c(mean(Pseudovalues[,2]), sqrt(var(Pseudovalues[,2])/n))) # b2
```

Output

```
summary(Out)
Formula: Y ~ b1 * (1 - exp(- b2 * Age))
Parameters:
numeric matrix: 2 rows, 3 columns.
        Value    Std. Error    t value
b1   96.77800    2.58364    37.45810
b2    1.11909    0.13374     8.36766
Residual standard error: 8.4837 on 23 degrees of freedom
Correlation of Parameter Estimates:
        b1
b2  -0.693
print(c(mean(Pseudovalues[, 1]), sqrt(var(Pseudovalues[, 1])/n)))
96.620969    2.871182
print(c(mean(Pseudovalues[, 2]), sqrt(var(Pseudovalues[, 2])/n)))
1.1034496    0.1563441
```

Alternative method using the jackknife routine

```
    PSEUDOVALUES <- matrix(0,n,2) # Create matrix to take pseudovalues
# Use jackknife routine to calculate required elements
    Out.jack <- jackknife(data =Data, nls(Y~b1*(1-exp(-b2*Age)),
    data=Data, start=list(b1=50, b2=.5))$parameters)
# Calculate pseudovalues
    for (i in 1:2) PSEUDOVALUES[,i] <- n*Out.jack$obs[i]-(n-1)*Out.jack
    $replicates[,i]
# Output results
    print(c(mean(PSEUDOVALUES[,1]), sqrt(var(PSEUDOVALUES[,1])/n),
    mean(PSEUDOVALUES[,2]), sqrt(var(PSEUDOVALUES[,2])/n)))
```

C.3.8 Analysis of parameter estimation of the von Bertalanffy function using the jackknife and MLE methods

```
# Generate data
    set.seed(1)                    # Set seed for random number generator
```

```
# Create 5 age groups with 5 individuals in each
    Age     <- rep(seq(1:5),5)
    n       <- length(Age)        # Find total number of individuals
    Ind     <- seq(1:n)           # Create a vector with individual names
    nreps   <- 1000                # Set number of replicates
    Output <- matrix(0, nreps,8)  # Create a matrix to store output
    for (irep in 1:nreps)         # Iterate across replicates
    {
    Error   <- rnorm(n,0,10)# Create vector of random normal errors N(0,10)
    Y       <- 100*(1-exp(-1*Age))+Error    # Generate length at age AGE
# Bind 3 vectors, convert to dataframe
    Data    <- data.frame(cbind(Ind,Age,Y))
# Jackknife estimation
    Pseudovalues <- matrix(0,n,2)   # Create matrix to take pseudovalues
# Use jackknife routine to calculate required elements
    Out.jack   <- jackknife(data=Data, nls(Y~b1*(1-exp(-b2*Age)),
    data=Data, start=list(b1=50, b2=.5))$parameters)
# Calculate pseudovalues
    for (i in 1:2)Pseudovalues[,i]   <- n*Out.jack$obs[i]-(n-1)*Out.jack
    $replicates[,i]
# Store means and SEs of the jackknife estimates
    Output[irep,1:2] <- c(mean(Pseudovalues[,1]),
    sqrt(var(Pseudovalues[,1])/n))
    Output[irep,3:4] <- c(mean(Pseudovalues[,2]),
    sqrt(var(Pseudovalues[,2])/n))
# Store means and MLE standard errors
# Here is one possible way to calculate SE
    D   <- matrix(Out$R,2,2)
    SE <-sqrt((sum(Out$residuals^2)/(n-2))*solve(t(D)%*%D))
# Here is an alternate method using summary(Out)
    x                   <- summary(Out)
    SE                  <- x$parameters[,2]
    Output[irep,5:8] <-c(B.obs[1],SE[1],B.obs[2],SE[2])
}
# Calculate coverage probabilities
# over is the proportion that lie above the upper confidence limit
# under is the proportion that lie below the lower confidence limit
# coverage is the proportion lying within the confidence limits
# M is the matrix containing the data
```

```
# Theta is the true value of the parameter
    CL.stats  <- function(M,I,t.value,Theta,n.cols,n.reps)
{
    over       <- length(M[(M[,I]+t.value*M[,I+1])<Theta,])/n.cols
    under      <- length(M[(M[,I]-t.value*M[,I+1])>Theta,])/n.cols
    coverage  <- (n.reps-(over+under))/n.reps
    return(c(mean(M[,I]),over/n.reps,under/n.reps,coverage))
}
# Print out stats. Results shown in Table 3.6
    TV <- qt(0.975, n-1)                        # Get t value
    CL.stats(Output,1,TV,100,8,nreps)  # Jackknife estimate of Lmax (=100)
    CL.stats(Output,3,TV,1,8,nreps)    # Jackknife estimate of k (=1)
    CL.stats(Output,5,TV,100,8,nreps)  # MLE estimate of Lmax
    CL.stats(Output,7,TV,1,8,nreps)    # MLE estimate of k
```

C.3.9 Generation of random data sets from the von Bertalanffy function using either a statistical model or bootstrapping an observed data set

Random generation using statistical model $Y = \theta_1\left(1 - e^{-\theta_2 * \text{Age}}\right) + \varepsilon$, where $\theta_1 = 100$, $\theta_2 = 1$, Age varies from 1 to 5, and ε is N(0, 1).

```
# Random generation using a particular statistical model
# set seed for random number generator
    set.seed(1)
# Generate 25 integer ages between 1 and 5 from uniform probability
# distribution
    Age    <- ceiling(runif(25, 0,5))
# Generate 25 random normal variables N(0,10)
    error <- rnorm(25,0,10)
    Y       <- 100*(1-exp(-1*Age))+error    # Generate lengths at age
# Concatenate to make a single file
    Data   <- data.frame(Age,Y)
```

Random generation using bootstrap approach. Suppose the observed data are in file Data

```
# Random generation using a bootstrap approach
# Observed data are in data file called Data.
# For simplicity I use the file generated above
```

```
# Generation using splus subroutine sample
   Data.random    <- sample(Data, replace=T)
```

The above generates a sample equal in size to the original. For different sized samples it is easiest to do it by separate columns, using the same set of random numbers

```
# Generate 100 bootstrap samples from data set Data
   set.seed(1)              # set seed for random number generator
   Age.sample     <- sample(Data[,1],100,replace=T)
   set.seed(1)              # reset seed for same set of random numbers
   Y.sample       <- sample(Data[,2],100,replace=T)
   Data.random    <- cbind(Age.sample, Y.sample)
```

C.4.1 Coding to generate 30 random normal values, generate 1000 bootstrap values and determine basic statistics

```
   set.seed(0)          # Set seed for random number generator
   x <- rnorm(30,0,1)   # Generate sample of 30 data points
# Call bootstrap routine, using routine mean to generate statistic "mean"
   boot.x <- bootstrap(x, mean, B=1000, trace=F)
# Output stats
   summary(boot.x)
# Calculate bias-corrected estimate
   Bias.corrected.estimate <-2*boot.x$observed-boot.x$estimate[2]
   Bias.corrected.estimate    # Print out estimate
```

Output

```
Number of Replications: 1000

Summary Statistics:
     Observed    Bias     Mean      SE
mean  -0.06844  -0.0022  -0.07064  0.186

Empirical Percentiles:
        2.5%      5%       95%      97.5%
mean   -0.4257  -0.376   0.2398   0.2971

BCa Confidence Limits:
        2.5%      5%       95%      97.5%
mean   -0.4179  -0.3541  0.2423   0.3105
```

```
Bias.corrected.estimate
         Mean
mean  -0.06623903
```

Notes: Observed = sample mean = $\hat{\theta}$

Bias = $\theta^* - \hat{\theta}$

Mean = θ^*

SE = $\sqrt{\frac{1}{999}\sum_{i=1}^{1000}(\theta_i^* - \theta^*)^2}$

BCa = Accelerated bias-corrected percentile method

C.4.2 Coding to generate 500 samples of size 30 from N(0,1) to test bootstrap method of estimating the mean

```
set.seed(0)              # Set seed for random number generator
nreps<- 500              # Set number of replicates
Out<- matrix(0,nreps,8)  # Create matrix to store output
for (i in 1:nreps)       # Iterate over replications
{
# Generate sample of 30 data points
x        <- rnorm(30,0,1)
boot.x  <- bootstrap(x,mean,B=250, trace=F)  # Call bootstrap routine
y        <- summary(boot.x)                    # Generate stats
# Store stats
# Bootstrap estimate of mean
Boot   <- as.numeric(unlist(boot.x$estimate[2]))
# Bootstrap estimate of SE
SE     <- as.numeric(unlist(boot.x$estimate[3]))
Out[i,1]    <- Boot-1.96*SE      # Lower 2.5% using SE
Out[i,2]    <- Boot+1.96*SE      # Upper 2.5% (97.5%) using SE
Out[i,3]    <- y$limits.emp[1]  # Lower 2.5% using empirical percentile
Out[i,4]    <- y$limits.emp[4]  # Upper 2.5% (97.5%)using Emp percentile
Out[i,5]    <- y$limits.bca[1]  # Lower 2.5% using BCa bootstrap
Out[i,6]    <- y$limits.bca[4]  # Upper 2.5% (97.5%) using BCa bootstrap
Out[i,7]    <- Boot              # Bootstrap estimate
Out[i,8]    <- as.numeric(2*unlist(boot.x$observed))-
as.numeric(unlist(boot.x$estimate[2]))     # Bias-corrected bootstrap
}
p1 <- length(Out[Out[,1]>0,])/(8*nreps)# L confidence limit excludes zero
p2 <- length(Out[Out[,2]<0,])/(8*nreps)# U confidence limit excludes zero
p3 <- length(Out[Out[,3]>0,])/(8*nreps)# L confidence limit excludes zero
```

```
p4 <- length(Out[Out[,4]<0,])/(8*nreps) # U confidence limit excludes zero
p5 <- length(Out[Out[,5]>0,])/(8*nreps) # L confidence limit excludes zero
p6 <- length(Out[Out[,6]<0,])/(8*nreps) # U confidence limit excludes zero

SE.Prob          <- p1+p2 # Overall confidence limit for SE method
Percentile.Prob <- p3+p4 # Overall confidence limit for percentile method
BCa.prob         <- p5+p6 # Overall confidence limit for BCa method

# One sample t test for difference from zero
t.test(Out[,7])
t.test(Out[,8])

print(c("SE method", p1,p2, SE.Prob))
print(c("E P method", p3,p4, Percentile.Prob))
print(c("BCa method", p5,p6, BCa.prob))
```

Output

```
One-sample t-Test
data: Out[, 7]
t = 0.8956,   df = 499,   p-value = 0.3709
One-sample  t-Test
data:    Out[, 8]
t = 0.8441,   df = 499,   p-value = 0.399
```

Method	Lower P	Upper P	Overall P
"SE method"	"0.032"	"0.02"	"0.052"
"EP method"	"0.04"	"0.022"	"0.062"
"BCa method"	"0.038"	"0.022"	"0.06"

C.4.3 Coding to bootstrap the Gini coefficient of inequality

Note that the vector z must be inside the function because the accelerated bootstrap routine uses the jackknife and hence the size of vector z is not constant.

```
set.seed(0)             # Set seed for random number generator
x  <- runif(25,0,19)  # Generate 25 data points from uniform distribution
# Function to calculate Gini coefficient
Gini   <- function(d)
{
g <- sort(d)
# Because of jackknife in BCa method it is necessary to have the following
# two lines within the function
```

```
n    <- length(g)   # Number of observations
z    <- seq(1:n)    # Generate vector of integers from 1 to n
return(2*sum(z*g)/(n^2*mean(g))-(n+1)/n) # Gini coefficient
}
boot.x   <- bootstrap(x,Gini,B=1000, trace=F)   # Call bootstrap routine
summary(boot.x)                                  # Generate stats
```

Output

```
Number of Replications: 1000

Summary Statistics:
     Observed        Bias      Mean        SE
Gini   0.2873   -0.005916   0.2814   0.03863

Empirical Percentiles:
numeric   matrix:   1 row,   4 columns.
          2.5%       5%      95%     97.5%
Gini   0.2125   0.222   0.3465   0.3579
```

C.4.4 Coding to generate data for linear regression with normal or gamma distribution of errors and estimate parameters using least squares, jackknife, and bootstrap

```
set.seed(0) # Set random number seed
# Function is y = 0.0 + 0.2*x
# Set up dataframe for data
    n                    <- 300                    # Sample size
    Data                 <- matrix(0,n,2)          # Create matrix
    dimnames(Data)       <- list(NULL,c("x","y"))  # Set up column names
    Data                 <- data.frame(Data)
# Generate regression data with x = 1,2,3 … 10 with 30 points for each x
    Data$x       <- rep(seq(1,10), times=30)
# Generate error terms with mean zero using a normal or gamma function
# Alternate error terms. 2nd used here
    # error   <- rgamma(n,shape=2,rate=2)-1    # gamma error with mean zero
      error   <- rnorm(n,0,0.5)                # normal distributed error
# Now generate y
    Data$y   <- 0+0.2*Data$x+error
# Fit model
    fit.lm <- lm(y~x,Data)
    summary(fit.lm)
```

```
# Generate jackknife estimates
    Jack.lm    <- jackknife(Data,coef(lm(y~x,Data)))
    summary(Jack.lm)
# Generate bootstrap estimates
    Boot.lm    <- bootstrap(Data, coef(lm(y~x,Data)), B=100, trace=F)
    summary(Boot.lm)
```

Output

```
> summary(fit.lm)
Call: lm(formula = y ~ x, data = Data)
Residuals:
   Min     1Q   Median    3Q     Max
  -1.61  -0.345  0.0192  0.368  1.51

Coefficients:
              Value  Std. Error  t value  Pr(>|t|)
(Intercept)  -0.009    0.067      -0.140    0.889
         x    0.199    0.011      18.568    0.000
Residual standard error: 0.533 on 298 degrees of freedom
Multiple R-Squared: 0.536
F-statistic: 345 on 1 and 298 degrees of freedom, the p-value is 0
Correlation of Coefficients:
     (Intercept)
 x    -0.886
> # Generate jackknife estimates
Jack.lm    <- jackknife(Data, coef(lm(y ~ x, Data)))
> summary(Jack.lm)
Call:
jackknife(data = Data, statistic = coef(lm(y ~ x, Data)))
Number of Replications: 300

Summary Statistics:
              Observed       Bias       Mean      SE
(Intercept)   -0.0093    0.00011905  -0.0093   0.06681
         x     0.1990   -0.00004197   0.1990   0.01061

Empirical Percentiles:
numeric matrix:   2 rows,   4 columns.
                  2.5%       5%        95%        97.5%
(Intercept)   -0.01835  -0.01508  -0.003727  -0.0003836
         x     0.19768   0.19797   0.200091   0.2003331
```

```
Correlation of Replicates:
              (Intercept)          x
(Intercept)    1.0000    -0.8872
        x     -0.8872     1.0000
> # Generate bootstrap estimates
Boot.lm   <- bootstrap(Data, coef(lm(y ~ x, Data)), B = 100, trace = F)
> summary(Boot.lm)
Call:
bootstrap(data = Data, statistic = coef(lm(y ~ x, Data)), B = 100, trace = F)

Number of Replications: 100

Summary Statistics:
              Observed        Bias       Mean        SE
(Intercept)   -0.0093    -0.008702    -0.0180    0.06622
        x      0.1990     0.002514     0.2015    0.01011

Empirical Percentiles:
numeric matrix:   2 rows,    4 columns.
                 2.5%          5%         95%      97.5%
(Intercept)    -0.1433    -0.1345    0.09437    0.1201
        x       0.1803     0.1838    0.21747    0.2201
BCa Confidence Limits:
numeric matrix:   2 rows,    4 columns.
                 2.5%          5%         95%      97.5%
(Intercept)    -0.1384    -0.1241    0.1166     0.1238
        x       0.1774     0.1785    0.2134     0.2174

Correlation of Replicates:
              (Intercept)          x
(Intercept)    1.0000    -0.8471
        x     -0.8471     1.0000
```

C.4.5 Coding to simulate 1000 linear regression data sets, analyze by least squares and test for 95% coverage

```
    set.seed(0)    # Set random number seed
# Function is y = 0.0 + 0.2*x
# Set up dataframe for data
    n                 <- 30                    # Sample size
    nreps             <- 1000                  # set up iterations
    Est               <- matrix(0,nreps,4)     # Output matrix for estimates
    Total.rows        <- n*nreps               # Total number of rows
```

```
    Data              <- matrix(0,Total.rows,3)
# Column names
    dimnames(Data)  <- list(NULL,c("x","y","Index"))
    Data              <- data.frame(Data)      # Convert to data frame
# Generate index
    Data$Index     <- rep(seq(1,nreps), times=n)
    Data$Index     <- sort(Data$Index)
# Generate regression data with x evenly distributed between 1 and 10
    Data$x     <- rep(seq(1,10), times=3*nreps)    # 3 data points per x value
# Generate error terms with mean zero using a normal or gamma function
    error        <-rgamma(Total.rows,shape=2,rate=2)-1
# error          <-rnorm(Total.rows,0,0.5)
# Now generate y
    Data$y     <- 0+0.2*Data$x+error
# Fit model using Index to split data set
Output  <- by(Data, Data$Index, function(Data)summary(lm(y~x,data=Data)))
Output  <- unlist(Output)      # Unlist for storage
# Assign values of Estimates to matrix Est
# Estimates start in positions 37,38,39,40 and then the next are +60 places
    J     <- 37-60
    for (i in 1:nreps)
    {
    J     <- J+60
    Est[i,1]    <- Output[J]      # Intercept
    Est[i,2]    <- Output[J+1]    # slope
    Est[i,3]    <- Output[J+2]    # SE Intercept
    Est[i,4]    <- Output[J+3]    # SE slope
    }
# Calculate coverage
    Est    <- data.frame(Est)            # Convert to data frame
# Convert cell entries from character to numeric
    Est    <- convert.col.type(target=Est,column.spec= "@ALL", column.type =
    "double")
# Calculate number that do not include true value
# Number of UC < 0
    Upper.intercept  <- nrow(Est[Est[,1]+ 2.048*Est[,3]<0,])
# Number of LC > 0
    Lower.intercept  <- nrow(Est[Est[,1]- 2.048*Est[,3]>0,])
# Number of UC < 0
    Upper.slope      <- nrow(Est[Est[,2]+ 2.048*Est[,4]<0.2,])
```

```
# Number of LC > 0
    Lower.slope        <- nrow(Est[Est[,2]- 2.048*Est[,4]>0.2,])
# Coverage
    1-sum(Upper.intercept+Lower.intercept)/nreps
    1-sum(Upper.slope+Lower.slope)/nreps
```

C.4.6 Coding to generate von Bertalanffy growth data and fit model by bootstrapping least squares estimates of parameters

```
# Generate data
    set.seed(1)                          # Set seed for random number generator
# Create 5 age groups with 5 individuals in each
    Age        <- rep(seq(1:5),5)
    n          <- length(Age)            # Find total number of individuals
# Create vector of random normal errors N(0,10)
    Error    <- rnorm(n,0,10)
    Y<- 100*(1-exp(-1*Age))+Error        # Generate length at age AGE
    Data     <- data.frame(Age,Y)        # Bind 2 vectors, convert to dataframe

# Set up function to Estimate parameter values by least squares
    VonBert   <- function(D)
    {
    Out   <- nls(Y~b1*(1-exp(-b2*Age)), data=D, start=list(b1=50, b2=.5))
    Ts    <- matrix(c(Out$parameters[1],Out$parameters[2]))
    # store parameters
    return(Ts)    # Return parameter values
    }
    Boot.Bert     <- bootstrap(Data, VonBert, B=1000, trace=F)
    summary(Boot.Bert)      # Output bootstrap results
```

Output

```
Number of Replications: 1000

Summary Statistics:
          Observed      Bias     Mean       SE
VonBert1.1   96.778   0.05060  96.829   2.7435
VonBert2.1    1.119   0.01665   1.136   0.1549

Empirical Percentiles:
             2.5%       5%      95%     97.5%
VonBert1.1   91.79   92.7469  101.635  102.746
VonBert2.1    0.88    0.9099    1.415    1.474
```

BCa Confidence Limits:

	2.5%	5%	95%	97.5%
VonBert1.1	92.0940	92.8333	101.822	102.903
VonBert2.1	0.8802	0.9115	1.417	1.477

Correlation of Replicates:

	VonBert1.1	VonBert2.1
VonBert1.1	1.000	-0.723
VonBert2.1	-0.723	1.000

C.4.7 Coding to generate multiple samples of the von Bertalanffy growth function and fit by bootstrap

```
# Initiate random number generator
   set.seed(1)
# Set up function to generate data set
   Growth.Data <- function()
   {
# Create 5 age groups with 5 individuals in each
   Age           <- rep(seq(1:5),times=5)
# Number of individuals in a sample
   n             <- 25
# Find total number of individuals
   Total.rows    <- length(Age)
# Create vector of random normal errors N(0,10)
   Error         <- rnorm(Total.rows,0,10)
   Y             <- 100*(1-exp(-1*Age))+Error  # Generate length at age AGE
# Bind 2 vectors, convert to dataframe
   D             <- data.frame(Age,Y)
   return(D)
   }
# Set up function to Estimate parameter values by least squares
   VonBert <- function(D)
   {
   Out    <- nls(Y~b1*(1-exp(-b2*Age)), data=D, start=list(b1=50, b2=.5))
# store parameters
   Ts     <- matrix(c(Out$parameters[1], Out$parameters[2]))
   return(Ts)                               # Return parameter values
   }
```

```
# Set up Bootstrap function
    Bootstrap.VonBert <- function(I)
    {
    D    <- Growth.Data()                    # Call routine to create growth data
# Call bootstrap routine
    b    <- bootstrap(D, VonBert, B=100, trace=F)
# Extract estimates and SEs
    Est  <- c(unlist(b$estimate[2]), unlist(b$estimate[3]))
    print(c(I,Est))    # Print output as simulation proceeds
    return(Est)        # Return estimates and SEs
    }
# Do Nreps runs passing output to text file Data.txt
    nreps <- 100
    for (Ith.rep in 1:nreps)
    {
    X <- Bootstrap.VonBert(Ith.rep)              # Do bootstrap and store in Out
# For safety write to text file
    write(t(X), file="Data.txt", ncolumns=4, append=T)
```

Coding to analyze Data.txt

```
# Read in data from text file called Data.txt
    Est <- read.table("Data.txt", row.names=NULL, header=F)
# Calculate number that do not include true value
    nreps                 <- nrow(Est)              # Number of replicates
# Number of UC < 100
    Upper.Theta.1         <- nrow(Est[Est[,1] +2.069*Est[,3]<100,])
# Number of LC > 100
    Lower.Theta.1         <- nrow(Est[Est[,1]- 2.069*Est[,3]>100,])
# Number of UC < 1
    Upper.Theta.2         <- nrow(Est[Est[,2]+ 2.069*Est[,4]<1,])
# Number of LC > 1
    Lower.Theta.2         <- nrow(Est[Est[,2]- 2.069*Est[,4]>1,])
# Coverage
    P.Theta.1 <- 1-(Upper.Theta.1+Lower.Theta.1)/nreps
    P.Theta.2 <- 1-(Upper.Theta.2+Lower.Theta.2)/nreps
    print(c(nreps,P.Theta.1,P.Theta.2))# Output results
```

Output

```
781.0000 0.9321 0.9398
```

C.5.1 A randomization test of a difference between two means

For an alternative approach see C.5.2.

```
# Generate two normally distributed sets of data
# Initialize the random number generator
    set.seed(20)
    n              <- 10                  # Set the number per sample
    M              <- 2*n                 # Total sample
    Data           <- rnorm(M,0,1)        # Generate M random normal deviates
# Create group indices
    Group.Index    <- matrix(c(rep(1,n), rep(2,n)),M,1)
# Calculate observed average absolute difference between the two groups
    Obs.abs.diff <- abs(mean(Data[Group.Index==1])−mean(Data
    [Group.Index!=1]))
# Routine to calculate the required statistic - here the diff between two means
# Group is the vector with the indexes for each group
# X is the vector of data
# Index is the value of one of the indexes
# Obs is the absolute observed difference
    Diff <- function(Group, X, Index, Obs)
    {
    R.Group   <- sample(Group,replace=F)    # Generate random randomization
# Mean difference
    d             <- mean(X[R.Group==Index])-mean(X[R.Group!=Index])
    d             <- abs(d)-Obs
    return(d)
    }
# Iterate over randomizations
    N                  <- 5000               # Number of randomizations
# Set up matrix to store differences
    Difference    <- matrix(0,N)
    for (Irep in 1:N)
    {
    Difference[Irep] <- Diff(Group.Index,Data,1,Obs.abs.diff)
    }
# Now calculate proportion greater than obs difference
    n.over    <- sum(Difference>=0)
    P         <- (n.over+1)/(N+1)           # Remember to add 1 for observed value
    print(c(Obs.abs.diff,P))
```

Output

```
print(c(Obs.abs.diff, P))
[1]  0.2341589  0.5006999
```

C.5.2 Using the S-PLUS bootstrap routine to do a randomization test

```
# Generate two normal distributed data
# Initialize the random number generator
   set.seed(20)
   n                 <- 10                        # Set the number per sample
   M                 <- 2*n                       # Total sample
# Generate N random normal deviates
   Data              <- rnorm(M,0,1)
# Create group indices
   Group.Index       <- matrix(c(rep(1,n), rep(2,,n)),M,1)
# Routine to calculate the required statistic — here the diff between two means
   Diff <- function(Group, X, Index){mean(X[Group==Index])—mean(X[Group!=
   Index])}
# samp.permute in "bootstrap" gives sampling without replacement
   N                 <- 5000     # Number of randomizations
   Meanboot    <- bootstrap(Group.Index, Diff(Group.Index,Data,1),
                     sampler= samp.permute, B=N)
# Calculate number of randomizations in which absolute difference >
# than observed
   n.over      <- sum(abs(Meanboot$replicates) >= abs(Meanboot$observed))
   P           <- (n.over+1)/(N+1)# Remember to add 1 for observed value
   P # Print P
```

C.5.3 Estimation of the required sample size for
a randomization test of a difference between two means

```
# Coding to compare two means
# Generate two normally distributed data sets
   set.seed(20)                          # Initialize random number generator
   n            <- 10                    # Set sample size for each group
   M            <- 2*n                   # Total sample size
   Data         <- rnorm(M,0,1)          # Generate M random normal deviates
   n1           <- n+1                   # Set starting row for group 2
   Data[n1:M]   <- Data[n1:M] + 1        # Add 1 to group 2
```

```
# Create group indices
    Group.Index  <- matrix(c(rep(1,n), rep(2,,n)),M,1)
# Calculate observed average absolute difference
    Obs.abs.diff    <- abs(mean(Data[Group.Index==1])-mean(Data
                        [Group.Index!=1]))
# Routine to calculate the required statistic - here the diff between two means
    Diff        <- function(Group, X, Index){mean(X[Group==Index])-
                    mean(X[Group!=Index])}
    N           <- 100     # Number of randomizations
    Meanboot    <-bootstrap(Group.Index, Diff(Group.Index,Data,1), sampler=
                    samp.permute, B=N)
    n.over      <- sum(abs(Meanboot$replicates) >= abs(Meanboot$observed))
    P           <- (n.over+1)/(N+1)     # Remember to add 1 for observed value
# Calculate required number
    N.req                   <- 0
    if (P<0.05) N.req     <- 4*P*(1-P)/(0.05-P)^2
    print(c(Obs.abs.diff, P, N.req))
```

Output

```
print(c(Obs.abs.diff,       P,        N.req))
        0.76584108   0.02970297   279.83343248
```

C.5.4 Estimation of the probabilities for differing values of High and Low for the jackal data. See Figure 5.2 for output

```
# Routine to calculate the required statistic - here the difference between
# two means
    Diff  <- function(Group, X, Index){mean(X[Group==Index])-
                mean(X[Group!=Index])}
# Put data into two vectors, one identifying the group the other with the data
    Group.Index  <- c(1,1,1,1,1,1,1,1,1,1,1,2,2,2,2,2,2,2,2,2,2)
    Data         <- c(107,110,111,112,113,114,114,116,117,120,105,106,
                    107,107, 108,110,110,111,111,111)
# Store the male values
    Data.store  <- matrix (Data[1:10],10)
    N           <- 10000             # Number of randomizations
    set.seed(0)                       # Initialize random number seed
    nreps       <- 10                # Set number of estimates to calculate
    High        <- 7.5               # Set initial value for upper value
    Low         <- 1.6               # Set initial value for lower value
```

```
PC              <- matrix(0,nreps,4) # Set up matrix to store output
for (i in 1:nreps)# Iterate over values of Low and High
{
High            <- High+0.05        # Increment High
PC[i,1]         <- High             # Store High in column 1
Low             <- Low+0.05         # Increment Low
PC[i,3]         <- Low              # Store Low in column 3
# Randomization test
Data[1:10]   <- Data.store[1:10]-High      # Subtract High from males
Meanboot     <- bootstrap(Group.Index,Diff(Group.Index,Data,1),
                    sampler=samp.permute,trace=F,B=N)
# Calculate number <= observed
n.over      <- sum(Meanboot$replicates <= Meanboot$observed)
PC[i,2]     <- (n.over+1)/(N+1)         # Store P for High in column 2
# Randomization test
Data[1:10] <- Data.store[1:10]-Low  # Subtract Low from males
Meanboot     <- bootstrap(Group.Index,Diff(Group.Index,Data,1),
                    sampler= samp.permute,trace=F,B=N)
# Calculate number <= observed
n.over      <- sum(Meanboot$replicates >= Meanboot$observed)
PC[i,4]    <- (n.over+1)/(N+1)               # Store P for Low in column 4
# Print P
print(PC[i,])
}
```

C.5.5 Estimation of the standard error, SE, using three approximate methods

```
# Create Lizard data set
Males          <- c(16.4,29.4,37.1,23,24.1,24.5,16.4,29.1,36.7,28.7,
                    30.2,21.8,37.1,20.3,28.3)
Females        <- c(22.2,34.8,42.1,32.9,26.4,30.6,32.9,37.5,18.4,27.5,
                    45.5,34,45.5,24.5,28.7)
n              <- length(Males)
Stamina        <- c(Males,Females)
Group          <- c(rep(1, times=n), rep(2, times=n))
Lizard.data    <- data.frame(Stamina, Group)
# First do randomization test on lizard data
Data             <- Lizard.data$Stamina   # Column containing data
Group.Index    <- Lizard.data$Group      # Column giving group number
# Routine to calculate the required statistic - here the diff between two means
```

```
Diff    <- function(Group, X, Index){mean(X[Group==Index])-mean(X
[Group!=Index])}
set.seed(20)       # Initialize random number
N    <- 10000      # Number of randomizations
# Randomization test
Meanboot       <- bootstrap(Group.Index,Diff(Group.Index,Data,1),
trace=F,sampler=samp.permute,B=N)
Lizard.Output  <- Meanboot$replicates        # Store replicates
# Calculating SE using normal approximation
                                   # Observed mean difference
xobs       <- abs(Meanboot$observed)
df         <-length(Data)-2            # degrees of freedom
# Output from randomizations
N          <- length(Lizard.Output)
n.over     <- sum(abs(Lizard.Output) >= xobs)  # Number exceeding xobs
P          <- (n.over+1)/(N+1)         # Estimated P
# Calculate x from t distribution
x.abscissa <- qt(P/2,df)
SE1        <- abs(xobs/x.abscissa)     # Estimate SE
t.value    <- abs(qt(0.025,df))        # Compute t value
U1         <- xobs+t.value*SE1         # Upper confidence bound
L1         <- xobs-t.value*SE1         # Lower confidence bound
print(c(xobs,P,SE1,L1,U1))            # Output
# Calculating SE using Average percentile method
# Absolute values of randomized values
D1         <- abs(Lizard.Output)
D.sorted   <- sort(D1)                # Sort into ascending order
Upper      <- 0.95*length(D1)         # Calculate upper 95% point
C          <- D.sorted[Upper]         # Find value at this point
t.value    <- abs(qt(0.025,df))       # Compute t value
SE2        <- C/t.value               # SE
U2         <- xobs+C                  # Upper confidence value
L2         <- xobs-C                  # Lower confidence value
print(c(C,SE2,L2,U2))                # Output
# Calculating SE using Percentile method
D.sorted   <- sort(Lizard.Output)    # Sort into ascending order
Upper      <- 0.975*length(D1)       # Calculate 97.5% point
CU         <- D.sorted[Upper]        # Find value at this point
Lower      <- 0.025*length(D1)       # Calculate 2.5% point
CL         <- D.sorted[Lower]        # Find value at this point
```

```
t.value     <- abs(qt(0.025,df))        # Compute T value
U3          <- CU+xobs                   # Upper confidence value
L3          <- CL+xobs                   # Lower confidence value
SE3         <- (U2-L2)/(2*t.value)       # SE
print(c(CL,CU,,SE3,L3,U3))              # Output
```

Output

```
print(c(xobs,          P              SE1,         L1,        U1))
          5.36    0.05649435      2.694143    -0.1587019   10.8787
print(c(  C,              SE2,           L2,                  U2))
      5.5066667       2.6882677     -0.1466667           10.8666667
print(c(CL,              CU,            SE3,         L3,        U3))
     -5.5066667     5.4666667      2.6785040   -0.1466667  10.8266667
```

C.5.6 Randomization of one-way analysis of variance of ant consumption data

```
set.seed(0)                            # Set seed for randomization
# Enter data
    X.data   <- c(13,242,105,8,59,20,2,245,515,488,88,233,50,600,82,40,
    52,1889,18,44,21, 5, 6, 0)
    Month    <- c("Jn","Jn","Jn","J","J","J","J","J","A","A","A","A",
    "A","A","A", "A","A","A","S","S","S","S","S","S")
    Group         <- factor(Month)     # Convert months to factor
    N             <- 1000              # Set number of randomizations
    F.replicate   <- matrix(0,N,1)     # Set up matrix to take permuted Fs
    for (Iperm in 1:N)                 # Iterate over N randomizations
    {
# Note that on the first pass the F stats for original data calculated
# Bind 2 variables into dataframe
    Data               <- data.frame(Group,X.data)
# One-way anova.
    Model              <- aov(Data[,2]~Data[,1], data=Data)
    Model.summary      <- summary(Model, SSType=3)   # Do analysis
    F.value            <- Model.summary$F[1]         # Extract F value
    F.replicate[Iperm] <- F.value                    # Store F value
# Permute set of observations
    X.data             <- sample(X.data)
    }
    P              <- mean(F.replicate >= F.replicate[1])    # Calculate P
    print(c(F.replicate[1],P))# Output original F and P
```

Output

```
print(c(F.replicate[1],  P))
          1.643906     0.196000
```

Output from analysis of variance

	Df	Sum of Sq	Mean Sq	F Value	Pr(F)
Month	3	726695	242231.6	1.643906	0.2110346
Residuals	20	2947024	147351.2		

C.5.7 Randomization of one-way analysis of variance using "by" routine of S-PLUS

```
set.seed(0)      # Initialize random number
X.data    <- c(13,242,105,8,59,20,2,245,515,488,88,233,50,600,82,40,
             52,1889,18,44,21, 5, 6, 0)
Month    <- c("Jn","Jn","Jn","J","J","J","J","J","A","A","A","A",
             "A","A","A", "A","A","A","S","S","S","S","S","S")
Group    <- factor(Month)      # Convert months to factor
Perm.data <- X.data            # Initiate collection of data
N        <- 999                # Number of randomizations
Total    <- N+1                # N + initial data results
# Produce N randomizations of data
    for (i in 1:N){Perm.data  <- c(Perm.data,sample(X.data))}
# Now add index and group membership
    Perm.Group    <- rep(Group,Total)             # Replicate N+1 times
    n             <- length(X.data))              # Size of data set
    Perm.Index    <- sort(rep(seq(1:Total),n))    # Produce N+1 indices
# Set up matrix to take permuted Fs
    F.replicate   <- matrix(0,Total,1)
# Bind three vectors together
    d             <- data.frame(Perm.Index,Perm.Group,Perm.data)
# Use by routine to do N+1 anovas and store result in object ANOVA
    ANOVA <- by(d, d$Perm.Index, function(d) summary(aov(Perm.data~
    Perm.Group, data=d)))
    F.replicate   <- matrix(0,Total,1)   # Set up matrix to take permuted Fs
# Extract F values from ANOVA object
    for (i in 1:Total)
    {
    a <- unlist(ANOVA[i])
    F.replicate[i] <- a[7]
    }
```

```
P        <- mean(F.replicate >= F.replicate[1])    # Calculate P
print(c(F.replicate[1],P))                         # Output original F and P
```

Output

```
print(c(F.replicate[1],  P))
        1.643906        0.175000
```

Output from analysis of variance

	Df	Sum of Sq	Mean Sq	F Value	Pr(F)
Month	3	726695	242231.6	1.643906	0.2110346
Residuals	20	2947024	147351.2		

C.5.8 Randomization testing a two-way analysis of variance

Data are in a matrix of three columns, the original results presented in the left-hand table below. To produce a randomized data set while keeping cell counts the same we randomize only the data column, X, giving, for example, the data set shown in the right-hand table.

Factor A	Factor B	X	Factor A	Factor B	X
1	0	50	1	0	110
1	0	57	1	0	50
1	1	57	1	1	85
1	1	71	1	1	94
1	1	85	1	1	105
2	0	91	2	0	57
2	0	94	2	0	120
2	0	102	2	0	102
2	0	110	2	0	71
2	1	105	2	1	91
2	1	120	2	1	110

```
# Coding to do a two-way anova
# Create data set
    Data        <- c(50,57,57,71,85,91,94,102,110,105,120)
    Factor.A    <- c(1,1,1,1,1,2,2,2,2,2,2)
    Factor.B    <- c(0,0,1,1,1,0,0,0,0,1,1)
    Groups      <- data.frame(Factor.A, Factor.B)
# Ensure that variables are factors
    Groups[,1]   <- factor(Groups[,1])
# Ensure that variables are factors
    Groups[,2]   <- factor(Groups[,2])
    Obs.model   <- aov(Data~Groups[,2] # Initial ANOVA *Groups[,1])
```

```
# Specify Type 3 SS
   Obs.results    <- summary(Obs.model, ssType=3)
# Function to get F statistics from two-way anova
   ANOVA<- function(Groups,Data)
   {
   Model   <- aov(Data~Groups[,2]*Groups[,1])
   Fs      <- summary(Model,ssType=3)$"F Value"    # Type III sums of squares
   return(c(Fs[1],Fs[2],Fs[3]))
   }
# Do randomization
   N            <- 1000          # Number of randomizations
   F.values   <- matrix(0,N,3)   # Set up matrix to take F values
   for (iperm in 1:N)            # Iterate through randomizations
   {
# Note that first pass is on original data
   F.values[iperm,]   <- ANOVA(Groups,Data)
   Data               <- sample(Data)          # Randomize data vector
   }
# Print out results
   Obs.results
   for (i in 1:3)
   {print(c("Random P for ",i," = ", mean(F.values[,i]>= F.values[1,i])))}
```

Output

Type III Sum of Squares

	Df	Sum of Sq	Mean Sq	F Value	Pr(F)
Groups[, 1]	1	4807.934	4807.934	45.00908	0.0002752
Groups[, 2]	1	597.197	597.197	5.59061	0.0500130
Groups[, 1]:Groups[, 2]	1	11.408	11.408	0.10679	0.7533784
Residuals	7	747.750	106.821		

```
"Random P for "    "1"        " = "           "0.001"
"Random P for "    "2"        " = "           "0.061"
"Random P for "    "3"        " = "           "0.77"
```

C.5.9 Levene's test for homogeneity of variances

```
# Calculating Levene's test
# Enter data
   Month   <- c("Jn","Jn","Jn","J","J","J","J","J","A","A","A","A","A",
               "A","A","A","A","A","S","S","S","S","S","S")
```

```
Ants.eaten    <- c(13,242,105,8,59,20,2,245,515,488,88,233,50,600,
                   82,40,52,1889,18,44,21,5,6,0)
   Ant.data <- data.frame(Month,Ants.eaten)
# Function to calculate absolute differences from means within groups
   Levene     <- function(x){abs(x-mean(unlist(x)))}

# Calculate absolute differences between group means and observations
   Abs.diffs   <- by(Ant.data$Ants.eaten, Ant.data$Month, Levene)
# Combine group designator "Group" with Absolute differences
   Abs.diffs   <- unlist(Abs.diffs)              # Remove list structure
# Sort Month to correspond with "by results"
   Month    <- sort(Month)
# Combine into columns
   Data    <- data.frame(Month, Abs.diffs)
# Do ANOVA
   summary(aov(Data$Abs.diffs~Data$Month, data=Data))
```

Output

	Df	Sum of Sq	Mean Sq	F Value	Pr(F)
Data$Month	3	639142	213047.4	2.857091	0.06284734
Residuals	20	1491359	74567.9		

Results from randomization test (Use "Data" file and coding in C.5.6 or C.5.7)

```
print(c(F.replicate[1],    P))
        2.857091         0.052600
```

C.5.10 χ^2 contingency analysis by randomization

```
   set.seed(3)                          # Initialize random number seed
# Get data and convert to matrix
   Data.File   <- numerical.matrix (Shad.data)
   n              <- sum(Data.File)       # Find total sample size
   nos.of.rows <- nrow(Data.File)        # Find number of rows
   nos.of.cols <- ncol(Data.File)        # Find number of columns
   rows           <- NULL                 # Set up vector for row entries
   cols           <- NULL                 # Set up vector for column entries
# Construct row and column vectors (= M matrix)
   for (irow in 1: nos.of.rows)                      # Iterate over rows
   {
   for(icol in 1:nos.of.cols)                        # Iterate over columns
   {
   rows  <- c(rows, rep(irow,Data.File[irow,icol])) # Row entries
```

```
    cols <- c(cols,rep(icol,Data.File[irow,icol]))  # Column entries
    }}
    obs.chi<- chisq.test(Data.File, correct=F)        # Test on original data
# Create function to reconstruct data matrix from M matrix and do test
# n.rows = number of rows, n.cols = number of columns,
# r.s, c.s = vectors corresponding to columns of M matrix, m = number of entries
    Chi.random    <- function(r.s,c.s,n.rows,n.cols,m)
    {
# Randomized data matrix
    M.random     <- matrix(0,n.rows,n.cols)
    for (i in 1:m){M.random[r.s[i],c.s[i]] <- M.random[r.s[i],c.s[i]]+1}
    chi <- chisq.test(M.random, correct=F)    # Chi-square test
    return(chi$statistic)                      # Return chi-square value

    }
    N                <- 1000                # Number of permutations
    Chi.replicate  <- matrix(0,N,1)        # Set up matrix to take permuted Fs
    for (Iperm in 1:N)                     # Iterate over N permutations
    {
# Note that on the first pass the Chi2 value for original data calculated
# Extract Chi value
    Chi.value          <- Chi.random(rows,cols,nos. of.rows,nos.of.cols,n)
    Chi.replicate[Iperm] <- Chi.value           # Store Chi value
# Randomize column vector
    cols               <- sample(cols)
    }
    P   <- mean(Chi.replicate >= Chi.replicate[1])  # Calculate P
    obs.chi                                          # Chi2 on original data
    print(c(Chi.replicate[1],P))                     # Output original F and P
```

Output

```
obs.chi
Pearson's chi-square test without Yates' continuity correction
data: Data.File
X-square = 236.4939,  df = 117,  p-value = 0
> # Chi2 on original data
    print(c(Chi.replicate[1],    P))
                236.4939         0.0010
    There were 11 warnings (use warnings() to see them)
```

Warnings are given from the routine chisq.test and refer to the small sample size in cells. These warnings can be ignored.

C.5.11 Randomization analysis of intercept and slope in linear regression on generated data

```
set.seed(4)                # Initialize random number seed
x    <- runif(20,0,1)      # Generate 20 uniform random numbers
y    <- x + rnorm(20,0,1)  # Generate 20 random normal, N(0,1) and add to x
# Routine to fit linear regression and extract coefficients
Lin.reg            <- function(x.data,y.data){coef(lm(y.data~x.data))}
# Calculate regression stats for observations
obs.regression   <- summary(lm(y~x))
# Use bootstrap routine to do permutations
N          <- 1000         # Number of permutations
Meanboot  <- bootstrap(x, Lin.reg(x,y), sampler=samp.permute, B=N,
               trace=F)
# Calculate number of permutations in which absolute difference > than
# observed intercept
n.over.a   <- sum(abs(Meanboot$replicates[,1]) >= abs(Meanboot
$observed[1]))
# Slope
n.over.b <- sum(abs(Meanboot$replicates[,2]) >= abs(Meanboot$observed
[2]))
Pa       <- (n.over.a+1)/(N+1)       # Remember to add 1 for observed value
Pb       <- (n.over.b+1)/(N+1)       # Remember to add 1 for observed value
obs.regression                       # Print observed regression stats
print(c(Pa,Pb))                      # Print P
```

Output

```
obs.regression
Call: lm(formula = y ~ x)
Residuals:
      Min        1Q     Median        3Q       Max
   -1.642    -0.6467    0.07469    0.6928     1.423
Coefficients:
              Value    Std. Error    t value    Pr(>|t|)
(Intercept)  0.2758    0.3814        0.7231     0.4789
        x    0.9142    0.7241        1.2625     0.2229

Residual standard error: 0.8685 on 18 degrees of freedom
Multiple R-Squared: 0.08135
F-statistic: 1.594 on 1 and 18 degrees of freedom, the p-value is 0.2229
```

```
Correlation of Coefficients:
    (Intercept)
  x -0.8607
> # Print observed regression stats
print(   c(Pa,        Pb))
         0.8811189    0.2147852
```

C.5.12 Coding to create distance and difference matrices shown in Figure 5.12

Spatial.data contains the following data

2	2	0	0	0
1	2	0	0	5
2	0	0	0	6
0	1	0	5	7
0	1	0	5	4
0	0	0	3	5

```
# First create three column matrix
# Col 1 contains "x" coordinate
# Col 2 contains "y" coordinate
# Col3 contains data
    Nrows       <- 6                              # Number of rows
    Ncols       <- 5                              # Number of columns
    N           <- Nrows*Ncols
    M           <- matrix(0,N,3)
    Row         <- 0                              # Set up row counter
    for (irow in 1:Nrows){                        # Iterate over rows
    for (icol in 1:Ncols){                        # Iterate over columns
    Row         <- Row+1                          # Increment row counter
    M[Row,1]    <- irow                           # Save x coordinate
    M[Row,2]    <- icol                           # Save y coordinate
    M[Row,3]    <- Spatial.data[irow,icol]        # Save data
}}
# Now form matrix of distances and differences using data in matrix
# Spatial.data
    Distance              <- matrix(0,N,N)     # Distance matrix
    Difference            <- matrix(0,N,N)     # Difference matrix
    for (irow in 1:N){                         # Iterate over x coordinates
    for (icol in irow:N){                      # Iterate over y coordinates
```

```
Distance[irow,icol]    <- sqrt((M[irow,1]-M[icol,1])^2+(M[irow,2]-
                           M[icol,2])^2)
Difference[irow,icol]  <- abs(M[irow,3]-M[icol,3])
}}
```

C.5.13 The Mantel Test

```
# Coding for Mantel Test using data shown in Figure 5.12 and C.5.12
    set.seed(1)                # Initialize random number
    Mx         <-Distance      # Enter X matrix
    My         <- Difference   # Enter Y matrix
# Function to Convert matrix into vector excluding duplicate elements
    Vector     <- function(M)
    {
    n          <- nrow(M)      # Number of rows and columns
    V          <- NULL         # Set up vector
# Iterate over cols
    for (i in 1:n)
    {
    # Iterate over rows
    for(j in i:n){if(i!=j)V    <- c(V,M[i,j])}   # Accumulate data
    }
# Note that diagonal is excluded. In some cases it may be included
    return(V)
    }
    Vx         <- Vector(Mx)   # Create vector x
    Vy         <- Vector(My)   # Create vector y
    N          <- 1000         # Number of permutations
# Use bootstrap routine to do permutations using cor
    Meanboot   <-bootstrap(Vx, cor(Vx,Vy), sampler=samp.permute, B=N,
    trace=F)
# Calculate number of permutations in which absolute difference > than
# observed
    n.over     <- sum(abs(Meanboot$replicates) >= abs(Meanboot$observed))
# Remember to add 1 for observed value
    P          <- (n.over+1)/(N+1)
    print(c(Meanboot$observed,P))     # Print observed correlation and P
```

Output
```
    print(c(Meanboot$observed,        P))
            0.1598842             0.003996004
```

C.6.1 Cross-validation of two multiple regression equations using 10% of the data set as the test set

```
set.seed(1)                          # Set random number seed
Data      <- Cricket.Data            # Pass data to file Data
Nreps     <- 1000                    # Number of randomizations
n         <- nrow(Data)              # Find number of rows in data set
# Create an index vector in ten parts
Index     <- rep(seq(1,10), length.out=n)
Data      <- cbind(Data,Index)       # Combine Data and index vector
Last.col  <- ncol(Data)              # Last column for index
# Function to determine residual sums of squares
# D=Data; I=Index value; K=col for Index; R=col for obs. value; Model=model
# object
SS                     <- function(D,I,K,R,Model)
{
Obs                    <- D[D[,K]==I,R]     # Observed value
# Predicted value using fitted model
Pred                   <- predict (Model,D[D[,K]==I,])
return(sum((Obs-Pred)^2))                   # Residual sums of squares
}
# Matrix for residual sums of squares
RSS                    <- matrix(0,Nreps,2)
for (i in 1:Nreps)                          # Iterate over randomizations
{
Index                  <- sample(Index)     # Randomize index vector
# Place index values in last col of Data
Data[,Last.col]        <- Index
# Compute model objects Note that Last.col is the column for the index
# values
Model1    <- lm(OVARY.WT~F.coef+MORPH+HEAD.WTH:MORPH, data=Data
             [Data[,Last.col]!=1,])
Model2    <- lm(OVARY.WT~HEAD.WTH*F.coef*MORPH, data=Data
             [Data[,Last.col]!=1,])
RSS[i,1]  <- SS(Data,1,Last.col,1,Model1)        # Store RSS for Model 1
RSS[i,2]  <- SS(Data,1,Last.col,1,Model2)        # Store RSS for Model 2
}
t.test(RSS[,1], y=RSS[,2],paired=T)        # Do paired t test
print(c(mean(RSS[,1]), mean(RSS[,2])))     # Print means
```

Output

```
data: RSS[, 1] and RSS[, 2]
t = -3.1672, df = 999, p-value = 0.0016
alternative hypothesis: mean of differences is not equal to 0
95 percent confidence interval:
  -0.04489097 -0.01054436
sample estimates:
mean of x - y
  -0.02771766
```

C.6.2 Coding to fit a smoothed function using the loess routine. Coding to generate plots given but output not shown

```
# Generate data for plots in Figure 6.12
set.seed(1)                                          # Set random number seed
n             <- 100                                 # Sample size
Curves        <- matrix(0,n,5)                       # Matrix for data
x             <- seq(5,20,length=n)                  # values of x
Curves[,1]    <- x                                   # Store x
error         <- rnorm(n,0,0.06)                     # Errors
Curves[,2]    <- dnorm(x, 10,1)+ dnorm(x,12,1)       # Curve
Curves[,3]    <- dnorm(x, 10,1)+ dnorm(x,12,1)+error # Add error to curve

# Fit function using loess
  SPAN      <- 0.3       # Set span value
  DEG       <- 2         # Set degrees (1 or 2)
# Fit loess model
  L.smoother1    <- loess(Curves[,3]~Curves[,1], span=SPAN, degree=DEG)
# Calculate predicted curve with standard errors
# Set range of x
  x.limits       <- seq(min(Curves[,1]), max(Curves[,1]),length=50)
# Prediction model
  P.model  <- predict.loess(L.smoother1, x.limits, se.fit=T)
  C.INT    <- pointwise(P.model, coverage=0.95)  # Calculate values
  Pred.C   <- C.INT$fit                          # Predicted y at x
  Upper    <- C.INT$upper                        # Plus 1 SE
  Lower    <- C.INT$lower                         # Minus 1 SE
  plot(Curves[,1], Curves[,3])                   # Plot points
  lines(Curves[,1], Curves[,2], lty=2)           # Plot true function
```

```
    lines(x.limits,Pred.C)                          # Plot loess prediction
    lines(x.limits,Upper,lty=4)                     # Plot plus 1 SE
    lines(x.limits,Lower,lty=4)                     # Plot minus 1 SE
    Fits<- fitted(L.smoother1)                      # Calculate fitted values
    Res<- residuals(L.smoother1)                    # Calculate residuals
# Plot residuals on fitted values with simple loess smoother
    scatter.smooth(fitted(L.smoother1),residuals(L.smoother1),
    span=1, degree=1)
# Output basic stats for smoothed function
    summary(L.smoother1)
```

Output

```
summary(L.smoother1)
Call:
loess(formula = Curves[, 3] ~ Curves[, 1], span = SPAN, degree = DEG)
    Number of Observations:                     100
    Equivalent Number of Parameters:            9.8
    Residual Standard Error:                    0.05831
    Multiple R-squared:                         0.92
    Residuals:
        min       1st Q      median      3rd     Q max
     -0.1357    -0.03365   -0.003501  0.03404   0.1517
```

C.6.3 Coding for 10-fold cross-validation of loess fit

```
# Generate data as shown in Figure 6.2
    set.seed(1)                          # initiate random number generator
    n              <- 100                # sample size
    Curves         <- matrix(0,n,2)      # Matrix for data
    x              <- seq(5,20,length=n) # values of x
    error          <- rnorm(n,0,0.06)    # errors
    Curves[,1]     <- x
# add error to curve
    Curves[,2]     <- dnorm(x, 10,1) + dnorm(x,12,1)+error
# Generate indexes for cross validation.
# Note that because data are created sequentially index is also randomized
    Index          <- sample(rep(seq(1,10), length.out=n))

# Do ten-fold cross validation
    for (i in 1:10)
```

```
   {
   Data          <- data.frame (Curves[Index!=i,])      # Select subset of data
   CV.data       <- data.frame (Curves[Index==i,])      # Store remainder
# Fit model
   L.smoother    <- loess (X1.2~X1.1, data=Data, span=0.3,degree=2)
# Multiple r for fitted values
   R2            <- summary(L.smoother) $ covariance
# Calculate predicted curve
   Predicted     <- predict.loess (L.smoother, CV.data)
# Calculate correlation between predicted and observed
   r             <- cor(CV.data[,2], Predicted, na.method="omit")
   print(c(i,r^2, R2))             # Print predicted and observed multiple R
   plot(CV.data[,2],Predicted)     # Plot not shown
   }
```

Output First column is index number, second column is r^2 for fit to excluded data, third column is fit for the original model

```
[1]    1.0000000  0.9441238  0.9223947
[1]    2.0000000  0.9317727  0.9236801
[1]    3.0000000  0.9173710  0.9226224
[1]    4.0000000  0.9558029  0.9225427
[1]    5.0000000  0.1852531  0.9328380
[1]    6.0000000  0.8964556  0.9292294
[1]    7.0000000  0.9436588  0.9231652
[1]    8.0000000  0.9102181  0.9250648
[1]    9.0000000  0.9553580  0.9211906
[1]   10.0000000  0.8396268  0.9322675
```

C.6.4 Coding to fit loess curve to multivariate data

```
# Generate data plotted in Figure 6.5
   set.seed(1)                      # Initialize random number
   N    <- 100                      # Number of data points
   X1   <-runif(N,15,19)            # Generate values of X1
   X2   <-runif(N,0,20)             # Generate values of X2
# Nest density at these sites
   Y    <- matrix(0,N)              # Set up matrix for Y values
   for (i in 1:N)
   {
   if(X1[i]<17)                     Y[i] <- 5 + rnorm(1,0,2)   # error N(0,2)
```

```
    if(X1[i]>=17 & X2[i] < 10)        Y[i] <- 10 + rnorm(1,0,4)    # error N(0,4)
    if (X1[i] >= 17 & X2[i] >= 10)    Y[i] <- 20 + rnorm(1,0,8)    # error N(0,8)
    }
# Plot perspective surface using interpolation (top row in Figure 6.5)
    persp(interp(X1,X2,Y), xlab="X1", ylab="X2", zlab="Y")
    Data                  <- data.frame(Y,X1,X2)        # create datafile
# Fit loess function using quadratic
    Density             <- loess(Y~X1*X2, data= Data, degree=2)
# Generate equally spaced grid (20x20) for plot of loess-generated surface
    X1.predict          <- rep(seq(from=min(X1), to=max(X1), length=20),
                             times=20)
    X2.predict   <- sort(rep(seq(from=min(X2), to=max(X2), length=20),
                        times=20))
# Convert to data frame
    X.predict           <- data.frame (X1.predict,X2.predict)
# Predict Y
# Concatenate X1,X2 values
    X.predict           <- cbind(X1.predict,X2.predict)
    dimnames(X.predict) <- list(NULL, c("X1","X2"))   # Add names to columns
    X.predict           <- data.frame(X.predict)       # Convert to data frame
    Density.predict     <- predict.loess (Density, X.predict)
# Set X1 for persp
    X1                  <- seq(from=min.X1, to=max.X1, length=20)
# Set X2 for persp
    X2                  <- seq(from=min.X2, to=max.X2, length=20)
# Convert predicted values into matrix
    Z                   <- matrix(Density.predict,20,20)
# Plot loess plot
    persp(X1.predict,X2.predict,Z,xlab="X1", ylab="X2", zlab="Y")
```

C.6.5 Comparison of two fitted loess surfaces using density data

```
# Generate data plotted in Figure 6.5
    set.seed(1)                     # Initialize random number
    N       <- 100                  # Number of data points
    X1      <-runif(N,15,19)        # Generate values of X1
    X2      <-runif(N,0,20)         # Generate values of X2
# Nest density at these sites
    Y       <- matrix(0,N)          # Set up matrix for Y values
```

```
    for (i in 1:N)
    {
    if (X1[i]<17)                    Y[i] <- 5 + rnorm(1,0,2)    # error N(0,2)
    if (X1[i]>=17 & X2[i] < 10)      Y[i] <- 10 + rnorm(1,0,4)   # error N(0,4)
    if (X1[i] >= 17 & X2[i] >= 10)   Y[i] <- 20 + rnorm(1,0,8)   # error N(0,8)
    }
    Data <- data.frame(Y,X1,X2) # Concatenate data into dataframe
# Do fits and compare
    Density1 <- loess(Y~X1*X2, data= Data, degree=1)  # Fitted with degree 1
    Density2 <- loess(Y~X1*X2, data= Data, degree=2)  # Fitted with degree 2
# Output result for 1st fit
    Density1
# Output result for 2nd fit
    Density2
    anova(Density1,Density2)                          # Compare with anova
```

Output

```
    Call:
    loess(formula = Y ~ X1 * X2, data = Data, degree = 1)

    Number of Observations:            100
    Equivalent Number of Parameters:   4.8
    Residual Standard Error:           5.043
    Multiple R-squared:                0.54
    Call:
    loess(formula = Y ~ X1 * X2, data = Data, degree = 2)
    Number of Observations:            100
    Equivalent Number of Parameters:   9.2
    Residual Standard Error:           4.458
    Multiple R-squared:                0.67

    > anova(Density1, Density2)
    Model 1:
    loess(formula = Y ~ X1 * X2, data = Data, degree = 1)
    Model 2:
    loess(formula = Y ~ X1 * X2, data = Data, degree = 2)
    Analysis of Variance Table
             ENP      RSS     Test    F Value      Pr(F)
    1        4.8     2358.0   1 vs 2   6.15      0.000074947
    2        9.2     1743.1
```

C.6.6 Coding to produce plots and fitted curves to the Chapman equation shown in Figure 6.7, with testing of fits. Text output but not plots shown

```
# Generate data
   set.seed(1)  # initiate random number generator
   n           <- 200                                # Sample size
   X           <- runif(n, 0.1,10)                   # values of X
   error       <- rnorm(n,0,20)                      # error terms
   Y           <- 5 + 95*(1-exp(-1*X))^5             # Chapman curve
   Y           <- Y + error                          # Add error term
# Deterministic curve (error = mean=0)
   X.zero      <- seq(min(X), max(X), length=n)      # Set X values
# Calculate deterministic value
   Y.zero      <- 5 + 95*(1-exp(-1*X.zero))^5
# Combine X and Y into data frame
   Data        <- data.frame(X,Y)

# Fit GAM to X, Y
   Fit.gam     <- gam(Y~lo(X), data=Data)
# Fit linear (Fit.lin) and then quadratic (Fit.quad) regressions
   Fit.lin     <- lm(formula = Y ~ X, data = Data, na.action = na.exclude)
   Fit.quad    <- lm(formula = Y ~ X + X^2, data = Data, na.action = na.exclude)
# Calculate predicted value using quadratic fit
   pred.y      <- predict(Fit.quad)
   d           <- data.frame(X, pred.y)              # Convert to data frame
# sort data to produce sequence for line plot
   newd        <- sort.col(target=d, columns.to.sort="@ALL", columns.to.
                    sort.by="X", ascending=T)

# Plot results with quadratic fit on first plot
   plot(X,Y)                                         # Original data
   lines(X.zero, Y.zero)                             # Deterministic curve
   lines(newd[,1], newd[,2])                         # Quadratic fit
   plot(Fit.gam, residuals=T, se=T, rug=F)           # GAM fit

# Output Results
   anova(Fit.gam)                     # Fit of GAM model
   summary(Fit.lin)                   # Fit of linear regression
   summary(Fit.quad)                  # Fit of quadratic
   anova(Fit.quad, Fit.gam)           # Compare quadratic and GAM models
```

Output (summarized)

```
>anova(Fit.gam)
DF for Terms and F-values for Nonparametric Effects
                Df   Npar Df    Npar F    Pr(F)

(Intercept)    1
     lo(X)     1      2.2      72.17551     0

>summary(Fit.lin)
Call: lm(formula = Y ~ X, data = Data, na.action = na.exclude)
Coefficients:
              Value   Std. Error   t value   Pr(>|t|)
(Intercept)  31.1040     3.6782     8.4564    0.0000
         X    8.8805     0.6428    13.8155    0.0000

Residual standard error: 27.51 on 198 degrees of freedom
Multiple R-Squared: 0.4908
F-statistic: 190.9 on 1 and 198 degrees of freedom, the p-value is 0
>summary(Fit.quad)
Call: lm(formula = Y ~ X + X^2, data = Data, na.action = na.exclude)
Coefficients:
              Value   Std. Error    t value   Pr(>|t|)
(Intercept)  -5.5812     4.1689     -1.3388    0.1822
         X   31.9984     2.0013     15.9887    0.0000
   I(X^2)    -2.3082     0.1937    -11.9160    0.0000

Residual standard error: 21.03 on 197 degrees of freedom
Multiple R-Squared: 0.7041
F-statistic: 234.4 on 2 and 197 degrees of freedom, the p-value is 0
>anova(Fit.lin, Fit.quad)
Analysis of Variance Table
Response: Y
    Terms   Resid. Df       RSS     Test   Df   Sum of Sq    F Value   Pr(F)
1     X          198   149894.1
2  X + X^2        197    87108.9   +I(X^2)   1   62785.15   141.9909     0

>anova(Fit.quad, Fit.gam)
Analysis of Variance Table
Response: Y
    Terms  Resid. Df       RSS   Test        Df   Sum of Sq   F Value        Pr(F)
1 X + X^2  197.0000  87108.94
2   lo(X)  195.7897  82594.88  1 vs. 2  1.210328   4514.059  8.840986  0.001830153
```

C.6.7 Coding to generate regression tree and perform cross-validation

```
# Coding to generate the data shown in Figure 6.5
    set.seed(1)              # Initialize random number
    N    <- 100              # Number of data points
    X1   <-runif(N,15,19)    # Generate values of X1
    X2   <-runif(N,0,20)     # Generate values of X2

# Nest density at these sites
    Y <- matrix(0,N) # Set up matrix for Y values
    for (i in 1:N)
    {
    if (X1[i]<17)                    Y[i] <- 5 + rnorm(1,0,2)    # error N(0,2)
    if (X1[i]>=17 & X2[i] < 10)      Y[i] <- 10 + rnorm(1,0,4)   # error N(0,4)
    if (X1[i] >= 17 & X2[i] >= 10)   Y[i] <- 20 + rnorm(1,0,8)   # error N(0,8)
    }
    Data.df <- data.frame(X1,X2,Y) # Concatenate into dataframe

# Coding to generate regression tree and perform cross-validation
    set.seed(1)                              # Initiate random number
    Tree            <- tree(Y~X1+X2,Data.df)  # Create tree
    plot(Tree); text(Tree)                    # Plot tree with text
    Pruned.Tree     <- prune.tree(Tree)       # Prune tree
    plot(Pruned.Tree)                         # Plots deviance against size
    Size            <- NULL                   # set up Size vector
    gtotal          <- NULL                   # set up matrix for output data

# Iterate over 10 cross-validations
    for (i in 1:10)
    {
    Tr        <-cv.tree(Tree, prune.tree)  # Apply cross-validation routine
    plot(Tr)                                # Plot results
# Make matrix with 2 cols, dev & size
    g        <- cbind(Tr$dev,Tr$size)
    gtotal   <- cbind(gtotal,g)             # Save data for later plotting
    g        <- data.frame(g)               # Convert to data frame and sort
    g1       <- sort.col(target=g, columns.to.sort="@ALL",
               columns.to.sort.by=list("g.1"),ascending=T)
# Store size for smallest deviance
    Size     <- c(Size,g1[1,2])
    }
```

```
    Size                                    # Print best Size for the ten runs
                                            # Get the integer value of mean Size
    Avg.Size      <- floor(mean(Size))
# Use Avg.Size to prune tree
    Tree.pruned    <- prune.tree (Tree, best=Avg.Size)
    summary(Tree.pruned)                    # Output results
    Tree.pruned$size                        # Output possible tree sizes
    plot(Tree.pruned);text(Tree.pruned)     # plot tree
```

Output (plots not shown)

```
# Print best Size for the ten runs
    Size
    3 3 3 3 3 4 4 4 3 3
# Use Avg.Size to prune tree
    summary(Tree.pruned)
    Regression tree:
    snip.tree(tree = Tree, nodes = c(2., 6., 7.))
    Number of terminal nodes: 3
    Residual mean deviance: 13.83 = 1342 / 97
```

C.6.8 Function to perform randomization test of a given regression tree

```
# Function to perform randomization
    Tree.Random <- function(formula, data, Ypos, Ibest, N.Rand=100)
    {
# Check that a tree of required size actually exists for the real data
    Random.tree    <- tree(formula, data)        # Calculate tree
# Prune tree
    R1             <- prune.tree (Random.tree, best=Ibest)
    R1.summary     <- summary(R1)                # Summary data
    if(R1.summary$size!=Ibest) stop("A tree of this size cannot be fitted
    to data")
# Matrix to store deviances
    Deviances      <- matrix(0,N.Rand)
# Matrix to store actual sizes used
    Sizes          <- matrix(0,N.Rand)
    nrows          <- nrow(data)                 # Number of rows in data
```

```
# Iterate over randomizations
    for (i in 1:N.Rand)
# Note that on first pass there is no randomization
    {
    Random.tree   <- tree(formula, data)         # Calculate tree
# Prune tree
    R1               <- prune.tree (Random.tree, best=Ibest)
    R1.summary       <- summary(R1)              # Summary data
    Deviances[i]    <- R1.summary$dev            # Store deviances
# Store number of terminal nodes
    Sizes[i]        <- R1.summary$size
    data[,Ypos]     <- data[sample(nrows),Ypos]   # Randomize y
    }
# Calculate P
    P               <- length(Deviances [Deviances<=Deviances[1]])/N.Rand
    SE              <- sqrt(P*(1-P)/N.Rand)       # Calculate SE
    print("Probability of random tree having smaller deviance (SE)")
    print(c(P, SE))                              # Output P and SE
    print("Summary of sizes actually used in randomization")
    print(summary(Sizes))                        # summary data of sizes
    }


# **********************************
# Call to function
# Description of parameters in order:
# Model statement, e.g., Y ~ X1 + X2
# Data file, e.g., Data.df, which is the dataframe created in C.6.7
# Column for Response variable, e.g., Ypos = 3
# Size of tree to compare, e.g., Ibest = 3
# Number of randomizations, e.g., N.Rand=100

Tree.R <-Tree.Random(Y~X1+X2, Data.df, Ypos=3, Ibest=3, N.Rand=100)
```

Output

```
[1] "Probability of random tree having smaller deviance (SE)"
[1] 0.010000000 0.009949874
[1] "Summary of sizes actually used in randomization"
   Min. 1st Qu. Median  Mean  3rd Qu.   Max.
   3.00   3.00    4.00   4.24   5.00   12.00
```

C.7.1 Calculation of posterior probabilities for a normal mean based on prior distributions of the mean and variance and a single observation, x

For clarity, I have used looping, though in most circumstances this will be inefficient.

```
# Set up parameter values
    mu0          <- c(0,0.5,1,3)                  # μ0
    Pmu0         <- c(.1,.2,.5,.2)                # p1(σ0)
    sigma0       <- c(.25,.3,.5,.75)              # μ0
    Psigma0      <- c(0.01,0.05,0.9,0.04)         # p2(σ0)
    x            <- 1.5                           # x
# Calculate matrix of prior probabilities for μ0 and σ0 based on x
    Px           <- matrix(0,4,4)                 # Set up matrix for data
# Iterate over the 16 combinations
    for (mu in 1:4)
    {
    for (sigma in 1:4)
    {
    Px[sigma,mu]      <- dnorm(x,mu0[mu],sigma0[sigma])*Pmu0[mu]*Psigma0
    [sigma]
    }
    }
    Denom        <- sum(Px)
    Px           <- Px/Denom

# Now iterate over values of theta
    Theta        <- seq(0,3,0.01)       # Vector of theta values
    n            <- length(Theta)       # Length of vector
    sd           <- 0.5                 # σ
    Posterior    <- matrix(0,n,1)       # Set up matrix for posterior
# Iterate over all combinations. Note that theta does not require a loop
    for (mu in 1:4)
    {
    for (sigma in 1:4)
    {
    sd1          <- 1/((1/sigma0[sigma]^2)+ (1/sd^2))        # σ1
# μ1
    mu1          <- sd1*((1/sigma0[sigma]^2) *mu0[mu]+(1/sd^2)*x)
    P            <- dnorm(Theta,mu1,sd1)*Px[sigma,mu]        # P given μ0,σ0
```

```
Posterior    <- Posterior + P                                    # Posterior
}}
sum(Posterior)
plot(Theta, Posterior)
```

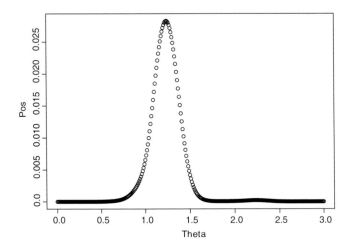

C.7.2 Bayesian analysis of binomial data

Note that the area under the curve is based on a frequency polygon with bin width 0.001.

```
# Function to calculate likelihood)
    Likelihood    <- function(theta,x,n) {choose(n,x)*theta^x*(1-theta)
                     ^(n-x)}

    Theta         <- seq(0,1,0.001)              # Vector of theta values
    L1            <- Likelihood(Theta,8,10)      # Likelihoods
# Approximate area under the curve
    Area          <- sum(L1)
    Posterior     <- L1/Area                     # Posterior probabilities
# Find maximum probability for scaling plot
    Max.Prob      <- max(Posterior)
                                                 # Plot scaled posterior probability
    plot(Theta, Posterior/Max.Prob)
# Calculate new posterior based on further observation of x=5, n=10
    Prior         <- Posterior                   # New prior
# Likelihoods x Prior
    L2            <- Likelihood(Theta,5,10)*Prior
# Approximate area under the curve
```

```
    Area           <- sum(L2)
    Posterior      <- L2/Area                    # Posterior probabilities
# Find maximum probability for scaling plot
    Max.Prob       <- max(Posterior)
# Plot scaled posterior probability
    plot(Theta, Posterior/Max.Prob)
```

C.7.3 Sequential Bayesian analysis of mark-recapture data

```
# Function to calculate probability of m marked in sample of n
    Recaptures    <- function(theta,n,m) {choose(n,m)*theta^m*(1-theta)
                      ^(n-m)}
# Get data elements
    n             <- c(34,42,43,40,32,56,42,44,56,44)
    M             <- c(50,84,125,168,207,239,294,335,375,428)
    m             <- c(0,1,0,1,0,1,1,4,3,1)
# Analyze First sample
    Nmin          <- 500                         # Lowest N
    Npop          <- seq(Nmin, 30000, by=100)    # Values of N used
    Nvalues       <- length(Npop)                # Find number of N
# Set up matrix to take posterior probabilities
    Posterior     <- matrix(0,Nvalues,10)
    Theta         <- M[1]/Npop                   # Vector of theta values
# Binomial probabilities
    Prob          <- Recaptures (Theta,n[1],m[1])
    Posterior[,1] <- Prob/sum(Prob)             # Posterior probabilities
# Now iterate over remaining nine samples
    for (i in 2:10)
    {
# Vector of theta values
    Theta         <- M[i]/Npop
# Binomial probabilities
    Prob          <- Recaptures(Theta,n[i],m[i])
# Posterior probabilities
    Posterior[,i] <- Prob*Posterior[,i-1]
# Posterior probabilities
    Posterior[,i] <- Posterior[,i]/sum(Posterior[,i])
    }
# Plot all 10 curves
    plot(rep(Npop,10),Posterior[,])
```

Appendix D

Solutions to exercises

Because of differences among computers, probabilities from randomization and bootstzap routines may vary slightly.

Solutions to Chapter 2

Question 2.1

The following is S-PLUS coding to do the exercise.

```
# The following two lines actually generated the X values
    set.seed(1)                          # Initialize random number generator
    x      <- rnorm(10,0,1)              # Generate 10 normal deviates
# The following is the coding for calculating LL and plotting
    Mean.X <- mean(X)                     # Calculate mean of data set
# Generate a sequence of mu values from -3 to +3
    mu     <- seq(from=-3, to=3, by=0.1)
    n      <- length(mu)                  # Get the length of mu
# Create a matrix to store log-likelihoods
    LL     <- matrix(0,n,1)
    constant <-log(1/sqrt(2*pi))         # Calculate constant
# Calculate log-likelihoods for each value of theta
    for (i in 1:n){LL[i] <- sum(constant-.5*(X-mu[i])^2)}
# Find maximum value and Concatenate mu and LL
    Out    <- matrix(c(mu,LL), nrow=n, ncol=2)
    LLmax  <- max(LL)                     # Find maximum LL
    mu.LL  <- Out[Out[,2]==LLmax]        # Find mu corresponding to this value
# Print results
    Mean.X
    mu.LL
# Plot data
    plot(mu,LL, xlab="mu", ylab="log-Likelihood", cex=1)
```

Output

Mean.X

0.2989654

mu.LL

0.30000 -13.13142

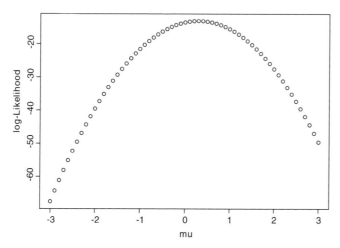

Question 2.2

$\sum_{i=1}^{n} (x_i - \bar{x})^2 = \sum_{i=1}^{n} x_i^2 - n\bar{x}^2$. As $\mu = 0$, then, letting $E(x)$ refer to the "expected value of x" we have $E(x_i) = 0$, $E(\bar{x}) = 0$, $E(x_i^2) = \sigma^2$, $E(\bar{x}^2) = \sigma^2/n$. Therefore, $E\left(\sum_{i=1}^{n} (x_i - \bar{x})^2\right) = n\sigma^2 - \sigma^2 = (n-1)\sigma^2$ and so $E\left((1/n)\sum_{i=1}^{n} (x_i - \bar{x})^2\right) = ((n-1)/n)\sigma^2$, which is less than σ^2 and hence is a biased estimate of σ^2. This bias can obviously be removed by dividing by $1/(n-1)$.

Question 2.3

The likelihood is $L = \prod_{i=1}^{m} e^{-\theta}(\theta^{r_i}/r_i!)$, where m is the number of sampling units, and r_i is the number observed in the ith sampling unit. Taking logs gives $LL = \sum_{i=1}^{m} (-\theta + r_i \ln \theta - \ln r_i)$. Differentiating, $(dLL/d\theta) = \sum_{i=1}^{m} (-1 + r_i/\theta) = -m + \sum_{i=1}^{m} (r_i/\theta)$, and setting the result to zero gives $(dLL/d\theta) = 0$ when $\theta = (1/m) \sum_{i=1}^{m} r_i$.

Question 2.4

The following coding is relatively slow but easy to follow. Note that there is a highly significant correlation between the estimated intercept and slope. To produce the same result each time, the program is run the random number generator is seeded using set.seed.

```
# Set up vectors for intercept A and slope B
    A <- matrix(0,20,1)
    B <- matrix(0,20,1)
```

```
# Construct X values evenly spaced from 1 to 10
    X <- seq(from=1, to=10)
    set.seed(1)                          # Set seed for random number generator
    for (irep in 1:20)                   # Iterate over 20 samples
    {
    Error   <- rnorm(10, mean=0, sd=1)   # Construct error term
    Y        <- X + Error                # Construct Y values
    Model   <- lm(Y~X)                   # Calculate regression coefficients
# Store coefficients
    A[irep]   <- Model$coefficients[1]
    B[irep]   <- Model$coefficients[2]
    }
    cor.test(A,B)                        # Test correlation between A and B
```

Output

```
cor.test(A, B)
    Pearson's product-moment correlation
data: A and B
t = -7.4259, df = 18, p-value = 0
alternative hypothesis: true coef is not equal to 0
sample estimates:
      cor
 -0.8682804
```

Question 2.5
Using the S-PLUS nonlinear regression dialog box

```
    *** Nonlinear Regression Model ***
Formula: Eggs ~ b1 * (1 - exp( - b2 * (Day - b3))) * exp( - b4 * Day)
Parameters:
      Value Std.        Error     t value
b1   107.613000   15.2729000    7.04598
b2     0.829793    0.2745190    3.02272
b3     1.173820    0.1019500   11.51370
b4     0.103955    0.0168184    6.18103
Residual standard error: 3.53335 on 5 degrees of freedom
Correlation of Parameter Estimates:
        b1      b2        b3
b2 -0.887
b3 -0.534    0.758
b4 0.965     -0.798   -0.455
```

Question 2.6

Equation 2.28 can be readily expanded to include two trials: $L = \prod_{i=1}^{2} \frac{n_i!}{r_i!(n_i-r_i)!} p^{r_i} (1-p)^{n_i-r_i}$. Taking log, $\ln(L) = \sum_{i=1}^{2} \left(\ln\left(\frac{n_i!}{r_i!(n_i-r_i)!}\right) + r_i \ln(p) + (n_i - r_i) \ln(1-p) \right)$. Differentiate and set the result to zero to find the MLE, $\frac{d \ln(L)}{dp} = \sum_{i=1}^{2} \frac{r_i}{p} - \sum_{i=1}^{2} \frac{n_i-r_i}{1-p} = \frac{1}{p} \sum_{i=1}^{2} r_i - \frac{1}{1-p} \left(\sum_{i=1}^{2} n_i - \sum_{i=1}^{2} r_i \right) = \frac{R}{p} - \frac{N-R}{p}$, where $R = r_1 + r_2$ and $N = n_1 + n_2$. Now $\frac{d \ln(L)}{dp} = 0$ when $\frac{R}{p} - \frac{N-R}{p} = 0$, which upon rearrangement gives $\hat{p} = R/N$, as required.

Question 2.7

The parameter to be estimated is μ and is designated as mu in the coding

```
# Generate 10 normally distributed random numbers
# Set seed to make runs repeatable
   set.seed(1)
   Xobs    <- rnorm(10,0,1)
   Mean.X  <- mean(Xobs)          # Calculate mean of data set
   SE.X    <- sqrt(var(Xobs)/10)  # Calculate SE by usual means
# Calculate lower and upper confidence values
   lower   <- Mean.X-2.262*SE.X
   upper   <- Mean.X+2.262*SE.X
# Generate a sequence of mu values from -2 to +2
   mu      <- seq(from=-2, to=2, by=0.01)
   n       <- length(mu)          # Get the length of mu
   L       <- matrix(0,n,1)       # Create a matrix to store likelihoods
# Calculate likelihoods for each value of mu
   for (i in 1:n){L[i] <- prod(exp(-.5*(Xobs-mu[i])^2))}
   Total <-sum(L)                 # Sum all likelihoods
# Divide by Total to make likelihoods sum to 1
   L       <- L/Total
   Cum.L   <- cumsum(L)           # Calculate vector of cumulative sums
# Concatenate mu and Cum.L for easy reading
   Out  <- matrix(c(mu,Cum.L), nrow=n, ncol=2)
# print out lower, mean and upper estimates of usual formula
   print(c(lower,Mean.X,upper))
```

Output

```
print(c(lower, Mean.X, upper))
-0.3705279   0.2989654   0.9684587
```

Examination of the file Out gives -0.330 and 0.910 for the lower and upper limits, which is reasonably close to the values estimated using the usual formula. The two sets of estimates will approach each other as the sample size is increased.

Question 2.8

Coding is based on that given in Appendix C.2.3

```
# Set up function to calculate negative of the log likelihood (omitting
    constants)
    LL <- function(mu)
    {
# Calculate log likelihood for the sample omitting constant
    L1 <- -(1/2)*sum((Xobs-mu)^2)
# Return negative of the log-likelihood
    return (-L1)
    }
# Main Program
    set.seed(1)
    Xobs       <- rnorm(10,0,1)
    mu         <- 0.0                  # Set initial estimates for Mean
    min.func   <- nlmin(LL,mu)         # Call minimization routine
    MLE.mu     <- min.func$x           # Save estimate
    Global.LL <- -LL(MLE.mu)           # Calculate Log-Likelihood at MLE
# Create a function to square Diff so that minima are at zero
    Limit  <- function(mu){(Global.LL+LL(mu)-0.5*3.841)^2 }
# Find lower limit by restricting upper value below MLE.mu
    mu <- -1
    min.func <- nlminb(mu, Limit, lower=-10, upper=MLE.mu-0.1)
    Lower.mu <- min.func$parameters  # Save estimate
# Find upper limit by restricting lower value above MLE.mu
    mu <- 1
    min.func <- nlminb(mu, Limit, lower=MLE.mu+0.01, upper=10)
    Upper.mu <- min.func$parameters  # Save estimate
# Print out results
    print(c(Lower.mu,MLE.mu,Upper.mu))
```

Output

```
print(c(Lower.mu, MLE.mu, Upper.mu))
-0.3207926   0.2989654   0.9187234
Using the standard error gives -0.37 to 0.97
```

Question 2.9

For simplicity let $C_t = (1 - e^{-k(t-t_0)})$. The log-likelihood function is $LL = -n \ln(\sigma\sqrt{2\pi}) - \frac{1}{2\sigma^2} \sum_{t=1}^{n} \theta C_t^2$. The second differential with respect to LL is $-\frac{1}{\sigma^2} \sum_{t=1}^{n} C_t^2$. Therefore the standard error of θ is $\sigma^2 \left(\sum_{t=1}^{2} C_t^2\right)^{-1}$.

Question 2.10

```
# Set up dataframe for data
    Age       <- seq(1:10)
    Length    <- c(23.61,43.10,57.54,68.24,76.16,82.03,86.38,89.60,91.99,
                  93.76)
    D         <- data.frame(matrix(c(Age,Length),nrow=10))
# Data are contained in dataframe D
    k         <- 0.3
    t0        <- 0.05
# Fit von Bertalanffy function
    Model     <- nls(D[,2]~b1*(1-exp(-k*(D[,1]-t0))), data=D, start=list
                  (b1=60))
# Save Estimate as Theta
    Theta     <- as.numeric(Model$parameters)
# Calculate predicted values
    D.fit     <- Theta*(1-exp(-k*(D[,1]-t0)))
# Calculate squared difference between observed and expected
    D.fit2    <- (D[,2]-D.fit)^2
# Find number of observations
    n         <- nrow(D)
# Calculate estimate of sigma (residual standard error)
    sigma.est <- sqrt(sum(D.fit2)/(n-1))
# Calculate estimate of standard error for Theta
    Sigma.Theta <- sqrt(sigma.est^2*(sum((1-exp(-k*(D[,1]-t0)))^2))^-1)
# Output results
    summary(Model)
    print(c(sigma.est, Theta, Sigma.Theta))
```

Output

```
summary(Model)
Formula: D[, 2] ~ b1 * (1 - exp( - k * (D[, 1] - t0)))
Parameters:
    Value Std.    Error   t value
b1  98.4992  0.152934  644.064

Residual standard error: 0.366283 on 9 degrees of freedom
```

```
> print(c(sigma.est, Theta, Sigma.Theta))
  0.3662833   98.4991562   0.1529339
```

Question 2.11

Using nls routine in S-PLUS (Appendix C.2.10)

```
# Enter data
    Eggs   <- c(54.8,73.5,78,71.4,75.6,73.2,65.4,61.9,61.7,60.1,55.1,50.4,44.3,42.3)
    Day    <- c(1,2,3,4,5,6,7,8,9,10,11,12,13,14)
    D      <- data.frame(Day,Eggs)
# Fit four parameter Drosophila model
    Model    <- nls(Eggs~b1*(1-exp(-b2*(Day-b3)))*exp(-b4*Day), data=D,
                 start=list(b1=100,b2=0.5,b3=1,b4=0.1))
# Store results
    Four.Parameter.Model <- Model
# Save residual sums of squares
    SS.4     <- sum(Model$residuals^2)
# Fit model assuming b3=0
    Model    <- nls(Eggs~b1*(1-exp(-b2*Day))*exp(-b4*Day), data=D,
                 start=list(b1=100,b2=0.5,b4=0.1))
# Store results
    Three.parameter.Model <- Model
# Save residual sums of squares
    SS.3     <- sum(Model$residuals^2)
# Get sample size n
    n        <- nrow(D)
# Compute F value
    F.value <- ((SS.3-SS.4)/(4-3) )/(SS.4/(n-4))
# Compute probability
    P        <- 1 - pf(F.value, 1, n-4) # p-value of stat
# Print out results
    summary(Four.Parameter.Model)
    summary(Three.parameter.Model)
    print(c(F.value, P))
```

Output

```
summary(Four.Parameter.Model)
Formula: Eggs ~ b1 * (1 - exp( - b2 * (Day - b3))) * exp( - b4 * Day)
Parameters:
        Value Std.       Error       t value
b1   103.8250000   7.40431000   14.022200
```

```
b2    0.6794940    0.20532000    3.309430
b3   -0.2321230    0.32006400   -0.725242
b4    0.0608795    0.00714721    8.517940
```

Residual standard error: 2.80399 on 10 degrees of freedom

Correlation of Parameter Estimates:

```
b1    b2    b3
b2   -0.873
b3   -0.670    0.922
b4    0.971   -0.808   -0.603
```

```
> summary(Three.parameter.Model)
```

Formula: Eggs ~ b1 * (1 - exp(- b2 * Day)) * exp(- b4 * Day)

Parameters:

```
          Value Std.        Error     t value
b1    100.0260000    4.32433000    23.13100
b2      0.8543150    0.09022010     9.46923
b4      0.0574262    0.00488552    11.75440
```

Residual standard error: 2.73403 on 11 degrees of freedom

Correlation of Parameter Estimates:

```
          b1          b2
b2    -0.837
b4     0.945    -0.757
```

```
> print(c(F.value, P))
0.4579748    0.5139206
```

Hypothesis that $\theta_3 = 0$ cannot be rejected. Note that the confidence range for θ_3 in the four parameter model considerably overlap zero.

Solutions to Chapter 3

Question 3.1

```
set.seed(0)                            # Set random number seed
n          <- 100                      # Number of replicates
x          <- rnorm(n,0,1)             # n random normal values
X.jack    <- jackknife(data=x, var(x))  # Jackknife the data
# Create pseudovalues see appendix C.3.2
   Pseudovalues <- n*X.jack$observed-(n-1)*X.jack$replicates
   shapiro.test(Pseudovalues)          # Test for normality
   hist(Pseudovalues, probability=T)   # Plot data
```

Output

```
Shapiro-Wilk Normality Test
data: Pseudovalues
W = 0.7305, p-value = 0
```

Pseudovalues

Question 3.2

The data were created using the model $y = x + \varepsilon$, where ε is $N(0, 1)$. Here are the lines used to create the data

```
set.seed(1)                          # Set seed for random number generator
n       <- 20                        # Number of points
x       <- runif(n,0,10)             # Construct uniform X values
error <- rnorm(n, mean=0, sd=1)      # Generate error term
y       <- x + error                 # Construct Y values
```

Next lines answer questions asked

```
# Calculate regression coefficients
Model           <- lm(y~x)
Obs.b           <- matrix(Model$coefficients,2)     # Store coefficients
# Create matrix to store pseudovalues
Pseudovalues    <- matrix(0,n,2)
```

```
# Create individual values
ind                    <- seq(1,n)
D                      <- data.frame(ind,x,y)       # concatenate data
    for (i in 1:n)
      {
      D.minus.i        <- D[D$ind!=i,]                  # Delete ith row
                                                        # Fit model
      Model.minus.i    <- lm(D.minus.i$y~D.minus.i$x)
# Pick out coefficients
      b.minus.i        <- matrix(Model.minus.i$coefficients,2)
      Pseudo.b         <- n*Obs.b-(n-1)*b.minus.i  # Create pseudovalue
      Pseudovalues[i,] <- Pseudo.b                      # Store pseudovalues
      }
    summary(Model)       # Print out results for least squares
# Print out Jackknife values
    print(c(mean(Pseudovalues[,1]), sqrt(var(Pseudovalues[,1])/n)))
    print(c(mean(Pseudovalues[,2]), sqrt(var(Pseudovalues[,2])/n)))
    t.test(Pseudovalues[,1],mu=0)                   # t test intercept = 0
    t.test(Pseudovalues[,2],mu=1)                   # t test slope = 1
    t.test(Pseudovalues[,2],mu=0)                   # t test slope = 0
```

Alternative coding using jackknife routine

```
# Next lines answer questions asked
    D                 <- data.frame(ind,x,y)       # concatenate data
# jackknife estimation
    Jack.Data     <- jackknife(data=D, lm(y~x,data=D)$coef)
# Extract delete-one replicates
    Replicates    <- matrix(Jack.Data$rep,n,2)
# Create Pseudovalues
    Pseudovalues  <- matrix(0,n,2)
    for ( i in 1:2){Pseudovalues[,i] <- n*Jack.Data$obs[i]-(n-1)
    *Replicates[,i]}
# Print out Jackknife values
    print(c(mean(Pseudovalues[,1]), sqrt(var(Pseudovalues[,1])/n)))
    print(c(mean(Pseudovalues[,2]), sqrt(var(Pseudovalues[,2])/n)))
    t.test(Pseudovalues[,1],mu=0)      # t test intercept = 0
    t.test(Pseudovalues[,2],mu=1)      # t test slope = 1
    t.test(Pseudovalues[,2],mu=0)      # t test slope = 0
```

Output (summarized)

```
summary(Model)
Coefficients:
            Value Std.    Error    t value   Pr(>|t|)
(Intercept)  0.0927   0.3111   0.2979    0.7692
         x   0.9332   0.0522  17.8877    0.0000
Residual standard error: 0.7338 on 18 degrees of freedom
Multiple R-Squared: 0.9467
F-statistic: 320 on 1 and 18 degrees of freedom, the p-value is 6.554e-013
# Print out Jackknife values
   [1] 0.07761845   0.32774423
   [1] 0.93522574   0.06224435
# t test intercept = 0
   One-sample t-Test
   data: Pseudovalues[, 1]
   t = 0.2368, df = 19, p-value = 0.8153
   alternative hypothesis: mean is not equal to 0
# t test intercept = 0
   One-sample t-Test
   data: Pseudovalues[, 2]
   t = -1.0406, df = 19, p-value = 0.3111
   alternative hypothesis: mean is not equal to 1
# t test slope = 1
   t.test(Pseudovalues[, 2], mu = 0)
   One-sample t-Test
   data: Pseudovalues[, 2]
   t = 15.0251, df = 19, p-value = 0
   alternative hypothesis: mean is not equal to 0
```

Question 3.3

```
# Lines that generated the tabulated data
# Set seed for random number generator
   set.seed(1)
   n        <- 20                        # Number of points
# Construct X values evenly spaced from 1 to 10
   x        <- runif(n,0,10)
   error    <- rnorm(n, mean=0, sd=1)    # Generate error term
```

```
y                <- x + error            # Construct Y values
xy               <- cbind(x,y)           # Data set to be examined
# Jackknife the correlation coefficient
Jack.r           <- jackknife(data=xy, cor(xy[,1],xy[,2]))
# Pseudovalues
Pseudovalues     <- n*Jack.r$observed-(n-1)*Jack.r$replicates
Jack.r                                   # Output
t.test(Pseudovalues, mu=0)               # t test for = 0
# Repeat using the z transformation
Jack.r           <- jackknife(data=xy, 0.5*log((1+cor(xy[,1],xy[,2]))/
                    (1-cor(xy[,1],xy[,2]))))
Pseudovalues     <- n*Jack.r$observed-(n-1)*Jack.r$replicates
Jack.r
t.test(Pseudovalues, mu=0)
```

Output
Untransformed estimate

```
Number of Replications: 20
Summary Statistics:
        Observed     Bias     Mean        SE
Param    0.973    0.001111   0.9731    0.008788
   One-sample t-Test
data: Pseudovalues = 110.5965, df = 19, p-value = 0
alternative hypothesis: mean is not equal to 0
95 percent confidence interval: 0.9535018   0.9902877
sample estimates: mean of x
   0.9718947
```

Transformed estimate

```
Number of Replications: 20
Summary Statistics:
        Observed     Bias    Mean       SE
Param    2.146    0.04788   2.148    0.1672
   One-sample t-Test
data: Pseudovalues = 12.5489, df = 19, p-value = 0
alternative hypothesis: mean is not equal to 0
95 percent confidence interval: 1.748053 2.447894
sample estimates: mean of x
   2.097973
```

Question 3.4

```
# Jackknife the correlation coefficient
    Jack.r <- jackknife(data=xy, cor(xy[,1],xy[,2]))
    Pseudovalues1 <- n*Jack.r$observed-(n-1)*Jack.r$replicates
    hist(Pseudovalues1) # Plot histogram
# Repeat using the z transformation
    Jack.r <- jackknife(data=xy, 0.5*log((1+cor(xy[,1],xy[,2]))/
    (1-cor(xy[,1],xy[,2]))))
    Pseudovalues2 <- n*Jack.r$observed-(n-1)*Jack.r$replicates
    hist(Pseudovalues2) # Plot histogram
# Concatenate two files
    Pseudovalues <- data.frame(cbind(Pseudovalues1,Pseudovalues2))
# Calculate basic statistics
# Note that this is the command issued by the dialog box and recovered from the
    history window
    menuDescribe(data = Pseudovalues, variables = "<ALL>", grouping.
    variables = "(None)", max.numeric.levels = 10, nbins = 6, min.p = T, first.
    quant.p = F, mean.p = T, median.p = T, third.quant.p = F, max.p = T, nobs.
    p = T, valid.n.p = T, var.p = T, stdev.p = T, sum.p = F, factors.too.p = T,
    print.p = T, se.mean.p = T, conf.lim.mean.p = F, conf.level.mean = 0.95,
    skewness.p = T, kurtosis.p = T)
# Test for normality
    shapiro.test(Pseudovalues[,1])
    shapiro.test(Pseudovalues[,2])
```

```
***  Summary Statistics for data in:   Pseudovalues ***
```

	X1.1	X1.2
Min:	6.932995e-002	-6.66044450
Mean:	9.462771e-001	1.79419576
Median:	9.521449e-001	1.85103918
Max:	1.106372e+000	3.32400503
Total N:	1.000000e+003	1000.00000000
Variance:	8.307094e-003	0.76286892
Std Dev.:	9.114326e-002	0.87342367
SE Mean:	2.882203e-003	0.02762008
Skewness:	-2.260647e+000	-2.28551313
Kurtosis:	1.309023e+001	13.34630679

```
# Test for normality
    Shapiro-Wilk Normality Test
data: Pseudovalues[, 1] W = 0.8599, p-value = 0
data: Pseudovalues[, 2] W = 0.8583, p-value = 0
```

The transformed values (right) are clearly more normally distributed than the untransformed values. Both are significantly different from normal according to the Shapiro–Wilk test. It is not clear if the deviation from normality is sufficient to discount the jackknife in this situation. Further simulations are required, particularly with respect to confidence intervals and hypothesis testing.

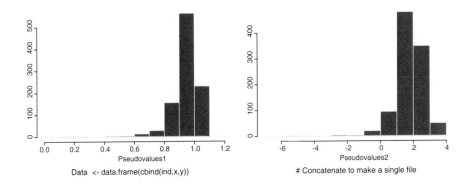

Data <- data.frame(cbind(ind,x,y))   # Concatenate to make a single file

Question 3.5

The actual model used was $Eggs = \theta_1(1-e^{-\theta_2 Age})e^{-\theta_3 Age}+\varepsilon$, where $\theta_1=100$, $\theta_2=1$, $\theta_3=0.1$ and ε is $N(0, 1)$. The data were generated using the following coding

```
# set seed for random number generator
    set.seed(1)
    n        <- 20                    # Number of observations
# Generate 20 integer ages using uniform probty dist
    x        <- ceiling(runif(n,0,5))
# Generate 20 random normal variables N(0,1)
    error    <- rnorm(n,0,1)
    y        <- 100*(1-exp(-1*x))*exp(-.1*x) +error   # Generate eggs
    y        <- floor(y+0.5)                           # set to nearest integer
# Generate individual identifiers
    ind      <- seq(1,n)
# Concatenate to make a single file
    Data     <- data.frame(cbind(ind,x,y))
# Coding not using the jackknife routine of S-PLUS to fit MLE and jackknife is
# Fit Drosophila model
    Model            <- nls(y~b1*(1-exp(-b2*x))*exp(-b3*x), data=Data,
                         start=list(b1=50,b2=0.5,b3=.5))
    Obs.Model        <- Model                    # Store results
# Store estimated parameters
    Obs.b            <- matrix(Model$parameters)
```

```
# Create matrix to store pseudovalues
    Pseudovalues  <- matrix(0,n,3)
    for (i in 1:n)
{
# Delete ith x and y values
    Data.minus.i      <- Data[Data$ind!=i,]
    Model   <- nls(y~b1*(1-exp(-b2*x))*exp(-b3*x),data=Data.minus.i,
    start=list(b1=50,b2=0.5,b3=.5))
    b.minus.i         <-matrix(Model$parameters,3)  # Pick out coefficients
    Pseudo.b          <- n*Obs.b-(n-1)*b.minus.i      # Create pseudovalue
    Pseudovalues[i,] <- Pseudo.b                       # Store pseudovalues

    }
# Print out results for least squares
    summary(Model)
# Print out Jackknife values
    print(c(mean(Pseudovalues[,1]), sqrt(var(Pseudovalues[,1])/n)))
    print(c(mean(Pseudovalues[,2]), sqrt(var(Pseudovalues[,2])/n)))
    print(c(mean(Pseudovalues[,3]), sqrt(var(Pseudovalues[,3])/n)))
```

Alternative coding using the jackknife routine of S-PLUS

```
# Create function to fit data to equation using nls
# Note that variables are assumed to be called x and y
    Model <-function(data)
{
    Fit <- nls(y~b1*(1-exp(-b2*x))*exp(-b3*x),data=data,start=list
    (b1=50,b2=0.5, b3=.5))
    return(Fit$param)

}
# jackknife estimation
    Jack.Data   <- jackknife(data=Data, Model(Data))
# Extract delete-one replicates
    Replicates <- matrix(Jack.Data$rep,n,3)
    for ( i in 1:3)
{
# Create Pseudovalues
    Pseudovalues <- n*Jack.Data$obs[i]-(n-1)*Replicates[,i]
    print(c(mean(Pseudovalues),
    sqrt(var(Pseudovalues)/n)) )                     # Print out mean and SE

}
    Model(Data)  # Output stats from model fit to all data
```

Output (order differs between coding alternatives)

```
summary(Model)
Formula: y ~ b1 * (1 - exp( - b2 * x)) * exp( - b3 * x)
Parameters:
            Value      Std. Error    t value
b1    101.948000      2.83085000    36.0131
b2      0.973440      0.04228580    23.0205
b3      0.105853      0.00598384    17.6898


# Print out Jackknife values
101.47614       2.88243
  0.97701871    0.04457281
  0.10479458    0.006226638
```

Question 3.6

```
# Generate Original data
    set.seed(1)                       # set seed for random number generator
    n       <- 20                     # Number of observations
# 20 integer ages from uniform probability distribution
    Age     <- ceiling(runif(n,0,5))
# Generate 20 random normal variables N(0,1)
    error   <- rnorm(n,0,1)
    Eggs    <- 100*(1-exp(-1*Age))*exp(-.1*Age) +error    # Generate eggs
    Eggs    <- floor(Eggs+0.5)                     # set to nearest integer
# Produce a bootstrap sample
    n       <- 10                     # Reset n
# set seed for random number generator
    set.seed(1)
    x    <- sample(Age,size=n,replace=T)  # Sample with replacement from Age
# reset seed for random number generator to get same run
    set.seed(1)
# Sample with replacement from Eggs
    y    <- sample(Eggs,size=n,replace=T)
    ind  <- seq(1:n)                        # Generate individual identifiers
    Data <- data.frame(cbind(ind,x,y))      # Concatenate to make a single file
# Coding not using the jackknife routine of S-PLUS to fit MLE and jackknife is
# Fit Drosophila model
    Model <- nls(y~b1*(1-exp(-b2*x))*exp(-b3*x), data=Data, start=list
    (b1=50,b2=0.5,b3=.5))
```

```
    Obs.Model    <- Model                        # Store results
# Store estimated parameters
    Obs.b        <- matrix(Model$parameters)
# Create matrix to store pseudovalues
    Pseudovalues <- matrix(0,n,3)
    for (i in 1:n)
      {
    Data.minus.i<- Data[Data$ind!=i,]        # Delete ith x and y values
    Model        <- nls(y~b1*(1-exp(-b2*x))*exp(-b3*x),data=Data.minus.i,
              start=list(b1=50,b2=0.5,b3=.5))
    b.minus.i    <- matrix(Model$parameters,3) # Pick out coefficients
    Pseudo.b     <- n*Obs.b-(n-1)*b.minus.i    # Create pseudovalue
    Pseudovalues[i,] <- Pseudo.b               # Store pseudovalues
      }
# Print out results for least squares
summary(Obs.Model)
print(c(mean(Pseudovalues[,1]), sqrt(var(Pseudovalues[,1])/n)))
# Print out Jackknife values
print(c(mean(Pseudovalues[,2]), sqrt(var(Pseudovalues[,2])/n)))
# Print out Jackknife values
print(c(mean(Pseudovalues[,3]), sqrt(var(Pseudovalues[,3])/n)))
# Print out Jackknife values
shapiro.test(Pseudovalues[,1])      # Test for normality
shapiro.test(Pseudovalues[,2])      # Test for normality
shapiro.test(Pseudovalues[,3])      # Test for normality
```

Alternative coding using the jackknife routine of S-PLUS

```
# Fit Drosophila model
# Create function to fit data to equation using nls
# Note that variables are assumed to be called x and y
    Model <- function(data)
{
    Fit <- nls(y~b1*(1-exp(-b2*x))*exp(-b3*x), data=data,start=list
    (b1=50,b2=0.5,b3=.5))
    return(Fit$param)
}

    Jack.Data    <- jackknife(data=Data, Model(Data))
    # jackknife estimation
    Replicates   <- matrix(Jack.Data$rep,n,3)
    # Extract delete-one replicates
```

```
    for ( i in 1:3)
{
    Pseudovalues <- n*Jack.Data$obs[i]-(n-1)
    *Replicates[,i]                           # Create Pseudovalues
# Print out mean and SE
    print(c(mean(Pseudovalues), sqrt(var(Pseudovalues)/n)) )
    print(shapiro.test(Pseudovalues))         # Test for normality
}
Model(Data)     # Parameter estimates for full data
```

Output (slightly different order for each alternative)

```
> summary(Model)
Formula: y ~ b1 * (1 - exp( - b2 * x)) * exp( - b3 * x)
Parameters:
             Value      Std. Error     t value
b1        103.104000    2.90365000    35.5084
b2          0.957489    0.03960900    24.1735
b3          0.107605    0.00652165    16.4997
# Print out Jackknife values
98.981087      7.299615
 0.98535813    0.07911545
 0.09784679    0.01672593
    Shapiro-Wilk Normality Test
data: Pseudovalues[, 1] W = 0.6508, p-value = 0.002
data: Pseudovalues[, 2] W = 0.7688, p-value = 0.006
data: Pseudovalues[, 3] W = 0.6044, p-value = 0.0001
```

The lack of normality arises from the conversion of Eggs to the nearest integer

Solutions to Chapter 4
Question 4.1

```
set.seed(0)                               # Set random number seed
n              <- 100                      # Number of replicates
x              <- rnorm(n,0,1)            # n random normal values
X.Boot         <- bootstrap(data=x, median(x)) # Bootstrap the data
# Create vector of replicates
Replicates     <- X.Boot$replicates
Replicates.df  <- data.frame(Replicates)  # Create data frame
summary(X.Boot)                           # Print results
shapiro.test(Replicates)                  # Test for normality
hist(Replicates, probability=T)           # Plot data
```

Output

```
Number of Replications: 1000
Summary Statistics:
         Observed       Bias     Mean       SE
Param      0.1517     0.0303    0.182   0.1197
Empirical Percentiles:
              2.5%         5%      95%    97.5%
Param     -0.02264   -0.01038   0.3441   0.4078
BCa Confidence Limits:
              2.5%         5%      95%    97.5%
Param     -0.03375   -0.02345    0.324   0.3405
shapiro.test(Replicates)
    Shapiro-Wilk Normality Test
data: Replicates
W = 0.9617, p-value = 0
```

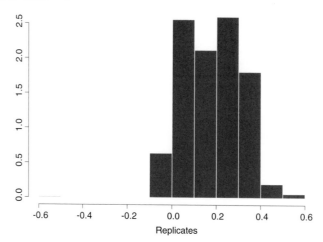

Question 4.2

The data were created using the model $y = x + \varepsilon$, where ε is $N(0, 1)$. Possible coding to solve problem:

```
# Data are in dataframe called Linear.data.df
  x   <- c(1.63,4.25,3.17,6.46,0.84,0.83,2.03,9.78,4.39,2.72,9.68,7.88,
          0.21,9.08,9.04,5.59,3.73,7.98,3.85,8.18)
  y   <- c(2.79,3.72,4.09,5.89,0.75,-0.13,1.76,8.44,5.15,2.16,9.88,
          6.95,0.03,7.50,9.92,5.37,3.79,7.18,3.37,7.81)
Linear.data.df <- data.frame(x,y)
```

```
# Bootstrap data
    Boot.LS          <- bootstrap(Linear.data.df, coef(lm(y~x,Linear.data.
                        df)),B=1000)
    Replicates       <- Boot.LS$replicates      # Store replicates
    Obs.intercept    <- Boot.LS$observed[1]     # Observed intercept
    Obs.slope        <- Boot.LS$observed[2]     # Observed slope
# Set up function for testing hypothesis
    P.Test      <- function(Predicted,Observed,Datafile,Col)
{
# Difference between obs and predicted
    Diff        <- abs(Observed-Predicted)
# differences between observed and predicted
    Diff.Boot   <- abs(Datafile[,Col]-Observed)
# Probability
    P           <- length(Diff.Boot[Diff.Boot Diff])/length(Diff.Boot)
    return(P)
}
# Test for intercept = 0
    P.intercept <- P.Test(0,Obs.intercept, Replicates,1)
    P.slope0    <- P.Test(0,Obs.slope,Replicates,2) # Test for slope = 0
    P.slope1    <- P.Test(1,Obs.slope,Replicates,2) # Test for slope = 1
# Output results
    summary(Boot.LS)
    print(c(P.intercept, P.slope0, P.slope1))
```

Output

Number of Replications: 1000

Summary Statistics:

	Observed	Bias	Mean	SE
(Intercept)	0.09087	0.014585	0.1055	0.31571
x	0.93370	-0.001995	0.9317	0.05794

Empirical Percentiles:

	2.5%	5%	95%	97.5%
(Intercept)	-0.4489	-0.3710	0.6742	0.794
x	0.8146	0.8317	1.0279	1.047

BCa Confidence Limits:

	2.5%	5%	95%	97.5%
(Intercept)	-0.4194	-0.3437	0.7181	0.8364
x	0.8174	0.8342	1.0308	1.0495

```
Correlation of Replicates:
            (Intercept)          x
(Intercept)     1.0000    -0.8649
        x      -0.8649     1.0000
> print(c(P.intercept, P.slope0, P.slope1))
          0.775      0.000      0.261
```

Parametric test

	Value Std.	Error	t value	Pr(>|t|)
(Intercept)	0.0909	0.3111	0.2921	0.7736
x	0.9337	0.0522	17.8943	0.0000

Question 4.3

```
# Data are in dataframe called Linear.data.df
# Do bootstrap
    Boot.Cor       <- bootstrap(Linear.data.df, cor(x,y),B=1000)
    Replicates     <- Boot.Cor$replicates       # Store replicates
# Create vector of transformed values
    z.replicates   <- 0.5*log((1+Replicates)/(1-Replicates))
    Obs.r          <- Boot.Cor$observed         # Observed correlation
# Set up function for testing hypothesis
    P.Test    <- function(Predicted,Observed,z.file)
{
# Transform observed values
    z.obs     <- 0.5*log((1+Observed)/(1-Observed))
# Transform predicted value
    z.pred    <- 0.5*log((1+Predicted)/(1-Predicted))
# Difference between observed and predicted
    Diff      <- abs(z.obs-z.pred)
# Differences between observed and predicted
    Diff.Boot <- abs(z.file-z.obs)
# Probability
    P         <- length(Diff.Boot Diff.Boot>Diff])/length(Diff.Boot)
    return(P)
}
# Test for r = 0.96
    P.r  <- P.Test(0.96,Obs.r,z.replicates)
# Do parametric test
    n        <- nrow(Linear.data.df)            # Sample size
```

```
   SE.z    <- sqrt(1/(n-3)  )              # Standard error of z
   z.pred  <- 0.5*log((1+0.96)/(1-0.96))   # Predicted z
   z.obs   <- 0.5*log((1+Obs.r)/(1-Obs.r)) # Observed z
   z       <- abs((z.obs-z.pred)/SE.z)     # z
   P.z     <- 1-pnorm(z, mean=0, sd=1)     # Proportion of normal above z
# Output results
   print(c(Obs.r, P.r, z, P.z))
```

Output

```
print(c(Obs.r,    P.r,          z,       P.z))
    0.9730251   0.221   0.8258512   0.2044443
```

Note excellent agreement between bootstrap results and parametric test.

Question 4.4

The data in Corr.df can be analyzed using the coding given in the previous question by replacing Linear.data.df with Corr.df and 0.96 with 0.0.

Output

```
print(c(Obs.r,    P.r,          z,       P.z))
    0.4325173   0.064   1.908953   0.02813406
```

There is considerable difference between the two probabilities. Reliability can only be assessed by simulation, as done in the following question.

Question 4.5

Coding is as follows:

```
# Set seed for random number generator
   set.seed(0)
   n              <- 20                      # Number of points
   z.values   <- matrix(0,nrow=10000)       # Set up matrix for z values
   Obs.r      <- 0.0                         # Set observed r
   Obs.z      <- 0.5*log((1+Obs.r)/(1-Obs.r)) # Calculate observed z
# Iterate over 10000 replicates
   for (i in 1:10000)
   {
   x              <- rnorm(n,0,1)            # Construct normal x values
   shape          <- 2                       # Set shape parameter
   rate           <- shape                   # Set rate parameter
   mu             <- shape/rate              # Set mean
# Generate error term with mean zero
```

```
    error             <- rgamma(n,shape,rate)-mu
    y                 <- 0.5*x+error                    # Construct Y values
    r                 <- cor(x,y)                       # Calculate r
    z.values[i]   <- 0.5*log((1+r)/(1-r))              # Calculate Fisher's z
}
    mean(z.values <Obs.z)                               # Find proportion less than Obs.z
```

Output

```
mean(z.values < Obs.z)
    0.005
```

The proportion of cases in which r is less than zero is 0.005 and hence, the tests in question 4 should give such a low probability. The parametric analysis actually gives a value closer to the correct one. Testing for r=0 on 10 data sets gave the following results

```
print(c(Obs.r,      P.r,        z,          P.z))        Closest test
      0.3463173    0.257   1.489484    0.06817992          bootstrap
      0.507786     0.002   2.307876    0.01050303          bootstrap
      0.4929078    0.085   2.22604     0.01300574          parametric
      0.4341699    0.074   1.917342    0.02759724          parametric
      0.6701658    0.002   3.344021    0.0004128681        parametric
      0.6194449    0.038   2.985557    0.001415314         parametric
      0.7875485    0       4.390874    5.644794e-006       bootstrap
      0.6127475    0.031   2.941048    0.001635521         parametric
      0.5172109    0.023   2.360579    0.009123207         parametric
      0.6551209    0.002   3.233412    0.0006116045        parametric
```

The parametric test P was closest to expected in seven of the ten runs and thus appears to be the better method in this case.

Question 4.6

The actual model used was $\text{Eggs} = \theta_1(1 - e^{-\theta_2 \text{Age}})e^{-\theta_3 \text{Age}} + \varepsilon$, where $\theta_1 = 100$, $\theta_2 = 1$, $\theta_3 = 0.1$ and ε is $N(0, 1)$. The data were generated using the following coding

```
    set.seed(1)                                  # set seed for generator
    n         <- 20                              # Number of observations
# 20 integer ages, uniform probty
    x         <- ceiling(runif(n,0,5))
# Generate 20 random normal variables N(0,1)
    error  <- rnorm(n,0,1)
```

```
# Generate eggs
    y        <- 100*(1-exp(-1*x))*exp(-.1*x)+error
    y        <- floor(y+0.5)                          # set to nearest integer
# Generate individual identifiers
    ind      <- seq(1,n)
# Concatenate to make a single file
    Drosophila.df <- data.frame(cbind(ind,x,y))

# The following coding does the bootstrap
# Set up function to estimate parameters
    Model.drosophila <- function(D)
{
    Model <-nls(y~b1*(1-exp(-b2*x))*exp(-b3*x), data=D,
    start = list(b1=50,b2=0.5,b3=.5))
    b        <-matrix(Model$parameters,3)        # Pick out coefficients
    return (b)
}
# Bootstrap
    Boot.LS <- bootstrap(Drosophila.df, Model.drosophila(Drosophila.df) ,
    B=1000)
    Boot.LS                                        # Print output
```

Output

```
Number of Replications: 1000

Summary Statistics:

          Observed        Bias        Mean          SE
Param1.1   101.6021   0.0962219   101.6983   2.680841
Param2.1     0.9778   0.0013379     0.9792   0.041636
Param3.1     0.1050   0.0001486     0.1051   0.005742
```

Method	b1(SE)	b2(SE)	b3(SE)
Bootstrap	101.70 (2.68)	0.979 (0.042)	0.105 (0.006)
MLE	101.95 (2.83)	0.973 (0.042)	0.106 (0.006)
Jackknife	101.48 (2.88)	0.977 (0.044)	0.105 (0.006)

Question 4.7

```
set.seed(1)
    n          <- 10
```

```
x           <- rnorm(n,4,1)

cv          <- sqrt(var(x))/mean(x)

cv

Boot.cv   <- bootstrap(x, sqrt(var(x))/mean(x),B=1000, trace=F)

Boot.cv

Boot.cv$parameters
# Generate 10,000 samples to get estimate of SE
nreps              <- 10000

CV.replicate       <- matrix(0,nreps)

for (i in 1:nreps)
{
x                  <- rnorm(n,4,1)

CV.replicate[i]    <- sqrt(var(x))/mean(x)
}
# CV
summary(Boot.cv)

sqrt(var(CV.replicate))

mean(CV.replicate)
```

Output

```
Summary Statistics:
       Observed       Bias      Mean         SE
Param   0.2177    -0.01205    0.2057    0.02855
Empirical Percentiles:
          2.5%        5%        95%       97.5%
Param   0.1358      0.1527    0.2454     0.2513
BCa Confidence Limits:
          2.5%        5%        95%       97.5%
Param   0.1747      0.1839    0.2607     0.2643
sqrt(var(CV.replicate))
       [,1]
[1,] 0.0624759
mean(CV.replicate)
[1] 0.2450466
```

Question 4.8
The data were generated using

```
set.seed(0)
n <- 10
```

```
a <- 1
b <- 19
Y <- runif(n,a,b)           # Generate n uniform random numbers
X <- floor(sort(Y+0.5))     # Sort Y into ascending sequence
```

The following coding from C.4.3 addresses the question

```
# Function to calculate Gini coefficient
    Gini <- function(d)
{
    g <- sort(d)
# Because of jackknife in BCa method it is necessary to have the following
# two lines within the function
    n <- length(g)                          # Number of observations
    z <- 2*seq(1:n)-n-1                      # Generate "numerator"
    return((n/(n-1))*sum(z*g)/(n^2*mean(g)))  # Gini coefficient
}
    boot.x <- bootstrap(X,Gini,B=1000, trace=F) # Call bootstrap routine
    summary(boot.x)                         # Generate stats
# Set up testing procedure
# Initial value of Gini coefficient
    H0      <- 0.05
    inc     <- .01                          # Increment
    Hmax    <- 0.4                          # Maximum value
    while(H0 <Hmax ) {                      # Increment over values
    b       <- unlist(boot.x$estimate[2])   # Get bootstrap estimate
    di      <- abs(boot.x$replicates-b)     # Calculate d_i vector
    d       <- abs(b-H0)                     # Calculate d
    dd      <- di-d                          # Compare values
    print(c(H0,length(dd[dd>0])/1000))      # print H0 and the probability
    H0      <- H0+inc                        # Increment H0
}
```

Output

```
    Number of Replications: 1000
Summary Statistics:
    Observed       Bias     Mean       SE
Gini   0.2155   -0.008927   0.2066   0.07537
Empirical Percentiles:
       2.5%      5%       95%       97.5%
Gini   0.0875   0.1036   0.3513    0.3637
```

```
BCa Confidence Limits:

          2.5%      5%       95%     97.5%
Gini    0.1185   0.1274   0.4262   0.4725
```

```
  H0      P
0.050   0.027
0.060   0.054
0.070   0.065
0.080   0.087
0.090   0.106
.................
0.350   0.057
0.360   0.032
0.370   0.017
0.380   0.014
0.390   0.014
```

Solutions to Chapter 5

Question 5.1

```
# Input data
  x  <- c(-0.79,0.79,-0.89,0.11,1.37,1.42,1.17,-0.53,0.92,-0.58)
  y  <- c(-0.88,-0.17,-1.16,-1.23,2.14,0.86,1.36,-1.46,0.74,-2.15)
  set.seed(1)                        # Initialize random number
  group   <- c(rep(1,10), rep(2,10))  # Set up group identity vector
  Data    <- c(x,y)                   # Concatenate x and y
# Do a t-test
  Test    <- t.test(Data[group==1], Data[group==2])
# Do N permutations of Data
  N <- 1000  # Number of permutations
  Meanboot <- bootstrap(group,t.test(Data[group==1], Data[group==2])
  $statistic, sampler=samp.permute, B=N, trace=F)
# Calculate number of permutations in which absolute difference > than
# observed
  n.over <- sum(abs(Meanboot$replicates) >= abs(Meanboot$observed))
  P <- (n.over+1)/(N+1)    # Remember to add 1 for observed value
  Test       # Print out results of paired t test on original data
  P          # Print P from randomizations
```

```
Standard Two-Sample t-Test
data: Data[group == 1] and Data[group == 2]
t = 0.9256, df = 18, p-value = 0.3669
alternative hypothesis: difference in means is not equal to 0
95 percent confidence interval:
  -0.6272868   1.6152868
sample estimates:
mean of x mean of y
    0.299   -0.195
> # Print out results of randomization estimate of P
          P
      0.3656344
```

To use the mean difference change call to bootstrap to

```
Meanboot <- bootstrap(group,t.test(Data[group==1], Data[group==2],
   paired=T)$estimate, sampler=samp.permute, B=N, trace=F)
```

The estimated P using this statistic is $P = 0.3876124$.

Question 5.2

Use the coding given in C.5.12 substituting given data for x and y.
Output for linear regression on data

```
Coefficients:
            Value Std.   Error   t value   Pr(>|t|)
(Intercept)  0.0909  0.3111   0.2921   0.7736
        x    0.9337  0.0522   17.8943   0.0000
Residual standard error: 0.7335 on 18 degrees of freedom
Multiple R-Squared: 0.9468
F-statistic: 320.2 on 1 and 18 degrees of freedom, the p-value is 6.513e-013
```

Randomization results based on 1,000 permutations

```
    print(c(Pa,      Pb))
    1.000000000   0.000999001
```

Testing for a slope of 1.

This is readily accomplished by creating a vector z, $z = y - x$ and changing the following two lines.

```
# Calculate regression stats for observations
obs.regression <- summary(lm(z~x))
Meanboot <-bootstrap(x, Lin.reg(x,z), sampler=samp.permute, B=N, trace=F)
Linear Regression results
Coefficients:
```

	Value Std.	Error	t value	Pr(>|t|)
(Intercept)	0.0909	0.3111	0.2921	0.7736
x	-0.0663	0.0522	-1.2706	0.2201

```
Residual standard error: 0.7335 on 18 degrees of freedom
Multiple R-Squared: 0.08231
F-statistic: 1.614 on 1 and 18 degrees of freedom, the p-value is 0.2201
Randomization statistics
        print(c(Pa,       Pb))
        0.8141858     0.2237762
```

Question 5.3

Coding to do F ratio test and randomization based on F ratio

```
x <- c(-0.06, -1.51, 1.78, 0.91, 0.05, 0.53, 0.92, 1.75, 0.73, 0.57, 0.17,
    0.31, 0.66, 0.01, 0.16)
y <- c(1.86, 0.44, 0.59, 0.18, -0.59, -1.16, 1.01, -1.49, 1.62, 1.89, 0.10,
    -0.44, -.06, 1.75, 1.74)
# For the purposes of randomization we must set means to common value, say zero
    x      <- x-mean(x)
    y      <- y-mean(y)
    Data  <- c(x,y)                        # Concatenate the two groups
    Group <- c(rep(1,15),rep(2,15))     # Create group membership
# Function to find F value in F ratio test
F.ratio <- function(Index,X, df1,df2)
{
    vx    <- var(X[Index==1])
    vy    <- var(X[Index==2])
    return( max(vx/vy,vy/vx))
}
# Maximum F from observed data
    obs.Fratio <- F.ratio(Group,Data,14,14)
    Pobs       <-(2*(1 - pf(obs.Fratio, df1, df2)))  # P from parametric test
    N          <- 1000                            # Number of permutations
    Meanboot  <- bootstrap(Group, F.ratio(Group,Data,14,14), sampler=samp.
    permute, B=N,trace=F)
```

```
# Calculate number of permutations in which absolute difference > than
# observed
   n.over  <- sum(abs(Meanboot$replicates) >= abs(Meanboot$observed))
   P        <- (n.over+1)/(N+1)   # Remember to add 1 for observed value
   print(c(Pobs,P)) # Print P from parametric test and randomization
print(c(      Pobs,           P       ))
            0.1916366       0.1688312
```

Using the difference between the two variances produces a probability of 0.1428571, which is close to that obtained using the F ratio.

Question 5.4

```
set.seed(5)       # Initiate random number
# The following lines generate the data given in the question
# The population follows logistic growth
   x       <- matrix(0,30,1)
   error <- runif(30,-40,40)
   for ( i in 1:30){x[i] <- trunc(100/(1+exp(1-0.5*i))+error[i])}
# Do Pollard and Lakhani test
   xlog <- log(x)      # Take logs
# calculate the differences
   d  <- c(x[2:30],1)-x   # Last entry is a dummy to keep lengths the same
   d  <- d[1:29]            # Discard dummy
   xd <- x[1:29]
# Use bootstrap routine to do permutations
   N  <- 1000               # Number of permutations
   Meanboot <-bootstrap(xd, cor(xd,d), sampler=samp.permute, B=N,
   trace=F)
# Calculate number of permutations in which difference > than observed
   n.over <- sum(Meanboot$replicates <= Meanboot$observed)
   P <- (n.over+1)/(N+1)                  # Remember to add 1 for observed value
# Parametric test
   cor.test(xd, d, alternative="less", method="pearson")
   P                                       # Print P
```

Output

```
Pearson's product-moment correlation
data: xd and d
t = -4.5061, df = 27, p-value = 0.0001
```

```
alternative hypothesis: coef is less than 0
sample estimates:
        cor
−0.6551566
Randomization        P
            0.000999001
```

Both methods of analysis give comparable results.

Question 5.5

The coding in the previous question is readily adapted to address this question.

```
set.seed(1)
clutch    <- c(1,1,2,2,2,3,4,4,4,5,5,6)
n.survs   <- c(1,1,1,1,1,2,2,2,3,1,2,3)
survival  <- n.survs/clutch
# Use bootstrap routine to do permutations
N         <- 1000                         # Number of permutations
Meanboot  <- bootstrap(clutch, cor(survival,clutch), sampler=samp.
permute, B=N, trace=F)
# Calculate number of permutations in which difference > than observed
n.over    <- sum(Meanboot$replicates <= Meanboot$observed)
# Remember to add 1 for observed value
P         <- (n.over+1)/(N+1)
cor.test(clutch, survival,
alternative="less", method="pearson")   # Parametric test
P                                          # Print P
```

Output

```
Pearson's product-moment correlation
data: clutch and survival
t = -2.7481,   df = 10,   p-value = 0.0103
alternative hypothesis: coef is less than 0
sample estimates:
        cor
−0.6559505
Randomization     P
            0.01298701
```

The results from both methods of analysis indicate that survival declines with clutch size.

Question 5.6

The data were constructed using

```
set.seed(5)
Group1        <- c(1,1,1,1,1,2,2,2,2,2,2)
Group2        <- c(1,1,2,2,2,1,1,1,1,2,2)
Group         <- cbind(Group1,Group2)
Groups        <- Group
Groups[,1]    <- factor(Groups[,1])
Groups[,2]    <- factor(Groups[,2])
Data          <- matrix(11,1)
Data          <- Group[,1]+5*Group[,2]+Group[,1]*Group[,2]+rexp(11)
Data          <- trunc(Data)
```

There is considerable imbalance in the data and the error term is exponentially distributed and hence, the anova results are suspect. Using the coding given in C.5.8 gives the following results for the data set

```
Type III Sum of Squares
```

	Df	Sum of Sq	Mean Sq	F Value	Pr(F)
Groups[, 2]	1	6.537415	6.537415	10.98286	0.0128699
Groups[, 1]	1	0.463768	0.463768	0.77913	0.4066967
Groups[, 2]:Groups[, 1]	1	0.280702	0.280702	0.47158	0.5143594
Residuals	7	4.166667	0.595238		

```
"Random P for " "1"    " = "    "0.017"
"Random P for " "2"    " = "    "0.379"
"Random P for " "3"    " = "    "0.474"
```

As found in the analysis given in the text, there is little difference between the anova and randomization tests.

Question 5.7

This can be done by randomizing one of the group vectors. It makes no difference if X is also randomized:

```
for (iperm in 1:N)
{
    F.values[iperm,]   <- ANOVA(Groups,Data)
    Data               <- sample(Data)
    Group1             <- sample(Group1)
    Group              <- cbind(Group1,Group2)
    Groups             <- Group
```

```
    Groups[,1]          <-factor(Groups[,1])
    Groups[,2]          <-factor(Groups[,2])
}
```

Output

```
Type III Sum of Squares
```

	Df	Sum of Sq	Mean Sq	F Value	Pr(F)
Groups[, 2]	1	6.537415	6.537415	10.98286	0.0128699
Groups[, 1]	1	0.463768	0.463768	0.77913	0.4066967
Groups[, 2]: Groups[, 1]	1	0.280702	0.280702	0.47158	0.5143594
Residuals	7	4.166667	0.595238		

```
[1] "Random P for " "1"    " = "      "0.01"
[1] "Random P for " "2"    " = "      "0.31"
[1] "Random P for " "3"    " = "      "0.39"
```

The results do change and appear to give a lower *P*-value in two cases. The overall conclusion is not changed and only more extensive simulation could show if the effect is consistent.

Question 5.8

```
# Create vector with habitat categories
Group          <- GammarusData$HABITAT
manova.model <- manova(cbind(OMMATIDI,EYE.L,EYE.W) ~ Group,
            data = GammarusData)
Obs.results  <- summary(manova.model)   # Results for observed data
# Create function to do manova and extract F value
    manova.F <- function(Group,GammarusData)
{
    manova.model <- manova(cbind(HEAD,OMMATIDI,EYE.L,EYE.W)~Group,
    data =GammarusData)
    summary(manova.model)$Stats[6]
}
# Do randomization
    N                <- 1000            # Number of permutations
    F.values         <- matrix(0,N,1)   # Set up matrix to take F values
    for (iperm in 1:N)                  # Iterate through permutations
{
# First pass is on original data
    F.values[iperm]   <- manova.F(Group, GammarusData)
```

```
    Group              <- sample(Group)      # Randomize data vector
}
# Print out results
    Obs.results
    print(c("Random P = ", mean(F.values>= F.values[1])))
```

Output

```
Obs.results
              Df    Pillai    Trace approx.   F num df   den df   P-value
    Group     1    0.85125         59.1333         3        31        0
Residuals 33
    print(c("Random P = ", mean(F.values >= F.values[1])))
    "Random P = "."0.001"
```

None of the *F*-statistics from the randomized data set exceeded the observed value.

Solutions to Chapter 6

Question 6.1

The data were generated using

```
set.seed(1)
x       <- runif(20,0,2)
error   <- rnorm(20,0,1)
y       <- x^2 +error
```

The analysis can be done as follows:

```
# Compare linear and quadratic fits
lin.fit     <- lm(y~x)                     # linear fit
quad.fit    <- lm(y~x+x^2)                 # Quadratic fit
anova(lin.fit,quad.fit, test="F")          # Comparison
```

Output

```
> anova(lin.fit, quad.fit, test = "F")
Analysis of Variance Table
Response: y
Terms   Resid.   Df      RSS    Test  Df   Sum of Sq    F Value      Pr(F)
1        x       18   10.29661
2  x + x^2       17   9.46891 +I(x^2)  1  0.8277026   1.486016   0.2394812
```

The simple linear regression model is sufficient, even though the actual model was quadratic.

Question 6.2

Possible coding is

```
# Generate data
    set.seed(1)
    x        <- runif(20,0,2)
    error    <- rnorm(20,0,1)
    y        <- x^2 + error
# Create index
    Index   <- seq(1:20)              # Set up index
    Data    <- cbind(x,y,Index)   # Combine data
    Data    <- data.frame(Data)   # Make into data frame
    RSS     <- matrix(0,20,2)      # Create matrix for residual sums of squares
    for (i in 1:20)                   # Iterate over index
{
    Data.training  <- Data[Index!=i,]               # Create training set
    lin.fit        <- lm(y~x, data=Data.training)     # Fit linear model
    quad.fit       <- lm(y~x^2, data=Data.training)   # Fit quadratic model
# Calculate residual sums of squares for the left-out data
    RSS[i,1]       <- (lin.fit$coeff[1]+lin.fit$coeff[2]*Data[i,1]
                       - Data[i,2])^2
    RSS[i,2]       <- (quad.fit$coeff[1]+quad.fit$coeff[2]*Data[i,1]^2
                       -Data[i,2])^2
}
    t.test(RSS[,1], y=RSS[,2], paired=T)       # Paired t test
    print(c(mean(RSS[,1]), mean(RSS[,2])))     # Output means
```

Output (only relevant portions shown)

```
    Paired t-Test
    data: SS[, 1] and SS[, 2]
    t = 0.4168, df = 19, p-value = 0.6815
print(c(  mean(SS[, 1]), mean(SS[, 2])))
         0.6546031     0.6193718
```

The quadratic fit gives a lower RSS but not significantly.

Question 6.3

Coding

```
set.seed(1)
n              <- 100
x              <- runif(n,0,2)
error          <- rnorm(n,0,1)
y              <- x^2 +error
Curves         <- matrix(0,n,2)      # Matrix for data
Curves[,1]     <- x
Curves[,2]     <- y
# Create index for cross validation.
# Note that because data is created sequentially index is also randomized
    Index          <- sample(rep(seq(1,10), length.out=n))
# Do ten-fold cross validation
# Set up matrix to store r^2 values for ( i in 1:10)
    Corr.store     <- matrix(0,10,2)
{
# Select subset of data
    Data           <- data.frame(Curves[Index!=i,])
    CV.data        <- data.frame(Curves[Index==i,])    # Store remainder
    Model          <- lm(X1.2 ~ X1.1+X1.1^2, data=Data)
# Multiple r for fitted values
    R2             <- summary(Model)$r.squared
    Corr.store[i,1]  <- R2                              # Store R2
# Calculate predicted curve
    Predicted      <- predict.lm(Model, CV.data)
# Calculate correlation between predicted and observed
    r                <- cor(CV.data[,2], Predicted, na.method="omit")
    Corr.store[i,2]  <- r^2       # Store r^2
    print(c(i,r^2, R2))          # Print predicted and observed multiple R
}
print (c(mean(Corr.store[,1]),mean(Corr.store[,2])))  # Print mean r^2
```

Output

```
[1]  1.0000000  0.8516866  0.5947113
[1]  2.0000000  0.5038258  0.6247513
[1]  3.0000000  0.8110234  0.6060138
[1]  4.0000000  0.5069817  0.6294819
[1]  5.0000000  0.5321204  0.6245366
```

```
[1]  6.0000000  0.5201748  0.6222061
[1]  7.0000000  0.8017032  0.5797304
[1]  8.0000000  0.7014673  0.6014567
[1]  9.0000000  0.5414010  0.6224561
[1] 10.0000000  0.4151990  0.6283841
> print(c(mean(Corr.store[, 1]), mean(Corr.store[, 2])))
[1]  0.6133728  0.6185583
```

In this example, there is an excellent correspondence between the two r^2 values, indicating that the proposed model does not overfit.

Question 6.4

To compare the two models using cross validation alter the coding of C.6.1, thus

```
set.seed(1)                                    # Set random number seed
Data      <- Multiple.regression.example       # Pass data to file Data
Nreps     <- 100                               # Number of randomizations
# Column of response (dependent) variable
obs.col   <- 1
Kfold     <- 5                                 # Set value for Kfold
.............................
.............................
.............................
.............................
Model1  <-lm(INTR.INDEX ~ FOREST + S.ABLE + S.LENGTH, data=Data
        [Data[,last.col]!=1,])
Model2  <-lm(INTR.INDEX ~ S.LENGTH + FOREST + S.ABLE + S.TEMP, data= Data
        [Data[,last.col]!=1,])
```

Output (modified)

```
Paired t-Test
data:  RSS[, 1] and RSS[, 2]
t = 1.6225, df = 99, p-value = 0.1079
alternative hypothesis:  mean of differences is not equal to 0
print(c(mean(RSS[, 1]), mean(RSS[, 2])))
[1] 0.1106308  0.1069008
```

There is no significant difference between the two models.

To examine each equation separately modify coding as follows (bold shows modifications)

```
set.seed(1)                              # Set random number seed
Data        <- Multiple.regression.example   # Pass data to file Data
last.col    <- ncol(Data)+1              # Find number of columns
obs.col     <- 1                         # Observed column
                                         # Number of randomizations
Nreps       <- 100
Kfold       <- 5                         # Set Kfold number
# Find number of rows in data set
   n            <- nrow(Data)
# Create an index vector in Kfold parts
   Index        <- rep(seq(1,Kfold), length.out=n)
# Combine Data and index vector
   Data         <- cbind(Data,Index)
# Function to determine correlations
# D=Data; I=Index value; K=col for Index; R=col for obs. value; Model=model
   object
   SS       <- function(D,I,K,R,Model)
{
   Obs    <- D[D[,K]==I,R]                 # Observed value
# Predicted value using fitted model
   Pred   <- predict(Model,D[D[,K]==I,])
   R2     <- cor(Obs,Pred)^2
   return(R2)                           # r^2 between prediction and observation
}
   Rsquare <- matrix(0,Nreps,2)            # Matrix for r^2
   for (i in 1:Nreps)                      # Iterate over randomizations
{
   Index           <- sample(Index)        # Randomize index vector
# Place index values in last col(last.col) of Data
   Data[,last.col]  <- Index
# Compute model objects note that last.col is the column for the index values
   Model            <- lm(INTR.INDEX ~ S.LENGTH+FOREST+S.ABLE+S.TEMP,
                       data=Data)
   Rsquare[i,1]     <- summary(Model)$"r.squared"    # r^2 for fitted model
# Store r^2 for pred vs obs
   Rsquare[i,2]     <- SS(Data,1,last.col,obs.col, Model)
}
   summary(Rsquare[,1]); summary(Rsquare[,2])
```

Output (for above model)

```
> summary(Rsquare[, 1])
    Min.   1st Qu.   Median     Mean   3rd Qu.     Max.
0.558015  0.558015  0.558015  0.558015  0.558015  0.558015
> summary(Rsquare[, 2])
     Min.   1st Qu.    Median      Mean    3rd Qu.      Max.
0.0004780  0.4213802  0.6046154  0.5679942  0.7396240  0.9538082
```

There is no substantial evidence that the four-parameter model overfits the data.

Question 6.5

Coding

```
# Generate data
  set.seed(1)
  x        <- runif(20,0,2)
  error    <- rnorm(20,0,1)
  y        <- x^2 + error
  Data     <- data.frame(x,y)       # Combine data & make into data frame
# Create function to plot data
# Note that data are assumed to be cols 1 and 2
Model.Plot  <- function(Loess.model, Data, xlimits)
{
  P.model   <- predict.loess(Loess.model, x.limits, se.fit=T)
  C.INT     <- pointwise(P.model, coverage=0.95)  # Calculate values
  Pred.C    <- C.INT$fit                          # Predicted y at x
  Upper     <- C.INT$upper                        # Plus 1 SE
  Lower     <- C.INT$lower                        # Minus 1 SE
  plot(Data[,1], Data[,2])                        # Plot points
  lines(x.limits,Pred.C)                          # Plot loess prediction
  lines(x.limits,Upper,lty=4)                     # Plot plus 1 SE
  lines(x.limits,Lower,lty=4)                     # Plot minus 1 SE
  Fits      <- fitted(Loess.model)                # Calculate fitted values
  Res       <- residuals(Loess.model)             # Calculate residuals
# Plot residuals on fitted values with simple loess smoother
  scatter.smooth(fitted(Loess.model),residuals(Loess.model),
  span=1, degree=1)
}
```

```
Lin1.smooth  <- loess(y~x, data=Data, span=.2, degree=1)
Lin2.smooth  <- loess(y~x, data=Data, span=1, degree=1)
summary(Lin1.smooth)
summary(Lin2.smooth)
anova(Lin1.smooth, Lin2.smooth)
```
`# Plot data`
```
x.limits <- seq(min(Data[,1]),max(Data[,1]),length=20)  # Set range of x
par(mfrow=c(2,2),pty="")
Model.Plot(Lin1.smooth,Data,xlimits)
Model.Plot(Lin2.smooth,Data,xlimits)
```

Output (summary output not included)

```
> anova(Lin1.smooth, Lin2.smooth)
Model 1:
loess(formula = y ~ x, data = Data, span = 0.2, degree = 1)
Model 2:
loess(formula = y ~ x, data = Data, span = 1, degree = 1)
Analysis of Variance Table
     ENP     RSS    Test  F Value   Pr(F)
1   12.1  4.3000  1 vs 2     0.4  0.9114
2    2.3  9.7171
```

There is no significant increase in variance accounted for decreasing the span to 0.2.

Question 6.6

```
# Generate data
   set.seed(1)
   x         <- runif(20,0,2)
   error    <- rnorm(20,0,1)
   y         <- x^2 + error
   Data    <- data.frame(x,y)        # Combine data & make into data frame
# Add index for cross validation.
# Note that because data is created sequentially index is also randomized
   Index   <- data.frame(sample(rep(seq(1,3), length.out=20)))
# Do three-fold cross validation
   for ( i in 1:3)
{
   Data.i      <- Data[Index!=i,]                      # Select subset of data
   CV.data     <- Data[Index==i,]                      # Store remainder
# Fit model
   L.smoother  <- loess(y~x, data=Data.i, span=1,degree=1)
# Multiple r for fitted values
   R2          <- summary(L.smoother) $covariance
# Calc predicted curve
   Predicted   <- predict.loess(L.smoother, newdata=CV.data)
# Calculate correlation between predicted and observed
   r            <- cor(CV.data[,2], Predicted, na.method="omit")
   print(c(i,r^2, R2))           # Print predicted and observed multiple R
}
```

Output

```
[1]  1.0000000  0.2064643  0.8032626
[1]  2.0000000  0.6640947  0.6813835
[1]  3.0000000  0.9060839  0.6559848
```

The correlation between predicted and observed is not very high and considerably less than suggested by the multiple R for the fitted values.

Question 6.7

Line changes indicated in bold.

```
# Generate data
   set.seed(1)
```

```
    x          <- runif(20,0,2)
    error      <- rnorm(20,0,1)
    y          <- x^2 + error
    Data       <- data.frame(x,y)        # Combine data & make into data frame
# Add index for cross validation.
# Note that because data is created sequentially index is also randomized
    Index      <- sample(rep(seq(1,3), length.out=20))
# Do three-fold cross validation
    MultR      <- matrix(0,100,2)
    for ( Irep in 1:100)
{
    Index         <- sample(Index)
    Data.i        <- Data[Index!=1,]        # Select subset of data
    CV.data       <- Data[Index==1,]        # Store remainder
# Fit model
    L.smoother    <- loess(y~x, data=Data.i, span=1, degree=1)
# Multiple r for fitted values
    R2            <- summary(L.smoother) covariance
# Calc predicted curve
    Predicted     <- predict.loess(L.smoother, newdata=CV.data)
# Calculate correlation between predicted and observed
    r             <- cor(CV.data[,2], Predicted, na.method="omit")
    MultR[Irep,   <- c(r^2,R2)
}
t.test(MultR[,1], MultR[,2],paired=T)
plot(MultR[,1], MultR[,2],cex=1.05, xlab="r^2 between predicted and
    observed", ylab="R^2 for training set")
```

Output

```
Paired t-Test
data:  MultR[, 1] and MultR[, 2]
t = -3.9504, df = 99, p-value = 0.0001
alternative hypothesis:  mean of differences is not equal to 0
95 percent confidence interval:
 -0.15021839 -0.04976847
sample estimates:
mean of x - y
 -0.09999343
```

There is a highly significant difference between the two values but the magnitude of the difference is slight. There is a curious negative relationship between the two correlations.

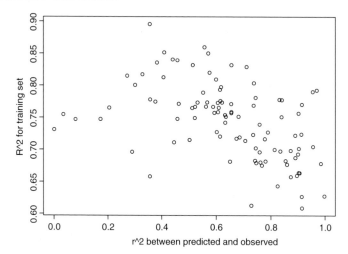

Question 6.8

The data were created using the coding

```
set.seed(1)
X1      <- floor(runif(30,0,10))
X2      <- floor(runif(30,0,10))
X3      <- floor(runif(30,0,10))
error   <- rnorm(30,0,10)
Y       <- 2*X1+floor(X2^2+exp(X2/10)+X3^3+error)
Data    <- cbind(X1,X2,X3,Y)
Data    <- data.frame(Data)
```

The appropriate sequence of gam functions is

```
Model.1  <- gam(Y~lo(X1)+lo(X2)+lo(X3), data=Data)
anova(Model.1)
Model.2  <- gam(Y~X1+lo(X2)+lo(X3), data=Data)
anova(Model.2, Model.1, test="F")
Model.3  <- gam(Y~lo(X2)+lo(X3), data=Data)
anova(Model.3, Model.2, test="F")
```

Output

```
> anova(Model.1)
```

```
DF for Terms and F-values for Nonparametric Effects
            Df  Npar Df    Npar F       Pr(F)
(Intercept)  1
    lo(X1)   1        3    0.7593   0.5322555
    lo(X2)   1        3    9.3432   0.0007132
    lo(X3)   1        3  528.1886   0.0000000
```

This analysis suggests that there is no non-linear component to *X1*.

```
> Model.2 <- gam(Y ~ X1 + lo(X2) + lo(X3), data = Data)
> anova(Model.2, Model.1, test = "F")
Analysis of Deviance Table
Response: Y
                      Terms   Resid. Df   Resid. DevTest  Df    F Value    Pr(F)
1     X1 + lo(X2) + lo(X3)    20.05509    3458.405
2  lo(X1) + lo(X2) + lo(X3)   17.05387    3143.790 1 vs. 2 3.001222 0.5686566 0.6432204
```

This analysis supports the former test and indicates that the non-linear component of *X2* is not significant.

```
> Model.3 <- gam(Y ~ lo(X2) + lo(X3), data = Data)
> anova(Model.3, Model.2, test = "F")
Analysis of Deviance Table
Response: Y
                   Terms   Resid. Df Resid. Dev Test Df Deviance   F Value        Pr(F)
1     lo(X2) + lo(X3)    21.05509   4355.890
2  X1 + lo(X2) + lo(X3)  20.05509   3458.405 +X1      1 897.4851  5.204463  0.03358977
```

The above analysis indicates that there is a significant contribution of *X1* to the model.

Question 6.9

Response variable as a numeric

Step 1: Create tree and plot deviance against possible tree sizes

```
Tree.1          <- tree(P~Wing+Egg+Body+Nest+Habitat, data=Q7.Data)
Tree.pruned     <- prune.tree(Tree)        # prune tree
plot(Tree.1); text(Tree.1)                 # Plot full tree
plot(Tree.pruned)                          # Plot deviance vs size
Tree.pruned$size                           # Text of sizes used in deviance plot
```

Output

```
Tree.pruned$size
[1]  38 37 34 32 29 28 24 21 19 18 17 13 12 9 8 5 4 3 1
```

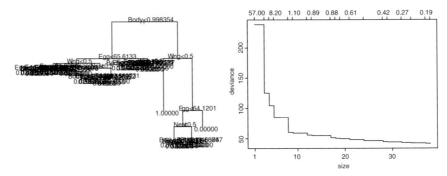

The above tree differs from the "true" tree in that the root node is split according to body size not flight capability. The deviance appears to stop declining markedly at about 10 leaves, which is considerably more than the 5 of the "true" tree. Note that a trees of size 2, 6, 7, etc. are not possible according to the cost-complexity measure. Also note that HABITAT never used.

Step 2: Cross-validation of tree to find optimal size (C.6.7)
Line changed from C.6.7

```
Tree <- tree(P~Wing+Egg+Body+Nest+Habitat, data=Q7.Data) # Create tree
```

Output (trimmed for presentation. Graphical output not shown.)

```
> # Print best Size for the ten runs
> Size
[1]  8 8 8 8 8 8 8 8 8 8
> # Output results
summary(Tree.pruned)
Regression tree:
snip.tree(tree = Tree, nodes = c(29., 18., 28., 5., 19., 8.))
Variables actually used in tree construction:
[1] "Body" "Egg" "Wing" "Nest"
Number of terminal nodes: 8
Residual mean deviance: 0.06066 = 60.18 / 992
```

```
Pruned.Tree$size
```
```
[1]  38 37 34 32 29 28 24 21 19 18 17 13 12 9 8 5 4 3 1
```
All 10 cross-validation runs give an optimal size of 8.

Step 3: Randomization test for tree with 8 leaves (C. 6.8)

The call to the function is

```
Tree.R  <-Tree.Random(P~Wing+Egg+Body+Nest+Habitat, Q7.Data, Ypos=6,
        Ibest=8, N.Rand=100)
```

Output

```
[1]  "Probability of random tree having smaller deviance (SE)"
[1]  0.010000000  0.009949874
[1]  "Summary of sizes actually used in randomization"
 Min.  1st Qu.  Median   Mean  3rd Qu.   Max.
 8.00    8.00   11.00   12.49   14.25   53.00
```

Only one (the observed tree) produced a deviance as small or smaller than the observed. Reject null hypothesis of no relationship. The final tree is

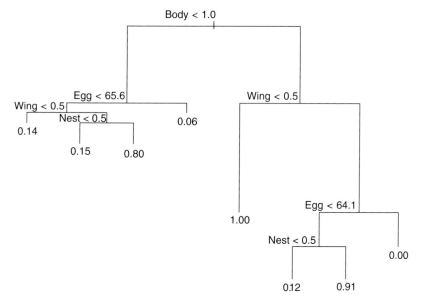

At first glance, the above tree looks quite different from the "true" tree. However, if we rewrite the "true" tree commencing with a body size split we get (branch lengths of arbitrary lengths)

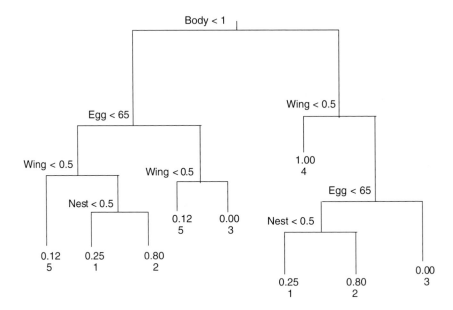

The above tree is very similar to that obtained using the regression tree model, the principle differences being (1) probabilities for leaf #1 are underestimated (0.15 and 0.12 when the correct value is 0.25) and (2) one node (where $P=0.06$) should be split according to wing morph. These results illustrate the fact that the same tree can be constructed in a variety of different manners.

Solutions to Chapter 7

Question 7.1

Using the terminology given in the section "A simple classification problem" we have

Probability of showing character, B, if infected $=p_A=0.97$

Probability of showing the character if non-infected $=p_{A^c}=0.67$

Proportion of infected snails in population $= P(A) = 1-0.83 = 0.17$

Proportion of non-infected snails in population $=1-P(A)=0.83$

Probability of being infected given that the snail shows the characteristic $=P(A|B)=((0.97)(0.17)/(0.97)(0.17)+(0.67)(0.83))=0.2287$

Question 7.2

$L(\theta|x) = (\theta^x e^{-\theta}/(x!c)/ \int_0^{.1} \theta^x e^{-\theta}/(x!c)\, d\theta) = (\theta^x e^{-\theta}/ \int_0^{.1} \theta^x e^{-\theta} d\theta)$. Note that the range of θ is restricted to 0–0.1.

Question 7.3

$$L(\theta|1) = \frac{\theta e^{-\theta}}{\int_0^{.1} \theta e^{-\theta} d\theta}$$

Coding to generate data is

```
theta  <- seq(from=0, to=.1, by=.001)
Prob   <- theta*exp(-theta)
Prob   <- Prob/sum(Prob)
```

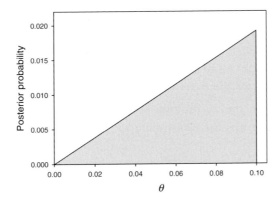

The probability distribution is very odd, with the probability increasing until 0.1 and then, because of the prior, dropping to zero. This strongly suggests that there is a problem with the prior probability.

Iterating from $\theta=0$ to $\theta=10$ produces a much more sensible curve, further suggesting that the initial prior is in error, or the most recent data are in error, or the two data sets represent different circumstances.

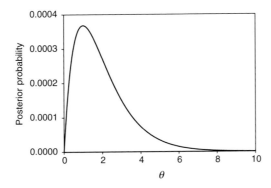

Question 7.4

$$\hat{\mu} = 0.52, \quad \sum_{i=1}^{5} (\hat{\mu}_i - \hat{\mu})^2 = 48.088$$

$$\mu_i = \hat{\mu} + (\hat{\mu}_i - \hat{\mu}) \left(1 - \frac{n-3}{\sum_{i=1}^{5} (\hat{\mu}_i - \hat{\mu})^2} \right) = 0.52 + (\hat{\mu}_i - 0.52) \left(1 - \frac{2}{48.09} \right)$$

$$\mu_i = 0.498 + 0.958\,\hat{\mu}_i$$

Species #	1	2	3	4	5	6	7	8	9	10
True value	−1.88	−1.02	−0.36	−0.13	−0.04	−0.03	0.00	0.01	0.34	1.21
Observed	−3.85	−1.74	1.74	0.32	4.10	−1.47	1.80	2.03	1.81	0.46
EB estimates	−3.22	−1.41	1.56	0.35	3.58	−1.18	1.61	1.81	1.63	0.47

There is a reduction in the sum of squared-errors (38.24 using the observed values and 28.29 using the EB estimators).

Index